THE SCIENCE OF AGRICULTURE: A BIOLOGICAL APPROACH

2nd Edition

**Delmar Publishers
is proud to support
FFA activities**

THE SCIENCE OF AGRICULTURE: A BIOLOGICAL APPROACH

2nd Edition

Ray V. Herren

DELMAR
™
THOMSON LEARNING

Australia Canada Mexico Singapore Spain United Kingdom United States

DELMAR

THOMSON LEARNING

The Science of Agriculture: A Biological Approach, 2nd Edition
by Ray V. Herren

Business Unit Director:
Susan L. Simpfenderfer

Editorial Assistant:
Elizabeth Gallagher

Executive Production Manager:
Wendy A. Troeger

Acquisitions Editor:
Zina M. Lawrence

Executive Marketing Manager:
Donna J. Lewis

Production Editor:
Carolyn Miller

Developmental Editor:
Andrea Edwards Myers

Channel Manager:
Nigar Hale

For permission to use material from this text or product, contact us by
Tel (800) 730-2214
Fax (800) 730-2215
www.thomsonrights.com

Library of Congress Cataloging-in-Publication Data

Herren, Ray V.
  The science of agriculture: a biological approach / Ray V. Herren.— 2nd ed.
    p. cm.
  ISBN 0-7668-1669-9
    1. Agriculture.

S495 .H625 2000
630—dc21   00-063846

NOTICE TO THE READER

CONTENTS

PREFACE

Constant change is a fact of our everyday lives. No segment of our society changes more rapidly than that of the sciences. In fact, it is *because* the changes in science bring about the other changes in our lives. Perhaps the most sensational changes occur in the biological sciences. Over the past decade our understanding of the life processes has dramatically expanded, and the rate of the understanding and change is increasing as new discoveries unlock the mysteries of life. All of the knowledge we have about biology has but three applications: medicine, ecology, and agriculture. By far the widest application is that of agriculture. Advances in medicine and ecology often come about as a result of agricultural research. The second edition of *The Science of Agriculture: A Biological Approach* is intended as a text that explains the scientific principles behind the production of food and fiber. All of modern agriculture is built on these principles, and it is through scientific inquiry that progress is made.

Traditionally, agricultural texts have concentrated on the *how* of production. *The Science of Agriculture: A Biological Approach, Second Edition* approaches the material from the *why* rather than the *how* aspect. It is written so that by completing an agriscience course based on the text, students will be able to receive science credit for the course.

Agricultural science is a very broad discipline. This text concentrates on the scientific principles of the central components of the agricultural industry. The production of plants and animals is central to fulfilling the need for food and fiber. Areas such as the environment, food safety, and the future of agriculture are addressed. In the second edition a chapter has been added on producing food and fiber through organic means. Also, a chapter on career opportunities in agricultural science is included.

The Science of Agriculture: A Biological Approach, Second Edition represents not just a text but a complete teaching package. Each chapter in the text ends with suggested student activities. These learning activities are designed to provide the students with exercises they can complete on their own. Laboratory activities are outlined in the *Laboratory Manual* designed for the course. Other instructional materials can be found on the ClassMaster CD-ROM. The ClassMaster contains lesson plans, answers to the text review exercises, a computerized test bank, and transparency masters. When used as an entire package, the text should provide an instructor with most of the teaching resources needed to conduct a high-quality course in agriscience.

This newly revised and exciting title from Delmar, a division of Thomson Learning, is written to be used as a core text for those courses that focus on a biological approach to teaching agriscience.

FEATURES

- Full-color photographs, illustrations, and design stimulates student interest and makes difficult concepts easier to understand.

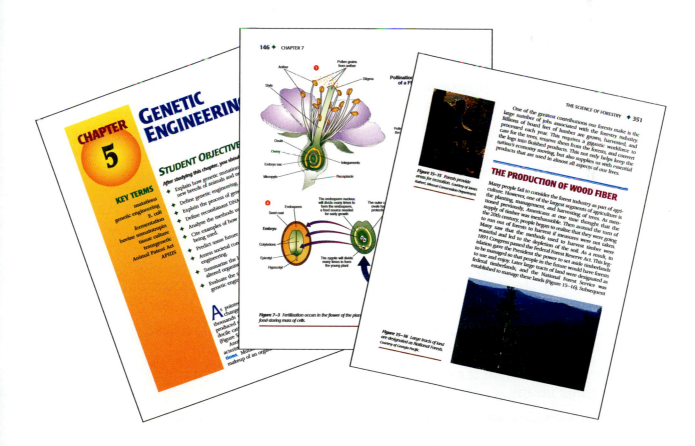

BIO ◆ BRIEF

What Does a Research Scientist Do?

Just how does a research scientist conduct research? The process involves several steps and takes a lot of time. Although many types of research investigate a wide variety of topics, most research projects follow a similar scheme.

The first step is that the scientist defines a definite problem to investigate. The problem has to be defined so that the scientist knows exactly what is to be accomplished. When this is established, the scientist begins a literature search. This involves reading all the research study summaries, books, and journal articles that have been written on the topic. This helps to provide a basis for designing the project.

After the literature search has been completed, the scientist carefully plans how the research project is to be conducted. Remember from Chapter 1 that many factors have to be considered to make the results of the research usable. Questions such as "Where will the research be conducted?"; "What steps will I take?"; "What materials will be needed?"; and "What will be the timeline?" all have to be answered.

Next, the scientist has to secure funding for the project. Quality research is very expensive, and plans have to be made to provide money. If the scientist works for a corporation, the money will come from the research and development funds of the corporation. If the scientist works for a university, funding will be applied for on a competitive basis. Only so much money is available, and researchers have to compete for it. The scientist may write a proposal for a grant. Grants come from foundations that sponsor research or from corporations that have an interest in a particular area. The grant proposal contains the rationale for doing the research, the benefits of the research, how the research will be conducted, how the results will be used, a timeline, and a budget. The proposals with the most potential are the ones that get funded.

When funds have been secured, the project begins. All the procedures outlined in the plant are followed. The research study is completed, and the data are gathered. The data are analyzed using appropriate formulas and tests. Conclusions are drawn based on the analysis of the data.

When the conclusions have been made, the research is still not complete. The scientist must decide what to do with the findings of the research. Recommendations have to be made as to what impact the research will have and how it should be used. The final step is to publish the results. The research is described in a written report that details how the project was done and what the results were. The report then can be written in the form of a journal article for a research publication or an oral presentation at a research meeting. By publicizing the results, other scientists can have access to the research and use the findings to conduct other research. Of course, the practical applications can be put into use.

A research scientist needs skills in communication almost as much as expertise in research methodology and technical know-how. Research is of not use if the results cannot be communicated to others.

Collecting data is only part of a research scientist's duties. Courtesy of USDA-ARS.

Examples of jobs requiring a PhD or doctorate of veterinary medicine are

Veterinarian

Meat inspector

Animal geneticist

Animal nutritionist

Reproductive physiologist

Microbiologist

Research scientist

College professor

CAREERS IN NATURAL RESOURCES

If you enjoy the beauty of nature and like to be outdoors, you might consider a career in natural resources. This area is expected to become increasingly more important. As the population

- Boxed articles highlight new developments in biotechnology and agriscience and introduce students to the latest advancements in agriscience.

Enhanced Content

- Strong science-based approach covering a broad range of agriscience subjects including: Plant Science, Animal Science, Genetic Engineering, Aquaculture, Environmental Science, Food Science and Technology, and more!

- Interdisciplinary connections to academic subject areas including: biology, chemistry, physics, math, history, and English are reinforced throughout the text.

- Complete chapter on careers in agriscience allows students to investigate career opportunities in the agriscience industry.

- New to the edition is a chapter on producing food and fiber through organic means.

- Developed in conjunction with science and agriculture teachers.

EXTENSIVE TEACHING/LEARNING PACKAGE

This supplement package was developed to achieve two goals:

1. To assist students in learning the essential information needed to continue their exploration into the exciting field of agriscience.

2. To assist instructors in planning and implementing their instructional program for the most efficient use of time and other resources.

Instructor's Guide to Text

The Instructor's Guide provides answers to the end-of-chapter questions and additional material to assist the instructor in the preparation of lesson plans.

Lab Manual

Order # 0-7668-1671-0

This comprehensive science-based lab manual reinforces science concepts in the text. It is recommended that students complete each lab to confirm their understanding of essential science content. Great care has been taken to provide instructors with low-cost, strong-content labs.

Lab Manual Instructor's Guide

The Instructor's Guide provides answers to lab manual exercises and additional guidance for the instructor.

ClassMaster CD-ROM

Order # 0-7668-1673-7

This new supplement provides the instructor with valuable resources to simplify the planning and implementation of the instructional program. It includes transparency masters, answers to questions in the text, lesson plans, and a computerized test book to provide the instructor with a cohesive plan for presenting each topic.

ACKNOWLEDGMENTS

The author gratefully acknowledges the following for their assistance in creating the second edition of this text:

Mary Herren for her help in compiling the Bio Briefs and Crystal Cummings for her assistance in creating review questions.

The author and Delmar wish to sincerely acknowledge the following reviewers for their invaluable input and for sharing their content expertise.

Alex Azcona
Lyman High School
Longwood, Florida

Roy Crawford
Lancaster High School
Lancaster, Texas

Murdock Leroy Gillis
Ponce de Leon High School
Westville, Florida

Heather McDowell
Ceres High School
Ceres, California

DEDICATION

This book is dedicated to my mother, Ethel Herren. She taught me from a very early age the value of an education. Her philosophy, "Where there is a will, there is a way," set the foundation for all my achievements.

THE SCIENCE OF AGRICULTURE

KEY TERMS

science
aquaculture
scientific method
basic research
applied research
genetics
cooperatives

STUDENT OBJECTIVES

After studying this chapter, you should be able to:

✦ Define science.

✦ Explain how agriculture helped develop civilization.

✦ Tell how humans first began to use science.

✦ Explain the concept of the scientific method.

✦ Distinguish between basic and applied science.

✦ Cite scientific discoveries that have made food better and less expensive for the consumer.

✦ Discuss how scientific research in agriculture developed in this country.

✦ Analyze the factors that allowed the development of the world's most efficient system of agriculture.

✦ Discuss the food needs of the average American family.

✦ Discuss the advances made by American agriculture.

✦ Analyze how agricultural research has benefited the consumer.

✦ Discuss how agricultural cooperatives have affected agriculture and consumers.

The **science** of agriculture is the basis on which all of civilization is built. Before humans began to devise ways to produce their own food, most of their lives were devoted to finding enough to eat (Figure 1–1). The only available food was the plants and animals that grew wild in the area. Hunting and gathering food was a process that was not only

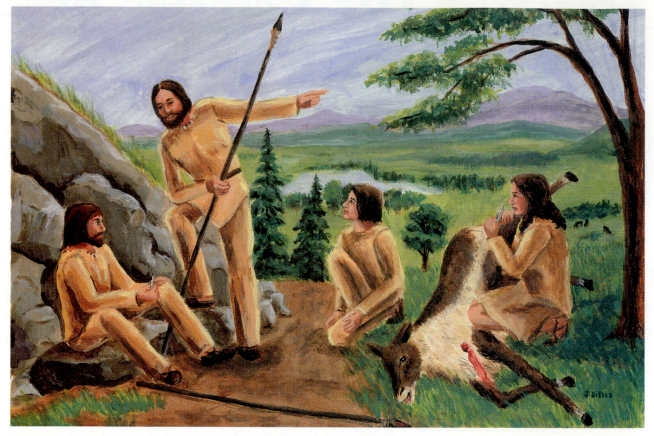

Figure 1–1 *Early humans spent most of their time searching for food.* Courtesy of June Di Pisa.

time-consuming but also prevented early humans from settling in one place. If a group of people stayed in one area very long, most of the wild game and wild plants that provided food would be exhausted. Because gathering food took so much time, these early humans had no time for such endeavors as building homes and cities or even in developing inventions that might make their lives better.

The very first science was agriculture. Science is knowledge obtained through a systematic study of naturally occurring phenomena. The first systematic study by humans was probably devising ways to obtain food, clothing, and shelter.

The systematic study of anything begins with observation. Early humans likely began to notice that plants sprouted from seeds and that by putting seeds in the ground, they could make the seeds come up where they wanted them. They also observed where the edible plants grew and the environment surrounding the growing plants. They noticed the time of year when the seeds, nuts, and fruits matured and returned to that area to harvest. By further observing, they saw that the seeds had to be planted at the right time of year; the plants had to have water and sunlight; and they had to be protected from animals. These early humans probably noticed which plants grew bigger and better than others and used these for obtaining seeds.

Similarly, humans observed the way animals developed patterns in where they ate, slept, and moved. They noticed that some animals were not as wild as others and would tolerate the presence of humans. Rather than following the herds of animals, they began to raise the animals in captivity and to live in one place. This provided a ready supply of food that required less time than hunting.

Most anthropologists agree that agriculture began about 10,000 years ago in what is now known as the Middle East. When humans began to grow their own food, they no longer needed to wander about in search of edible plants and animals. This allowed them to settle down in one place and to develop villages where they could live together as one society (Figure 1–2).

Figure 1–2 Agriculture allowed people to settle down in one place and develop villages. *Courtesy of Ray Donahue.*

As humans grew more of their own plants and animals for food, they began to search for better ways to produce food. These ways were discovered through trial and error and passed down from parents to children. All modern agricultural crops and livestock were developed from the plants and animals tamed and cultured by early humans (Figure 1–3).

Figure 1–3 *For centuries agriculture relied on animal power.* Courtesy of James Strawser, The University of Georgia.

As more efficient ways of growing plants and animals were developed, food could be produced in less time. As soon as people could supply enough to feed themselves and have some left over, the surplus food was traded to other people. Because food could be obtained through trading, time was spent developing skills in building, engineering, literature, and art that led to the great civilizations. When everyone had to find food every day, there were few inventions because of the lack of time and energy. Without the development of agriculture, humans would still be hunting and foraging for their food.

As people began to raise their own food, the necessity arose to invent implements to open the soil to plant seed or to dig out weeds. These first crude tools were made of wood or stone and later evolved into metal implements. The more tools they made, the more efficient they became at growing food. The more food they grew, the more time they had for inventing and making tools. After the technology was developed to make agricultural tools, humans discovered that the tools could have other uses also. These uses might include carving stone for buildings or statues.

Some scientists think that counting and writing developed from agriculture. As people began to harvest crops and had surplus left over, bins and storage areas had to be built. To indicate ownership, contents, and the amount in the containers, a system of marking had to be developed. As more and more containers had to be marked, a system of written language developed. This allowed the expansion of trade and barter so that the excess food could be traded to other people.

AMERICAN AGRICULTURE

When the first Europeans came to the New World, they found a system of agriculture already in existence. Native Americans in North America planted, cultivated, and harvested such crops as corn, squash, okra, and pumpkins. In Central and South America, civilizations such as the Incas, the Mayas, and the Aztecs had elaborate systems of fields and irrigation. They also developed tools of flint and wood to dig irrigation canals and drainage ditches, plant, cultivate, and harvest crops. Their system of agriculture allowed them to feed their people and have enough time and energy to develop cities, roads, and works of art.

Historical accounts tell of how Native Americans helped the settlers understand the methods used in the New World to produce food. Most of the Europeans who emigrated to America in the early years grew their own food. They learned from the Native Americans the plants and crops that grew best in the area and also the techniques of planting, cultivating, and harvesting the crops. Until about 100 years ago, the vast majority of Americans were involved in growing crops and animals (Figure 1–4). Advancements in agriculture have increased production, allowing more people to leave the farm to pursue careers in other areas.

Today, the American agricultural system is the envy of the world. According to the United States Department of Agriculture, one American farmer feeds more than 100 people. Not only do people in this country benefit from our agriculture, but much of the world is fed and clothed by American producers. Estimates are that of the more than 100 people each producer feeds, around half of them live in other countries.

Americans enjoy an abundance of food (Figure 1–5). According to the American Farm Bureau Federation, in one year the average American consumes about 2.5 tons of food. Each of us consumes 194.1 pounds of flour and cereal products, 66 pounds of fats and oils, 212.7 pounds of beverage milk, 237 eggs, 114.7 pounds of red meat, 19.3 pounds of rice, 26.8 pounds of cheese, 63.3 pounds of poultry, 175.9 pounds of fresh vegetables, and 126.5 pounds of fresh fruit. If you multiply this by the number of people in the United States, it adds up to a tremendous amount of farm production!

Not only do Americans have a lot of food, but they can buy it at a relatively low cost. People in many countries

Figure 1–4 *At one time the vast majority of Americans made their living through production agriculture.* Courtesy of PhotoDisc, The Palma Collection.

Figure 1–5 *Americans enjoy an abundance of a wide array of foods. Courtesy of Fleming Companies, Inc.*

Figure 1–6 *Flowers are part of agriculture that brightens our lives. Courtesy of USDA-ARS/K5547-19.*

spend the bulk of their income on food. Even in advanced countries people may spend half of their paychecks at the grocery store. In the United States, the average family spends only 10.9 percent of its income for food. This compares to 15.2 percent in France, 17.7 percent in Germany, and 51.4 percent in India.

The American agricultural industry is the most efficient the world has ever seen. No one in the world can come even close to matching us in the power of our output. This powerful, efficient system has evolved for three basic reasons.

The Climate and Soil

Crops and livestock are grown in all 50 states. Climate and soil conditions vary widely across the country, and this allows the cultivation of a tremendous variety of crops and animals. More than 200 different commodities are produced by American farmers, and this does not include some of the highly specialized crops. American agriculture produces such familiar crops as cotton, corn, soybeans, wheat, rice, apples, citrus fruits, green beans, grapes, and cherries. Think of all the sugar beets grown to sweeten our food or the mint grown to flavor our chewing gum. Thousands of acres of flowers are produced to brighten our lives (Figure 1–6). No other country in the world grows such a wide variety of agricultural products.

Almost everyone knows that livestock producers raise beef cattle, hogs, and sheep. Fewer people realize that these producers raise other animals as well. A growing industry of **aquaculture** is developing in this country. Animals such as fish, crawfish, shrimp, and oysters are raised for food and are just as much a part of agriculture as beef cattle (Figure 1–7). Bees are raised to pollinate crops and to produce honey. Millions of dollars each year are earned by producers who raise bait for fishermen. Producers are always coming up with new ideas of how animals can be used for profit. Some examples include ostriches and alligators that are raised on farms for meat. Only a country with a wide variety of climatic conditions could produce such a wide variety of agricultural products (Figure 1–8).

Our country lies in a temperate zone. This means that although certain parts of the country have extreme cold or heat, the majority of the country enjoys mild temperatures that make ideal growing conditions for crops and livestock. Many countries, Norway and Russia for example, have most of their land so far to the north that many crops are not suitable

Figure 1–7 *Animals grown in the water for food are as much a part of agriculture as plants and animals grown on the land.* Courtesy of Progressive Farmer.

for growth there. In the United States, we even have areas in the southern parts of the country that are subtropical, with temperatures warm enough to allow the cultivation of crops in the winter. The climate there also allows the production of such tropical crops as citrus, avocados, and mangos. Many regions in the country get adequate amounts of rainfall to produce crops. Other parts have irrigation systems that supplement rainfall and provide all the moisture necessary to grow crops.

Although many parts of the country have soils that are productive, the Midwest section has the most productive soils in the nation. These deep, rich soils combined with adequate rainfall and near ideal growing conditions have earned this area the title of "breadbasket of the country" because of all the grain that is produced there. Few countries in the world can match the United States for a combination of climate and productive soils for agriculture (Figure 1–9).

Our Economic System

American economics is based on the free enterprise system. This means that people are free to own and operate businesses as they see fit without interference from the government beyond that necessary to protect the economy and public interest. American producers have always been hard workers because they could see the benefits of their labor. Higher yields usually meant higher profits, and the higher profits

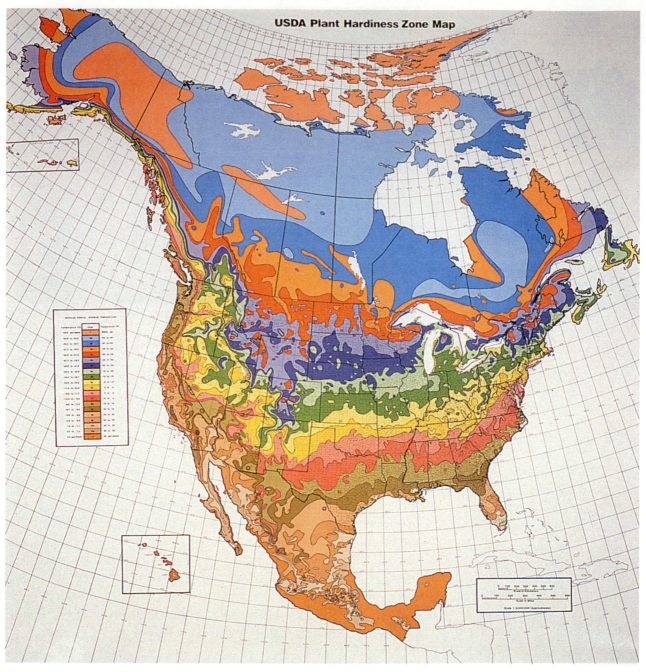

Figure 1–8 *Many different climate zones in the United States allow us to produce a wide variety of crops.* Courtesy of USDA Natural Resources Conservation Service.

meant a better living. In addition, producers have always taken a great deal of pride in the crops and animals they raise. Just think of all the livestock and produce exhibits that can be seen at shows and fairs across the country. Government-run farms

Figure 1–9 *Few countries in the world can match the United States for a combination of climate and rproductive soils suited to agriculture.* Courtesy of California Agriculture, University of California.

in other countries have never been very successful because of the lack of incentive for producing for the government. Historically our agriculturists have kept moving westward in search of new land to expand their operations and produce more and better crops and animals. Government programs have always been designed to help and encourage the free enterprise system of production.

Scientific Research

The creation of new knowledge about the growing of living organisms has allowed us to take advantage of the fertile soils and ideal climate in this country. The knowledge gained over the past 150 years has had a profound effect on the efficiency of our agricultural system. The accumulation of knowledge has allowed us to increase our yields dramatically. Through the use of trial and error, the best ways of caring for crops and animals were discovered and passed along from parents to children.

Progressive scientific research began in this country about the middle of the 1800s. At that time, the universities in the United States taught a curriculum known as the classics in which the main subjects studied were Latin, Greek, history, philosophy, and mathematics. People began to realize that there was a need for institutions where students could study areas that had practical applications. The nation was emerging as an industrial and agricultural economy. To make progress in these areas, young people needed to be taught how to produce food and manufacture goods in a more efficient manner.

In the late 1850s, a senator from Vermont, Justin Morrill, introduced a bill that would provide public land and funds for universities to teach practical methods of manufacturing and food and fiber production. The bill passed in 1862 and became known as the Land Grant Act or the Morrill Act. During that same year, President Lincoln signed into law a bill that established the United States Department of Agriculture (USDA). Soon almost all of the states in the country established land grant colleges.

As students enrolled and classes began, people began to realize that there was little scientific knowledge about agriculture. Most of the knowledge about growing plants and animals had been passed from generation to generation and represented peoples' beliefs rather than proven facts. To solve this problem, in 1872 Congress passed the Hatch Act, which authorized the establishment of experiment stations in all the states with land grant colleges. The purpose was to create new knowledge through a systematic process of scientific investigation. These experiment stations used the **scientific method** of investigation.

Later, in 1914, Congress passed the Smith Lever Act. This act founded the Cooperative Extension Service to disseminate information learned from the research (Figure 1–10). Consequently, this completed a system known as the Land Grant

Figure 1–10 *The Cooperative Extension Service began in 1914 to teach people about agriculture. Agents began to teach farmers about new production methods.*
Courtesy of James Strawser,
The University of Georgia.

Concept, which held that the purpose of the Land Grant universities was to teach, conduct research, and carry the new information to people through the extension service.

In 1917, the Smith Hughes Act established Vocational Agriculture in the public high schools to teach new methods of agriculture.

The Scientific Method

The scientific method helps to ensure that conclusions reached through study are valid and reliable. This process involves formulating a hypothesis, designing a study, collecting data, and drawing conclusions based on an analysis of the data (Figure 1–11).

A scientist begins by identifying a problem that needs to be solved. He or she may have an idea about what causes the problem or what might solve it. This is called a *hypothesis* and is the basis for investigating a problem. The hypothesis is subjected to a test called an *experiment* that attempts to isolate the important factors.

Seven Steps in the Scientific Method

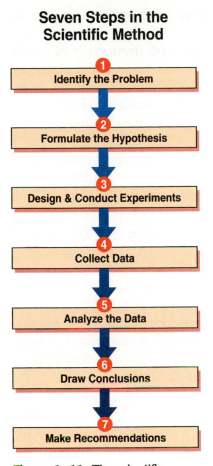

1. Identify the Problem
2. Formulate the Hypothesis
3. Design & Conduct Experiments
4. Collect Data
5. Analyze the Data
6. Draw Conclusions
7. Make Recommendations

Figure 1–11 *The scientific method is a standard procedure for conducting research.*

For instance, if a scientist wanted to know whether a newly formulated fertilizer would increase wheat yields, he or she would conduct a study using the scientific method. The scientist would formulate the hypothesis that the new fertilizer will produce higher yields of wheat than the old fertilizer. He or she then would search for other studies that had been done on this topic to learn what others had discovered. This information might help the scientist plan the study.

Plots of wheat would be planted using the new fertilizer. These plots would be called the treatment plots. Other plots would be planted with the same type of wheat using the same procedures as the treatment plots except that the old fertilizer would be used. These plots would be called the *control plots* (Figure 1–12). Certain conditions must be met to ensure that the data collected in the experiment are valid. The wheat must be of the same type, the same variety, and planted at the same time; and the same methods of production also must be used. The number of plots should be large enough so that differences in the soil and other factors would average out. All the plots must be harvested using the same machine and weighed on the same scales.

The weights from the control and treatment would be compared. If the plots that received the new fertilizer yielded more wheat, the scientist might conclude that the new fertilizer was responsible. After this has been decided, other areas of agricultural research would become involved. Agricultural economists would determine whether the yields gained by using the new fertilizer would offset any increased cost. Nutritionists might analyze the wheat to determine the value

Figure 1–12 *A scientist might use one plot for an experimental treatment and another plot for a control. All conditions are as closely controlled as possible.*
Courtesy of USDA-ARS.

Figure 1–13 *Basic research deals with how or why processes occur in plants and animals.* Courtesy of USDA-ARS.

of the wheat grown using the new fertilizer as opposed to wheat grown with the old.

The scientific method has been used thousands of times to develop ways to produce agricultural crops and animals. As more knowledge was gained, the experiments became more involved and more complicated. Today scientific research is classified into two broad areas: **basic research** and **applied research**.

Basic research investigates why or how processes occur in plants and animals (Figure 1–13). For instance, basic research was used to discover the specific hormones that control growth in animals. Applied research uses discoveries made in basic research to help in practical ways (Figure 1–14). If basic research discovers growth-regulating hormones, applied research uses this knowledge to develop ways of using hormones to increase the growth efficiency of agricultural animals.

Agriculture applies almost all of the research and knowledge associated with plants and animals. Aside from the field of medicine, about the only application of basic research in the life sciences is that found in the field of agriculture.

As a result of scientific research and the application of research findings, gigantic strides have been made in the efficient production of food from plants and animals. According to the Council for Agricultural Science and Technology (CAST), since 1925 beef cattle live weight marketed per breeding female has increased from 220 pounds to 482 pounds. This means that for every cow raised, we are now selling more than twice the beef we did in 1925 (Figure 1–15).

Figure 1–14 *Applied research deals with using the results of basic research to improve the way plants and animals are grown.* Courtesy of USDA.

Figure 1–15 *As a result of research, twice the amound of beef can be produced from a cow than in 1925. Courtesy of* Progressive Farmer.

This increase has come about as a result of the scientific selection of breeding animals, a better understanding of beef cattle nutrition, and better control of parasites and diseases.

A better understanding of all phases of the lives of the animals has led to higher quality, less expensive beef for the consumer. To a large degree, the lower cost accounts for the tremendous increase in the annual per capita consumption of beef. Since 1925, consumption has doubled from 60 pounds of carcass weight equivalent to 120 pounds. Americans now have a more nutritious diet at less cost.

According to CAST, sheep live weight marketed per breeding female has increased from 60 to 130 pounds. This, too, represents an increase of 100 percent over the production in 1925. Around the turn of the century, sheep were raised primarily for wool. Then through research in selection, sheep began to be produced that were raised mainly for meat, and research efforts concentrated on raising better meat-type animals.

Milk marketed per dairy cow has increased from 4,189 to almost 15,000 pounds, an increase of 300 percent (Figure 1–16). In the period from 1950 to 1975, the number of dairy cows was reduced by about half and the amount of milk produced remained about the same. This greater efficiency has resulted in a bargain for consumers. A gallon of milk now costs 36 to 84 cents less than it would using 1950s technology.

Swine live weight marketed per breeding female has increased from 1,600 to 2,850 pounds. Since 1950, the amount of feed required to produce a 220-pound market hog has been reduced by about 50 pounds, and the average time to produce it has been reduced from 170 to 157 days. Because

B I O   B R I E F

Who Does Agricultural Research?

Most of the advances made in agriculture have come about as a result of scientific research in the multitude of different areas related to the discipline. Each year, new knowledge is added that helps us understand how organisms grow and reproduce. It takes a lot of money and a lot of people to conduct the experimentation and draw conclusions based on the findings. Have you ever wondered where these research studies are conducted? The answer is that they are conducted in a variety of settings.

The first major agricultural research was conducted by Land Grant universities. When the experiment stations were added to the colleges, research began in a big way. For many years, almost all of the research was conducted at these research stations connected to the universities. Even to this day, a major portion of the research is carried out at the Land Grant universities. Every state has at least one university that conducts agricultural research. Faculty are hired to do research, and the same faculty might also teach classes.

The United States Department of Agriculture (USDA) also sponsors a lot of research. All across the country, the USDA has stations where scientists make discoveries pertaining to agriculture. These are usually independent of the universities, although the two may cooperate on projects. This branch of the USDA is called the Agricultural Research Service (ARS). Many of the research studies that are highlighted in this book are reports from the ARS.

In recent years, large corporations have conducted quite a lot of agricultural research. The advantages to these companies are that they make the discovery and that they have exclusive rights for many years to market any resulting product. They may use the results of university or ARS research as the basis for doing studies that result in the development of a new product. These products are thoroughly evaluated

pigs can be raised in a shorter time on less feed, the cost of production is less. This savings is passed along to the consumer in less expensive pork.

No segment of agriculture has made more dramatic advancements than the poultry industry. Since 1925, the time for broiler chickens to reach market weight has been cut in half, from 15 weeks to 7.5 weeks, CAST reports. The amount of feed required per pound of live weight gain for broiler chickens was cut in half from 4 pounds to around 2 pounds. The weight of broilers at marketing increased from 2.8 to 3.38 pounds. This means that we can now produce a

and regulated by government agencies before they can be marketed. This makes the development of new agricultural products an expensive endeavor. Companies such as Monsanto, DowElanco, Pioneer, and many others have made a tremendous contribution to agricultural research.

A common arrangement is that of cooperation between a university and a company. The company may provide money to a university to conduct research on a topic. Depending on the terms of the agreement, the company may or may not have exclusive rights to any product that comes from the research. The university benefits from the funding for the research, and the company benefits from the results of the research.

Agricultural research will continue because a lot of problems still need to be solved. As agriculture expands and more knowledge is gained, a variety of settings probably will be available for this research. Perhaps you might be one of the scientists who works for the USDA, a university, or a large company.

This scientist works for the Agricultural Research Service (ARS) of the United States Department of Agriculture (USDA). Courtesy of USDA-ARS.

heavier broiler in half the time on half the feed compared to 1925 (Figure 1–17). Not only do consumers get a better broiler at less cost, but they also get a bird that is completely dressed and ready to cook. These advances have come about as a result of breeding and nutrition research.

The annual production per laying hen has doubled from 112 to 232 eggs, and the feed required has decreased from 8.0 to 4.2 pounds. Advances made through scientific research have made the production of eggs in enclosed buildings possible. This allows the hens to be much better managed because the producer can control the environment in which

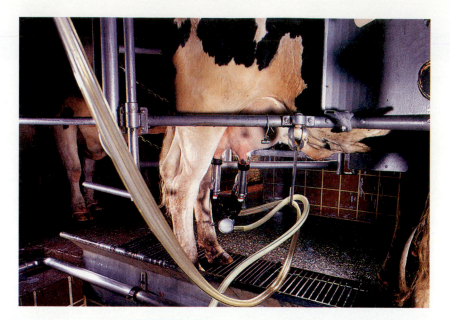

Figure 1–16 *Milk production has tripled as a result of research.* Courtesy of James Strawser, The University of Georgia.

the eggs are produced. The weight of marketed turkeys has increased from 13.0 to 18.4 pounds. This increase was achieved on less feed (from 5.5 to 3.1 pounds) and in nearly half the time (from 34 to 19 weeks).

Crop production has also advanced dramatically. Since 1950, corn yields have increased by 125 percent, wheat yields have increased by 38 percent, and cotton has advanced 51 percent. An interesting note is that soybeans were first

Figure 1–17 *We can now produce a heavier broiler in half the time on half the feed than we could in 1925.* Courtesy of James Strawser, The University of Georgia.

Figure 1–18 *More than two billion bushels of soybeans are produced a year. We now export soybeans to China, the country where soybeans originated.* Courtesy *of* Progressive Farmer.

brought to the United States from China in the early 1900s. In 1924 only 5 million bushels were produced in this country. Today over 2 billion bushels are produced. The United States now exports soybeans to China (Figure 1–18).

The countries in North and Central America, Europe, and Oceania (Australia and New Zealand) have only 29.9 percent of all the world's cattle; yet these countries produce 68 percent of the world's beef and veal. Similar statistics are true for other areas of agriculture. These countries are where most of the scientific knowledge about agricultural plants and animals has been discovered. People in these areas are the best fed and enjoy the highest living standards of any people in the world as a result of basic and applied research. Before countries can develop and prosper, a sound basis for producing food must first be achieved. Many poorer countries are attempting to make their agricultural systems more productive.

MILESTONES IN AGRICULTURAL RESEARCH

Through the years, many discoveries and developments have aided the advancement of agriculture. Some progress has been the result of many small discoveries that combine to provide greater efficiency in agricultural production. Other advancements have been the result of milestone breakthroughs. The following are some of the milestones that have revolutionized agriculture.

Animal Immunization

Until the last half of the 1800s, diseases devastated all types of agricultural animals all over the world. When a disease started in an area, all the animals in the surrounding countryside contracted the disease because no method of preventing the spread existed.

During the 1870s and 1880s, a French scientist named Louis Pasteur developed a means of vaccinating animals to make them immune to disease (Figure 1–19). Using the scientific method, Pasteur hypothesized that animals that had contracted a disease and survived must have built an immunity to the disease. Using the blood from sheep that had contracted and survived the deadly disease of anthrax, he separated a serum from the blood of the sheep.

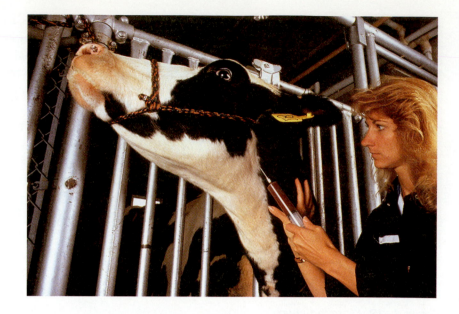

Figure 1–19 *Louis Pasteur discovered the concept of vaccination. As a result, agricultural animals can be protected from disease.* Courtesy of USDA-ARS.

His experiment consisted of using two groups of healthy sheep. One group was injected with the serum and was later injected with anthrax organisms. The other group received only an injection of the anthrax organisms. The group that had received the serum remained healthy, and the group that did not receive the serum died.

Using the discovery of Pasteur, other scientists began to conduct research on other diseases. During the next century, a host of new vaccines were developed to control most of the diseases that are contracted by agricultural animals. Animals in a disease-free environment can be raised at a much lower cost and at less risk to producers.

Canning and Refrigeration

A problem that has plagued producers since the time humans first began raising crops is how to preserve the food. When a large animal was slaughtered, not all of the meat could be eaten at once. Particularly in the summer months, the meat spoiled quickly, and most would go to waste. In colder climates, the animals would be killed in the winter and frozen to preserve the meat until the spring thaws. Similarly, fresh fruits and vegetables that were ripe in the summer soon spoiled. About the only other way of preserving the produce was to salt or dry the food. Both of these methods were time-consuming and did not produce a very palatable product. Another problem was getting the food to market. Until the turn of the century, live animals had to be delivered to population centers. This meant driving the animals to market or driving them to a railhead where they could be transported live to the market, slaughtered, and sold as fresh meat (Figure 1–20). If the meat did not sell quickly, it was lost to spoilage. Fresh fruits and vegetables could not be transported long distances to markets and arrive in an edible condition.

Figure 1–20 *At one time cattle had to be driven to market because the meat would spoil during shipment.* Courtesy of Texas and Southwestern Cattle Raisers Foundation, Ft. Worth, Texas.

The first real breakthrough in food preservation came around 1800. Feeding troops in the field during warfare had always been a severe problem. Feeding an entire army of thousands of men in one location presented a difficulty no one knew how to solve. The problem was particularly bad if the army was a long way from home and their supply base. An even greater need was that of sailors who had to stay at sea for months at a time with no opportunity to get fresh food. In 1795, in an attempt to solve the problem, the French government offered a prize to anyone who could come up with a way to preserve food long enough to be useful in feeding the army. A Paris chef named Nicholas Appert developed a process of placing food in glass bottles and heating the bottles in a hot water bath (Figure 1–21). As soon as the food in the bottles was heated to a sufficient temperature, Appert sealed the tops with corks and wired them closed. Although

Figure 1–21 *The process developed by Nicholas Appert allowed food to be stored for long periods. This can be done at home.* Courtesy of James Strawser, The University of Georgia.

he never understood why the process worked, the heating killed the microorganisms, and the seal prevented others from entering. Food that had undergone this process stayed edible for months. The method was used widely by the navy, and in 1810, Appert received 12,000 francs from Napoleon Bonaparte. After the English invented a tin-coated can and used it for this process, *canning* came to be the term for this method of food preservation.

People had known since prehistoric times that cold preserved foods. Meat could be hung in a storage room when the weather stayed below freezing, and the meat would be preserved. The problem was that this temperature occurred only during part of the year. The first attempt at cooling meat and other produce involved the use of ice cut from frozen lakes during the winter and stored in ice houses. The ice blocks were suspended from the ceilings in the storage rooms in an effort to produce cool. This effort was not very successful. During the 1880s, mechanical refrigeration was developed and used in slaughterhouses to store meat.

A few years later the refrigerated boxcar was invented. This innovation allowed the transportation of meat and fresh produce anywhere in the country anytime during the year. Now, not only could animals be slaughtered anytime of the year, but the meat could be stored for a long period of time. Also, fruits and vegetables grown in the warmer areas could be transported long distances to cities in the North. This meant that fresh meat and produce could be distributed to everyone in the country (Figure 1–22). Later methods of freezing meats, fruits, and vegetables ensured that people could have ready access to the foods they needed and wanted.

Agricultural Mechanization

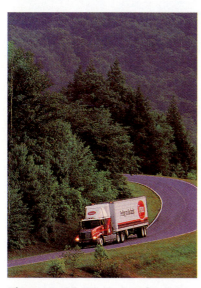

Figure 1–22 *Modern refrigerated trucks allow food to be transported all across the country.* Courtesy of Tyson Foods.

As mentioned earlier, as humans developed agriculture, they invented tools to make the job easier. These early tools of deer antler, stone, or wood helped, but the tasks were still time-consuming and laborious. Plows of sharpened wood were pulled through the ground by the people themselves. Later, the use of animals made the task a lot easier. Wooden plows worked on light soils but were ineffective on ground with a lot of clay content or ground covered with vegetation. The Greeks and Romans developed plows with iron points, and this helped to solve the problem. Later, during the Middle Ages, moldboards were developed that turned the soil to one side and allowed the turning of heavy clay soils.

In this country, John Deere developed a plow that would handle the heavy, sod-laden soils of the Midwest (Figure 1–23). Wooden plows would not turn the heavy sod, and iron plows weighed so much they could not be pulled through the soil. Deere invented a steel plow. Because steel is much stronger than iron, plowshares could be thinner than those made of iron. John Deere's plow was light and effective in slicing through the sod. This allowed the working of the highly fertile soil and opened up the area to become the breadbasket of the country.

Big advancements in mechanization came during the industrial revolution, when machines were invented that made other machines. Interchangeable parts enabled people to repair machinery without having parts custom-made by hand. If a machine broke down, the broken part could be replaced quickly, and the operation continued without long delays.

In 1831 Cyrus McCormick developed a machine for reaping wheat. The wheel-drawn machine cut wheat that previously had to be cut by hand (Figure 1–24). Later, a threshing machine was developed to separate the grain from the rest of the plant. This machine was run by steam power. People fed wheat shocks into the machine, and it threshed the wheat grain out. After the invention of the internal combustion engine, cutting wheat and threshing it was combined into one operation with a machine that came to be known as a combine.

Figure 1–23 *John Deere invented a plow that would cut through the heavy sod of the Midwest.*
Courtesy of John Deere & Company.

Figure 1–24 Cyrus McCormick's reaper increased the efficiency of harvesting grain. One machine could do the work of 12 people. *Courtesy of Case I-H.*

The development of the internal combustion engine had a revolutionary effect on agriculture. After its development, work that once took days using human or animal power could now be done in a matter of minutes and with much better results. Today almost all operations involved in the production of agricultural products are mechanized (Figure 1–25). This has come about through the use of scientific research in mechanization.

Figure 1–25 Modern harvesting machines are technological wonders. Their efficiency has had a tremendous effect on agriculture. *Courtesy of James Strawser, The University of Georgia.*

Figure 1–26 *Insects pests have always threatened our food supply.*
Courtesy of University of California.

Pesticides

Pesticides are substances used to kill pests. Pests may be insects, rodents, weeds, or other organisms that destroy or damage crops. Since people first began growing their own food, they have had to fight pests in order to have sufficient amounts left to eat (Figure 1–26). Insects can destroy an entire crop within a few days. Weeds can totally crowd out the desired plants. Until relatively recently, producers had to accept the fact that pests were going to get a high percentage of their crops.

The invention of modern chemical pesticides has been a tremendous boon to agriculture. If applied properly, they protect crops and livestock from pests and pose little threat to humans (Figure 1–27). Because of modern pesticides, people no longer see wormy apples or vegetables with insect damage.

Pesticides have a bad reputation for damaging the environment. In the past, these chemicals have been used to excess, but under modern regulations environmental damage is minimal.

Genetics

Many advancements in agriculture are due to research in **genetics**. Genetics is the study of how organisms pass on characteristics from one generation to the next. New and improved varieties of crops have been developed over the years that have doubled or tripled yields. Through the selection of superior parents, animal production has gained in the efficiency and quality of animals produced (Figure 1–28).

Figure 1–27 *Modern pesticides protect crops from insect damage. If properly applied, they are safe.*
Courtesy of James Strawser, The University of Georgia.

Figure 1–28 *Scientific research in genetics has brought about superior agricultural animals.*
Courtesy of Progressive Farmer.

The study of genetics started with an Austrian monk named Gregor Mendel, who lived from 1822 to 1884. Mendel developed the theory that organisms inherited their characteristics by what he termed "hereditary factors" (later called genes). His hypothesis was that an organism received factors from both parents and that each characteristic was inherited separately. He tested his hypothesis using garden peas. After his death, his theories were accepted as sound scientific research. Most modern research on genetics is based on Mendel's theories. Using these theories, innovations such as hybrid seed have been developed. Hybrids are crosses between unlike parents. The offspring usually outproduce the parents. Around 95 percent of all corn is produced from hybrid seeds (Figure 1–29).

Artificial Insemination

Superior animals are the product of superior parents. With the advent of artificial insemination in the 1930s, the transfer of genes from superior sires was greatly multiplied. Through modern techniques of semen collection, storage, and distribution, almost any producer can have access to the very best genes in the industry (Figure 1–30). This innovation is one of the reasons for the phenomenal advancements of the dairy industry. Most of the dairy animals born in this country result from artificial insemination.

Figure 1–29 *Hybrid seed developed through research has profoundly increased corn yields.*
Courtesy of James Strawser, The University of Georgia.

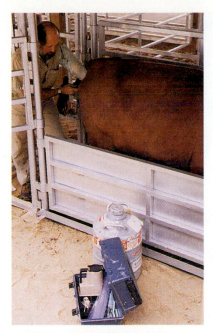

Figure 1–30 *Artificial insemination gives producers access to superior sires from all across the nation.* Courtesy of James Strawser, The University of Georgia.

Embryo Transfer

Although artificial insemination increased access to superior sires, improving herds through the use of superior dams was slow because of the gestation period of the female. With the development of the embryo transfer process, now one superior dam can produce many offspring each year (Figure 1–31). This process combined with artificial insemination allows producers to make an extremely rapid gain in the quality of their herds at a relatively low cost.

The Use of Computers

Computers were developed during the 1940s, but it was in the 1980s that the use of the computer became such an integral part of our lives. By the end of the 1990s, almost 84 percent of all American farms used computers. In agriculture, the use of computers has had a profound effect.

The computer has made all areas of research move more rapidly (Figure 1–32). Data that once took days and even weeks to analyze now can be computed in seconds. Computer-simulated experiments and models have helped to decrease the cost and time involved in scientific research.

Figure 1–31 *Embryo transplant has allowed superior females to produce multiple offspring. All these calves are from this cow's eggs.* Courtesy of Robert Newcomb, The University of Georgia.

Figure 1–32 *Computers have made all areas of research move more rapidly.* Courtesy of PhotoDisc.

The selection of superior dams and sires is made more convenient and accurate through computerized production records of progeny. Sires for artificial insemination can be matched with dams for embryo transfer through the use of computers. Many breed associations keep detailed records of all the animals in their registry on computer files. Much more rapid improvement of a herd can be made using computer sire and dam selection.

Feed formulation is now done by computer, including balancing nutrients and controlling the mixing of ingredients. This allows a more accurate blending of the nutrients.

Crop producers use computers to determine the optimum time for running irrigation systems. Computers are used to analyze soil tests and to balance the proper amount and type of fertilizer to increase yields and profits. Greenhouse operators rely on computers to control irrigation and lighting systems. Heat and humidity are also controlled using computers. Computers also give agriculturists ready access to the latest information on market conditions, weather forecasts, and an almost unlimited amount of information on methods and techniques.

Agricultural Cooperatives

Although they did not come about as a direct result of scientific research, agricultural **cooperatives** have had a major impact on the development of American agriculture.

Cooperatives go back to Colonial times when Benjamin Franklin organized the first cooperative for insurance. The principle is that many people can organize together to buy and sell goods and also to provide services. Basically, three types of agricultural cooperatives exist: marketing, supply, and service.

Marketing. Most farmers do not sell enough produce to sell directly to wholesalers and retailers, but if they organize and pool their products, enough volume can be generated to ensure efficient marketing. Modern cooperatives often process agricultural products. For example, they may shell pecans, dry soybeans, make juice concentrate from fruit, freeze vegetables, or complete many other tasks involved with processing agricultural products.

Supply. A basic principle of buying is that purchasing supplies in volume cuts the costs. If producers band together in large numbers through cooperatives, better prices can be obtained when buying supplies (Figure 1–33). Such needed inputs as seed, fertilizer, pesticides, and fencing can be bought in large volume and distributed through cooperative retail outlets. Most often, a cooperative is the most efficient way to purchase supplies.

Service. Producers need services such as fertilizer and pesticide applications, crop harvesting, and feed mixing. Cooperatives can purchase and operate machines and services that individual producers cannot afford on their own. Often the cooperative can supply services much cheaper than the producer can supply for him/herself.

Perhaps the best example of service cooperatives are the rural electric cooperatives. Agricultural cooperatives began to flourish in the 1920s when Congress passed the Capper-Volstead Act (1922) that allowed producers to band together to sell their products without being in violation of antitrust laws. Electrical conveniences were not available to the rural areas because of the cost involved in running lines to isolated areas. Rural electric cooperatives began to bring electricity to those isolated rural areas, and farms began to have the comforts and efficiency provided by electrical power. Today these cooperatives still operate more than half the electric distribution lines and provide electricity for more than 25 million people.

Figure 1–33 *Cooperatives can buy supplies in bulk and sell to producers at a lower cost.* Courtesy of USDA-ARS/Fred S. Witte.

SUMMARY

No branch of science touches our lives more than the science of agriculture. It has revolutionized every aspect of our lives. Other countries are fierce competitors in electronics, automotives, and manufacturing, but no one in the world even comes close to competing with American agriculture. Our research, combined with our free enterprise system, soils, and climate, have made us the envy of the world in agricultural production.

CHAPTER 1 REVIEW

Student Learning Activities

1. Write out what you consider to be an area of agriculture in which more knowledge is needed. Formulate a hypothesis and outline how you would conduct a scientific research study to gain knowledge in this area.

2. Interview someone involved in agriculture to determine what he or she considers to be the greatest advances in agriculture. Go to the library and find information about these developments.

Define the Following Terms

1. science
2. aquaculture
3. scientific method
4. basic research
5. applied research
6. genetics
7. cooperatives

True/False

1. The average American family spends 50 percent of their income on food.

2. The average American family spends a smaller percentage of their income for food than 40 years ago.

3. A hybrid is a cross between like parents.

4. Pesticides are substances used to kill pests.

5. The systematic study of anything begins with observation.

6. Government-run farms in other countries are very successful.

Fill in the Blank

1. The first science was _____.

2. _____, an Austrian monk, began the study of genetics, and his theories became known as _____ theories.

3. _____ developed the first reaping machine.

4. _____ developed the process of preserving food now called canning.

5. _____ is used to immunize animals to disease. _____ developed a means of vaccinating animals.

6. An average American family of four consumes about _____ tons of food in one year.

7. The _____ was established to make research information available to people.

8. The _____ established vocational agriculture programs in public high schools.

Discussion

1. When and where did agriculture begin?

2. What was the purpose of establishing experiment stations at the land grant universities?

3. Name the four steps involved in the scientific method.

4. Explain why advances in agriculture have helped society to grow.

5. Explain why scientific research is important in agriculture.

6. Explain the difference between basic and applied research.

7. Describe some of the advancements in agriculture since 1925 as compiled by the Council for Agricultural Science and Technology (CAST).

8. Why are interchangeable machine parts important in agriculture?

9. List some of the ways the computer has helped agriculture.

10. Why is the preservation of food so important to agriculture?

SOIL: THE SOURCE OF LIFE

STUDENT OBJECTIVES

After studying this chapter, you should be able to:

✦ Explain why agriculture and all of life depends on the soil.

✦ Explain the difference between organic and inorganic materials as it relates to soil.

✦ Discuss the carbon cycle.

✦ Discuss the different ways soils are formed.

✦ Discuss the role of ions in soil fertility.

✦ Define the difference between acidity and alkalinity and the effect of each on the soil.

✦ Analyze how soils are scientifically classified.

✦ Explain the concept of an ecosystem.

✦ Describe the balance of organisms in the soil.

✦ Explain how physical properties affect soil fertility.

✦ Name and list the characteristics of the soil classes.

Soil is defined by the Soil Conservation Service of the United States Department of Agriculture (USDA) as "the collection of natural bodies on the earth's surface, in places modified or even made by humans of earthy materials, containing living matter and supporting or capable of supporting plants out-of-doors." In other words, soil is the material covering the face of the earth that supports the growth of plants, including soil built up artificially by people who modify it or move it.

The most precious natural resource we have is the soil. Although it is only a thin layer of material that covers the earth, all life on our planet depends on it. If you can imagine the earth as an apple, the soil would be the peel. Although the thickness of the soil on the earth is not uniform, like the peel of an apple, the soil is a very thin layer compared to the thickness of the earth. It is a fragile natural resource that needs a lot of care. Without the soil, few plants could survive. Without plants, animal life on this planet would be nonexistent because the **food chain** of land animals begins with the soil (Figure 2–1).

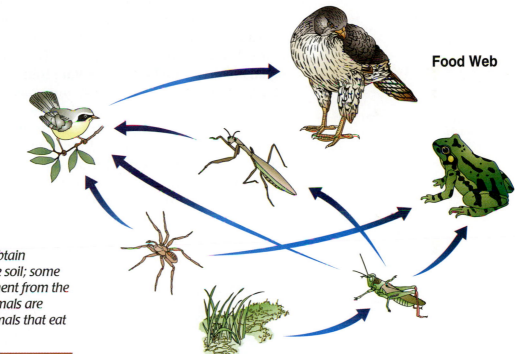

Food Web

Figure 2–1 *Plants obtain nourishment from the soil; some animals get nourishment from the plants, and other animals are nourished by the animals that eat the plants.*

The food chain is the progression of food energy from one species to the next. For example, grass grows from the soil, taking in nutrients and water and using light from the sun to convert these nutrients into plant material. A grasshopper then may eat the grass and convert it into animal tissues. A praying mantis eats the grasshopper and is then eaten by a hawk; and the energy is transferred to the hawk. Eventually the hawk dies; the body of the hawk is broken down by microorganisms; and nutrients are returned to the soil.

Figure 2–2 *All of the protein humans receive comes from plants either directly or indirectly. Courtesy of Progressive Farmer.*

Most of the food we eat is dependent on the soil. All of our fruits, vegetables, and grains originate from the soil. All of the protein we get from animals grown on the land depends on the soil, because most agricultural animals eat grass and grain that come from the soil (Figure 2–2). All of the systems and organs of an animal's body contain minerals such as calcium, phosphorus, and others. These minerals come from the soil either through direct contact with the soil or through plants that have extracted these minerals from the soil.

The soil is made up of four components: minerals, air, water, and organic matter (humus) (Figure 2–3). Minerals are the substances in the soil that are inorganic (derived from nonliving sources) that have chemical and physical properties

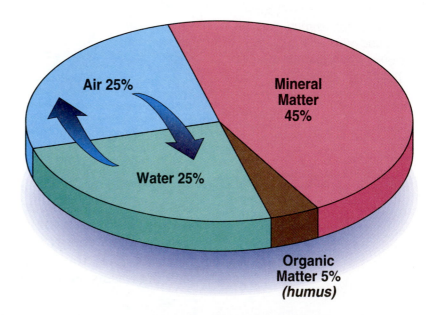

Figure 2–3 *Soil is composed of minerals, water, air, and organic matter.*

that go together to form rock. Air and water fill the gaps left in the soil by the grouping together of the mineral particles. These gaps or spaces are called pores. Humus is the organic material in the soil that comes from living sources. Carbon is the common element in organic substances and identifies a material as being organic. All organic material comes from living sources and contains carbon. All inorganic material comes from nonliving sources and does not contain carbon. The organic material in soil can be either living or nonliving.

SOIL ORIGINS

All soil comes from matter known as parent material. Parent material can be either organic or inorganic, and most soils are a combination of the two. One way of classifying soil is to term the soil organic or inorganic. **Organic soils** were derived from parent material that was at one time living. **Inorganic soils** were made up from parent materials that were of mineral or inorganic origin.

Organic Soils

In some places, the grass and other vegetation has grown and died for thousands of years. As the plants died, they fell to the ground and began to decay. Century after century, the layers of decayed plant materials accumulated and formed soil that might be several feet thick. Because these soils derive almost entirely from plant material, they are called organic soils. These soils are black and productive but not very abundant (Figure 2–4). In the United States, the largest deposits of these soils are in southern Florida in land that has been reclaimed from the Everglades (Figure 2–5). For thousands and perhaps millions of years, the grass growing there has been dying and decaying into soil. When the water was drained off, all of these very fertile deposits resulted in farm land that is tremendously productive. Other parts of the North and Pacific Northwest have deposits of organic soil that are the result of reclaiming peat bogs.

Inorganic Soils

Soils that originated from minerals are the result of the decomposition of rock materials. This process is the result of the wearing away of rock by actions of the weather. As wind

Figure 2–4 *Organic soils are black in color and are very productive. Courtesy of USDA/Bruce Fritz.*

Areas with Organic Soils

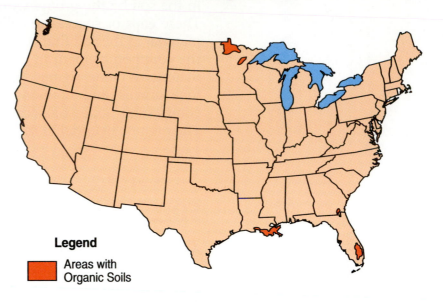

Legend

▮ Areas with
Organic Soils

Figure 2–5 *The largest deposits or organic soils are found in southern Florida.* Courtesy of USDA Natural Resources Conservation Service.

blows against a rock, particles are loosened and removed. Likewise, rain, sleet, and snow remove particles that become soil. The process of wetting and drying as well as the dissolving of minerals by the water can also decompose stone. The freezing and thawing of water in cracks and crevices of rocks can wear away tiny particles and can open other cracks for the accumulation of more water. The weathering of rocks is a slow process that occurs over thousands and even millions of years.

Thousands of years ago, great rivers of slowly moving ice called glaciers extended across the northern portions of America, Europe, and Asia. As these huge masses of ice crept southward, they carried along rocks that were in their paths. The rocks were crushed together with enough force that some were pulverized and made into particles small enough to become soils. With the change of the seasons, part of the ice would melt and freeze again. During particularly mild periods, the front of the ice mass would melt and drop all the debris it had picked up along the way. When the weather again turned cold, the ice resumed its forward movement and picked up other rocks and soil particles. This mass of boulders, rocks, sand, silt, and clay is called **till**. The landmass built up by a glacier dropping till is called a moraine. The type of debris picked up by the glacier determines the type of soil deposited. Moraines vary in the type and texture of the soil. Some contain a lot of sand and are coarser while some contain finer silt.

Water-Deposited Soil

As mentioned earlier, moving water wears away rock and forms soil particles. Soil that has been transported and deposited by moving water is called **alluvial soil**. Some of the world's most productive soils are the result of alluvial action. Water is a major way that soil is moved from one place and deposited in another. Water that moves in a stream picks up soil from the ground it flows over, so when rains wash soil particles into the stream, the currents carry the material down the stream (Figure 2–6). As long as the water is moving rapidly, the soil particles remain suspended in the water. As soon as the stream bed levels out and the water slows down, the

Figure 2–6 Soil becomes suspended in water and is moved along down the stream. *Courtesy of USDA-ARS.*

soil particles settle to the bottom. If the stream flows out of its bank, deposits are left in the area covered by the water. These areas are called **flood plains**. Also, as the stream nears its mouth, the water slows down and the deposits settle to the bottom. These areas are called **deltas**. Gradually over the years, the soil builds to a great depth. The ancient Egyptians fed a large population from the soil of the Nile River Valley. Every year the river would flood and bring not only water but fresh deposits of soil. Egyptian farmers timed planting, cultivating, and harvesting to coincide with the flooding of the Nile (Figure 2–7). They developed accurate calendars that predicted when the flooding would occur and made preparations for working the soil at the proper time.

In this country, the Mississippi River has carried deposits south for thousands of years and deposited them. The result has been the rich farmland of the Mississippi Delta, which has produced cotton and other valuable crops. It should be noted, however, that the soils of the delta were once in the Midwest,

Figure 2–7 *The flood plain of the Nile River has been cultivated for thousands of years.* Courtesy of PhotoDisc, Hisham F. Ibrahim.

and the soils in that area are less productive because of the erosion.

Soil may also be deposited into a lake. As a stream empties into a lake, the soil deposits are dropped, and they sink to the bottom. Over thousands of years, the lake becomes filled with soil deposits and no longer holds water. When the lake disappears a deep, fertile soil is left in place of the water. These soils are called **lacustrine deposits**. Areas around the Great Lakes, parts of North Dakota, Minnesota, Nevada, and the Pacific Northwest have large areas with lacustrine soils.

Likewise, oceans may become filled with deposits moved in by streams that feed them. The shoreline recedes and leaves the deposits as dry land. These soils are known as **marine sediments**. Areas of marine sediment in this country are the coastal areas of the Gulf of Mexico, parts of California, and parts of the Great Plains. Obviously, beach deposits are the result of marine sediments, but large areas of the United States that were once covered by the ocean are made up of land developed by marine sediments.

Soil Deposited by Wind

The wind also creates and transports soil. In many areas of the world tall outcroppings of rock have been carved by the wind. As the wind ate away the stone, the particles worn loose formed deposits known as **aeolian soils** (Figure 2–8). If the particles are large, the soil is called sand and the deposits are called dunes. In Saudi Arabia and many other places, vast areas are covered in sand dunes that continually move and

Figure 2–8 *As the wind carves stone, soil is formed.* Courtesy of PhotoDisc, Bruce Heinemann.

change shape. Sand dunes may be found in arid areas of the western part of the United States. Because the sand is constantly moving, little vegetation can grow on the dunes (Figure 2–9).

Fine soil particles like silt and clay deposited by the wind create **loess soils**. These soils may be found through the Mississippi Valley, parts of the Midwest, Washington, and Idaho. Much of these soils were formed when the glaciers

Figure 2–9 *Constantly moving sand dunes are difficult places for plants to grow.* Courtesy of PhotoDisc, Bruce Heinemann.

Figure 2–10 *Volcanic ash creates some of the world's most productive soil. Courtesy of PhotoDisc, John Wang.*

melted and climatic conditions turned dry. When the soil dried out, the fine particles were blown by the wind and accumulated in new areas.

In many of the mountainous regions of the world, soil was (and still is) formed by the action of volcanoes. Volcanoes bring deposits from deep within the earth and spew the material on the surface. Much of this material is in the form of molten rock (called lava) that flows out over the surface and cools into solid rock. Other deposits are spewed high into the air. The finest of these particles are called volcanic ash and may be carried many miles by the wind. Near the volcano, many feet of volcanic ash may be built up in a matter of days. Areas as far away as 100 miles may receive measurable deposits of ash. Soils from volcanic ash are generally very productive soils (Figure 2–10). Volcanic soils can be found in Hawaii, Oregon, and Washington.

PHYSICAL PROPERTIES OF SOILS

The physical properties or characteristics of the soil determine to a large degree how usable and productive the soil will be. How quickly water penetrates, how well water stays in the soil, how well the soil holds up under machinery, the ease of root penetration, and the aeration of the soil are all greatly influenced by the physical characteristics of the soil. Texture, structure, and consistency are all physical characteristics of the soil.

Soil Texture

The **texture** of the soil refers to the size of the individual soil particles, called soil separates. The larger the size of the soil separates, the coarser the soil feels when rubbed between the fingers. The largest of the soil particles is sand. Sand particles are soil separates that are between 2 mm and .05 mm in diameter. Silt particles are smaller than sand and have diameters between .05 mm and .002 mm. The smallest of the particles are less than .002 mm in diameter and are referred to as clay. One way to illustrate the relative sizes is to imagine that if a clay particle was the size of a BB, a silt particle would be the size of a golf ball, and a sand particle would be the size of a basketball (Figure 2–11).

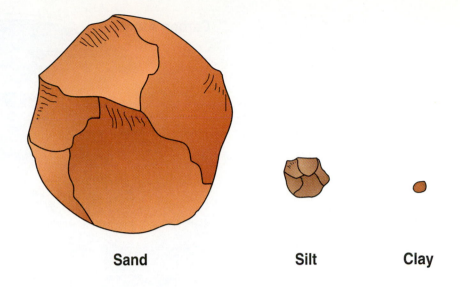

Sand   **Silt**   **Clay**

Figure 2–11 *If a particle of sand were as large as a basketball, a particle of silt would be the size of a golf ball, and a particle of clay would be the size of a BB.*

Soils are seldom composed of pure sand, pure silt, or pure clay, but rather a combination of the three. Soil that contains less than 52 percent sand, 28 to 50 percent silt, and 7 to 27 percent clay is called loam soil and is nearly ideal for growing most crops. Soils that have a large concentration of sand are coarse and do not hold water very well. Because the particles are so large, the water from a rain or irrigation system passes through the soil and little is retained for use by plants. Clays and silts slow the water down as it is absorbed into the ground and hold a portion of it that the plants can use. If the soil has a large percentage of clay, the ground will not let enough water through and may either cause the water to run off or may hold the water too long and cause problems for the plant.

Clay also plays an important role in soil fertility. The productivity of a soil depends to a large degree on the amount of essential nutrients that are contained in the soil and that are available for the plants to use. Clay particles usually have what is called a net negative charge. This is brought about by the chemical structure of the molecules that make up the clay particles. Molecules are composed of atoms that are chemically bound together. Atoms are composed of a nucleus consisting of particles known as neutrons, protons, and electrons that orbit around the nucleus (Figure 2–12). The neutrons in the nucleus are neutral—that is, they have no electrical charge. The protons in the nucleus have a positive charge, and the electrons have a negative charge.

Ordinarily, an atom has the same number of electrons as it does protons. Differences in elements are mostly due to the

Carbon Atom

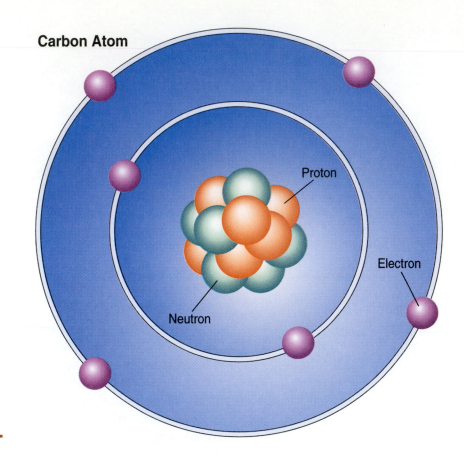

Figure 2–12 An atom is composed of negatively charged electrons, positively charged protons, and neutral neutrons.

number of electrons and protons that the atoms contain. Electrons sometimes move from one atom to another. If this happens the atom losing the electron becomes positively (+) charged, and the atom gaining the electron becomes negatively (–) charged. These atoms are called ions. When the electrons are exchanged, the two atoms are attracted to each other and form a molecule. The molecule may have a positive or negative charge depending on the nature of the electron exchange.

The chemical makeup of clay usually results in the particles having a negative charge. The chemical makeup of many essential plant nutrients results in a positive charge. For example, calcium (Ca++), magnesium (Mg++), and potassium (K++) all have a positive charge. The differences in the charge of the nutrients and the clay molecules result in the attraction of the nutrients to the clay particles in much the same way that opposite ends of magnets attract each other. The positively charged nutrients are referred to as **cations**. They are held until they are released in a process known as **cation exchange**. When water surrounds the soil particles, this is known as the soil solution, and nutrients are suspended or dissolved within the liquid. In a process called osmosis, the cations from the soil particles are pulled into the soil solution. It is from here that the plant roots take in the nutrients through their root systems. Soils are sometimes measured by the amount of cations they are able to exchange. This is known as the cation exchange capacity of the soil.

The coarser the soil (the higher the sand content), the fewer soil nutrients that will be in the soil and available to the plants. Coarse, sandy soils have a lower cation exchange rate because they are not negatively charged as are the finer clay particles. Also, most plant nutrients are water soluble and move out of the soil if the soil particles are too large. The loss of soil nutrients by water movement is called **leaching**. Coarse or sandy soils are less productive because they do not hold either water or nutrients as well as the finer soils.

In addition to having a higher cation exchange capacity, finer textured silts and clays slow down the water, and the dissolved nutrients are there longer for the plants. Nutrients adhere to clay because the fine texture of the clay has more surface area to which nutrients can cling.

Soil pH

The **pH** of the soil has to do with how acid the soil is. An acid is a substance containing hydrogen that forms hydrogen ions when dissolved in water. A hydrogen atom has one proton and one neutron. During the ionization process, the electron from one hydrogen atom may be lost to another hydrogen atom leaving only a single proton. This is known as the hydrogen ion. The atom with two electrons is known as the hydroxide ion. The pH scale is a measure of the relative strength of the hydrogen ion or hydroxide ion activity in a substance (Figure 2–13). A substance that is neutral has as many hydrogen as hydroxide ions and ranks a 7.0 on the scale. Substances with more hydrogen ions than hydroxide ions are said to be acid and score lower than 7 on the pH scale. The higher the relative concentration of hydrogen ions (the more acid), the lower the number. The higher the relative concentration of hydroxide ions (the more alkaline), the higher the number.

Certain crops may have problems growing in soils that are too acid or too alkaline. Producers frequently adjust their soils to the right pH by having soil samples tested and applying the correct amount of lime to acid soils or sulfur-based

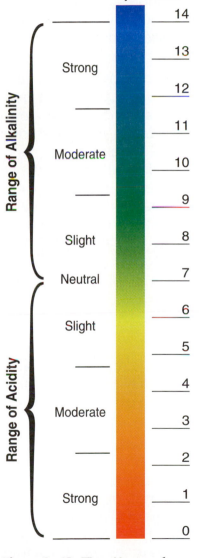

Figure 2–13 The pH range for soils generally runs from 4 to 10 on the pH scale.

B I O   B R I E F

Erosion Control with Power Plant Byproducts

In an effort to improve air quality, coal-fired power plants are required to have scrubbers installed in their systems to remove sulfur dioxide from the flue gases. This practice has resulted in more than 100 million tons of gypsiferous material, high in calcium and sulfur, being produced yearly as a byproduct. Scientists with the Agricultural Research Service (ARS) have discovered a way to use the gypsum byproducts to help control soil erosion. Soil scientist Darrell Norton, with the USDA-ARS National Soil Erosion Research Lab, believes that recycling these low-cost byproducts of industry to control erosion and increase yield is a win-win situation.

More than two hundred years ago, Benjamin Franklin demonstrated the advantages of using gypsum as a soil amendment, but the plentiful supply, currently estimated at enough to supply a ton per acre to a quarter of U.S. farmland, makes it even more appealing. Norton says, "Using gypsiferous byproducts would give farmers a low-cost remedy for acid, sodic, and erosion-prone soils."

Studies conducted in West Virginia and Maryland have shown that calcium from gypsum applied to acidic soils gets down into the subsoil where it is needed, so crop roots can grow deeper and access more water. In further studies, NSERL found gypsum byproducts releases electrolytes that keep clay particles clumped together, thereby reducing crusting, which improves water entry and reduces erosion. Tests of a byproduct of almost pure gypsum from a special coal-burning technique in power

---

materials to alkaline soils (Figure 2–14). Crops such as blueberries grow well in soils that are relatively acid (pH of 4.0 to 5.0), while crops such as asparagus do well in soils that are less acid (pH of 6.0 to 8.0).

Soil Structure

Freezing and thawing, the movement of roots, earthworms and other life forms, and variations of moisture content cause soil particles to cling together. These groups of clinging soil particles are called **peds** or soil aggregates. Peds are held together by the clay and humus (organic matter) in the soil. The rate at which water flows through the soil is greatly influenced by the structure of the soil. Peds are categorized by type (shape), class (size), and grade (strength).

plants increased water filtration and reduced soil loss by about one-fourth. Gypsum byproducts from fertilizer manufacturing did almost as well in reducing soil erosion between rows.

Further tests are showing that not only will gypsum fight soil erosion, but it will increase crop yields. Cooperating farmers report encouraging results. In one of more than 50,000 acres of field tests, Ken Curtis of Prairie City, Illinois, used high-purity gypsum, a scrubber byproduct from a coal-fired unit of City Water, Light, and Power of Springfield, Illinois. He applied three tons of gypsum per acre to a 20-acre field of no-till soybeans, randomly applying various amounts. The treated soybeans yielded four more bushels per acre than nongypsum treated soybean plots.

Source: Agricultural Research Service

To evaluate the effects of gypsum byproducts on soil erosion, scientists collect runoff water to measure sediment.

Figure 2–14 *A sample of the soil down to about 6 inches is taken for testing pH.* Courtesy of James Strawser, The University of Georgia.

Kinds of Soil Structure

Figure 2–15 *Groups of clinging soil particles are called peds.*

Types of peds are spherical (round), plate-like (flat and thin), block-like (cubic), and prism-like (long, with several sides). These four basic shapes are then divided into seven common shapes as shown in Figure 2–15.

Classes or sizes of soil structure are very fine, fine, medium, coarse, and very coarse.

The grade indicates the strength of the ped or how stable the structure is. Some soils have no real structure and are known as structureless soils. These soils are either single grains, such as sand in a dune, or massive, such as clays that stick together with no distinguishable peds. Structured soils have three grades.

Weak structure–The peds are hard to distinguish and only a few can be separated in moist soil.

Moderate structure–The peds are visible and can be handled without breaking up.

Strong structure–Most of the soil is formed into peds and can be handled without breaking up.

Figure 2–16 *The soil horizon can be examined by digging a pit and looking at the side of the hole.* Courtesy of James Strawser, The University of Georgia.

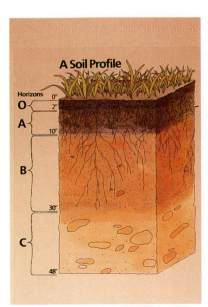

Figure 2–17 *The different soil horizons.* Courtesy of USDA Natural Resources Conservation Service.

SOIL HORIZONS

Soil from rock or from organic material forms in layers that lie parallel to the surface of the Earth. As layers are added they begin to take on different characteristics. These are called **soil horizons**. If a pit is dug into the ground, the side of the hole gives a picture of the different layers or horizons of soil (Figures 2–16 and 2–17). This picture is called the soil profile.

The uppermost layer is called the O horizon. This layer contains undecomposed and decomposed organic matter.

The next layer is the A horizon, composed of topsoil that contains organic matter along with minerals. It is generally darker than the horizons below and may be from a few inches to several feet thick.

The region under the A horizon is the E horizon, the area of the topsoil that has had all of the organic material leached out. Leaching is the removal of materials by dissolving in and being carried out by water. This horizon is light in color because the clays and humus have been leached out.

The next region, the B horizon or subsoil, is the region where materials leached from the other horizons accumulate and form. This area is generally high in clay content. The B horizon combined with the O, A, and E horizons make up a region of the soil known as the solum, which means true soil. Most of the roots of plants grow in the solum.

Underneath the B horizon is the C horizon, made up of parent material. Parent material is the material from which the soil originated. This horizon has not been affected by the soil-making process and does not have the properties of the soil in the other horizons.

The last horizon in the soil profile is the R horizon, the bedrock upon which the other soil horizons rest.

SOIL TAXONOMY

The word taxonomy is derived from the word taxis which means to arrange or organize. Plants and animals are arranged and classified beginning with kingdom and ending with species. Likewise, soils are arranged to group those that are alike together. The broadest grouping is called soil orders. There are 11 soil orders that are broken down into 47 subgroups. The subgroups are divided into 185 great groups, and these are divided into 970 subgroups. Subgroups are divided into 4,500 families that are then subdivided into 12,620 soil series. Many different kinds of soils are in the world, and this scheme of classification is needed in order for scientists to communicate about the different types (Figure 2–18).

Orders of Soils

Entisols. These soils were relatively recently created and have no real horizon development. Soils from recent floods or recent accumulations of volcanic ash would be classified as entisols.

Inceptisols. These soils have a weak B horizon and are relatively young. They are generally moist.

Andisols. These soils are black mineral soils containing glass. They derive from volcanic sources.

Histisols. These soils are organic and result from many years of plant decomposition. They are known as peat and mucks. These soils can be found in southern Florida and in the peat bogs of the North and Pacific Northwest.

Aridisols. These soils are found in arid desert areas. They often have high alkaline content with salt in the horizons. The deserts of the southwestern United States have these soils.

Figure 2–18 *Examples of the soil orders. (A) Entisols, (B) Aridisols, (C) Inceptisols, (D) Alfisols.* Courtesy of USDA Natural Resources Conservation Service.

Mollisols. These soils are from grasslands or hardwood forests. These soils are dark in color and are deep in the A horizon. This order of soil is present in much of the Great Plains.

Vertisols. These soils are high in clay content and have deep cracks when the soil dries. Vertisols are slightly to moderately acid and are found in the southern part of the United States.

Alfisols. These soils are found beneath forest growths. The climate where these soils are found is usually cool and moist.

Spodosols. These soils are coarse and high in sand content. They are generally found beneath coniferous forests. These soils are found in the northeastern United States.

Ultisols. These soils are highly acid and are found in warm tropical or subtropical areas. They can be found in the southeastern United States.

Oxisols. These soils are in tropical and subtropical regions and are highly weathered. The only state in this country with oxisols is Hawaii.

THE SOIL ECOSYSTEM

Soil might look lifeless, but actually it is abounding with life. In fact, the soil is an **ecosystem** (Figure 2–19). An ecosystem is all of the plant and animal life that lives in an area. The life-forms depend on each other for the proper balance of food and other environmental factors. Within this ecosystem are many different forms of plant, animal, and microbial life. Most of these forms of life depend on each other for their existence or for a maintenance of the proper balance in nature.

Plant Life

Most of the plants on earth depend on the soil for their existence. The soil supports their root systems and supplies nutrients. The roots of many plants reach as far as several feet into the soil (Figure 2–20). In fact, such common crops as alfalfa have been known to grow 25 feet or more into the ground. The area of the soil that contains the roots of plants is called the rhizosphere. In this zone of the soil, the plant receives water and the nutrients it needs to live and grow.

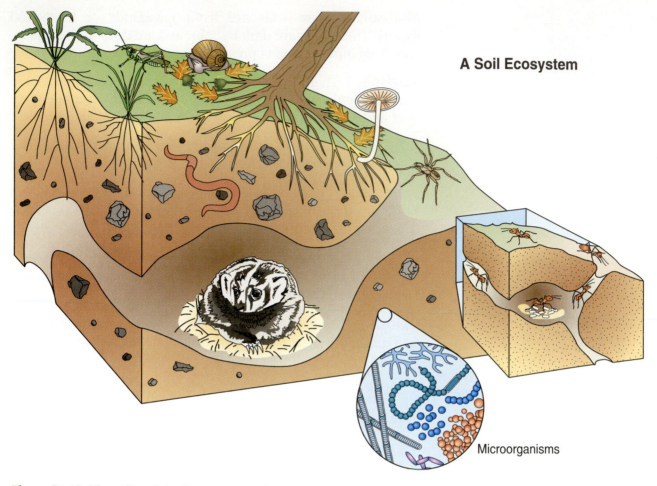

A Soil Ecosystem

Microorganisms

Figure 2–19 *The soil contains its own ecosystem.*

Microorganisms

Within the rhizosphere live billions of microorganisms of different types. Many of these organisms live off the roots of plants. As plants live and grow, they are constantly oozing materials through the roots. This material contains protein and other nutrients that the microorganisms feed on. In addition, as root cells mature and die, the microorganisms decompose them. When the entire plant dies, decomposition returns the nutrients the plant took from the soil back to the soil. This is known as the carbon cycle (Figure 2–21).

The most common soil microorganisms are bacteria. They are so abundant that in only one teaspoon of soil there can be as many as 500 million bacteria. Several types of these bacteria live in what is known as a symbiotic relationship with

Figure 2–20 *Plant roots can penetrate several feet into the soil.* Courtesy of Keith Karnack, The University of Georgia.

plants. This means that organisms of different types live together for mutual benefit. The most common type of symbiotic bacteria are the nitrogen-fixing bacteria, rhizobia, that live in the roots of certain plants (Figure 2-22). Plants that host these organisms are called legumes and include such plants as beans, clovers, peanuts, alfalfa, and peas. Rhizobia live in lumps on the roots called nodules (Figure 2–23). The bacteria live in the soil and attach themselves to the root hairs of the legume shortly after the plant sprouts. As a reaction to the bacteria attaching to the root, the nodules form. The bacteria receive all the nutrients they need to live and reproduce from the host plant. In return the bacteria convert nitrogen from the air in the soil into a form of nitrogen that the plant can use.

Many different types or strains of rhizobia exist, and those that live in one type of legume will not necessarily live in another. At one time, if a field had never grown a type of

Figure 2–21 *Through the carbon cycle, living matter decays and helps put nutrients back into the soil.*

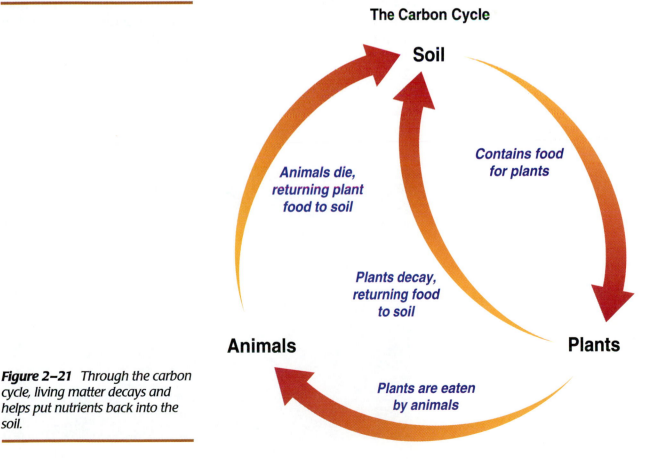

The Carbon Cycle

Soil

Contains food for plants

Animals die, returning plant food to soil

Plants decay, returning food to soil

Animals

Plants

Plants are eaten by animals

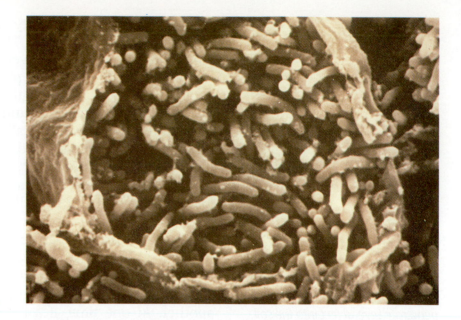

Figure 2–22 *Rhizobia are bacteria that fix nitrogen in the soil.* Courtesy of Rob Maier, Johns Hopkins University.

legume before, soil from a field that the legumes grew in was transferred to the new field. This introduced the rhizobia bacteria to the soil in the new field. Today bacteria are raised in a laboratory and are sold commercially to coat the legume seed in order to inoculate the soil with the bacteria (Figure 2–24). In fact, scientists select and reproduce superior types of rhizobia by using those found in the most rapidly growing, darkest green (nitrogen causes plants to be dark green) plants they can find!

Figure 2–23 *Rhizobia live in nodules on the roots of leguminous plants.* Courtesy of USDA-ARS.

Figure 2–24 *Seeds of legumes such as soybeans (A) may be inoculated with rhizobia bacteria before they are planted. Other seeds such as corn (B) are not affected by innoculation.* Courtesy of PhotoDisc, Siede Preis.

Fungi are plant-like organisms that contain no chlorophyll. They range in size from microscopic to the large mushroom fungi that grow on the surface of the soil or on decaying **plant material**. These organisms are as abundant as bacteria in the soil and play an important role in the breakdown and decay of plant materials. Fungi are particularly important in forest soils because they break down lignin, a primary component of wood (Figure 2–25). Trees that die and fall to the ground are returned to the soil through the action of fungi, bacteria, and other organisms.

Protozoa are one-celled organisms that live in moist soil (Figure 2–26). They are aquatic organisms because they live in particles of water in the soil. When the soil becomes dry, the protozoa change into an inactive state until the soil becomes moist again. Protozoa feed on bacteria in the soil. Through this process a balance of bacterial life in the soil is better maintained.

One of the most important groups of microscopic animals in the soil is nematodes. Nematodes are worms in the class Nematoda that have smooth round bodies that are not segmented (Figure 2–27). They move about by means of the high pressure of the fluid in their body cavity that acts to move the body parts as a result of fluid pressure. Although many nematodes are not microscopic, most of those that live in the soil are so minute they cannot be seen with the naked eye. These

Figure 2–25 *Soil fungi break down lignin and helps wood to decay and return to the soil.* Courtesy of Peter Hartel, The University of Georgia.

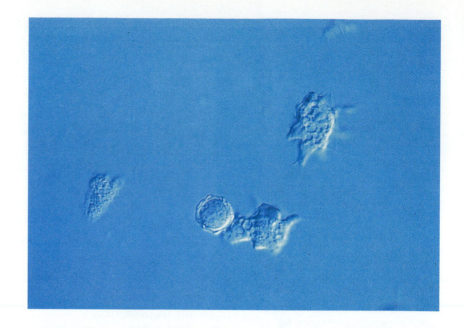

Figure 2–26 *Protozoa are one-celled organisms that live in the soil.* Courtesy of J. F. McClellan, Colorado State University.

tiny worms are very abundant. In a typical spade full of moist soil, there may be more than a million nematodes. In fact, they are the most abundant multicelled animals in the soil. Soil nematodes fall into three basic groups: those that consume decaying organic matter, those that eat other microorganisms, and those that are plant parasites. Of the three groups, those that are parasitic to plants are by far the most important because of the damage they do. Nematodes have a needle-like projection on the front end of their bodies that is

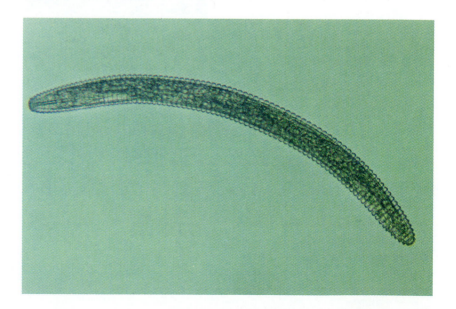

Figure 2–27 *Nematodes are tiny worms that live in the soil. Many feed on the roots of plants.* Courtesy of Ed Brown, The University of Georgia.

used to open a tiny hole in the cells of a plant root. Once the hole is opened, the contents of the cell are sucked out, and the plant is damaged. Not only is the plant damaged by the destruction of the cells, but also the opening lets in disease organisms that may attack the plant. The plant creates a gall or knot on the root where the damage occurred, so these nematodes are known as root-knot nematodes. Certain crops are very susceptible to nematodes, and producers use pesticides to help control them.

MACROORGANISMS IN THE SOIL

Macroorganisms are organisms that are visible to the unaided eye. In other words, you do not need a magnifying glass or microscope to see them. Many of these organisms live in the soil, and they all play a part in the soil ecosystem.

Earthworms are among the most abundant animals that live in the soil. In an acre of good land, as many as 50,000 to 100,000 may live in the topsoil. These worms enhance the soil by burrowing through the layers of soil. These burrows allow the passage of air and water deep into the soil. Some types of earthworms drag organic material into the soil where it adds to the richness of the soil. The worms feed on organic material that passes through their system and is deposited on the surface of the soil in droppings called castings. Earthworms deposit several tons of castings per acre each year and may add several inches of topsoil over a period of 10 to 20 years. Many agriculturists have added earthworms to their garden plots to enhance the ability of the soil to produce plants.

Many species of insects live in the soil. Some live their entire lives in the soil while others live only a portion of their lives there. During the warm months, a spadeful of moist soil usually yields several "grub worms" (Figure 2–28). These are the larval stage of beetles that have laid their eggs in the soil. A good example is the June bug (actually a beetle). These white grubs are about an inch long and are about as thick as a finger. Most grubs feed on plant roots, and some species, such as the wireworm, can be very destructive to the crops growing in the soil.

Larger animals also live in the soil. Prairie dogs, moles, shrews, armadillos, groundhogs, and chipmunks dig in the ground to make nests or to find food. In doing so, they open

Figure 2–28 *Grub worms live in the soil and can be very destructive by feeding on plant roots. Courtesy of Dekalb Genetics Corporation.*

passages for air and water to get into the soil and also turn the soil over so that organic matter from the surface is carried to the lower portions of the soil. Some of these animals may be destructive to plants growing in the soil (Figure 2–29). A large

Figure 2–29 *Animals also burrow into the ground and may damage living plants. These ridges in the ground are caused by mole crickets.* Courtesy of James Strawser, The University of Georgia.

burrowing groundhog (woodchuck) can cause considerable damage to roots and leave holes in the ground that cause problems for machinery. Others, such as shrews, are beneficial because of the large number of insects they consume.

The soil is full of living organisms that depend on each other in an ecological system. The more these organisms are in balance, the more productive the soil. Soil management uses principles derived from scientific research to keep the ecosystem in balance.

SUMMARY

Soil is the basis of all life. This thin, fragile layer that surrounds the Earth provides humans with food. Our huge agricultural industry depends on our fertile soils, and scientists are constantly seeking ways to protect the soil and improve its productivity.

CHAPTER 2 REVIEW

Student Learning Activities

1. Collect several soil samples from different areas around your school. Determine the physical characteristics of each sample and try to determine the origin of the soil.

2. On a warm day, dig down about a foot deep in a moist area and collect about 1 gallon of soil. Carefully go through the soil and with the aid of a large magnifying glass, find all the organisms you can that are living in the soil. Determine the role of each in the soil's ecosystem.

Define the Following Terms

1. soil
2. food chain
2. organic soils
4. inorganic soils
5. till
6. alluvial soil
7. flood plains
8. deltas
9. lacustrine deposits
10. marine sediments
11. aeolian soils
12. loess soils
13. texture
14. cation
15. cation exchange
16. leaching
17. pH
18. peds
19. soil horizons
20. ecosystem
21. plant material

True/False

1. The most precious natural resource we have is the soil.
2. The food chain of land animals begins with the soil.
3. Humus is the organic material in the soil that comes from living sources.
4. The landmass built up by the dropping of till by a glacier is called alluvial.
5. Soil particles settle to the bottom of rivers and streams when the water is moving rapidly.
6. Physical properties or characteristics of the soil determine to a large degree how usable and productive the soil will be.
7. Silt particles are smaller than .002 mm in diameter.
8. Protons and neutrons of an atom both have a positive electrical charge.
9. A soil that is considered neutral will score lower than 7 on the pH scale.
10. The layer of soil that contains the most organic material is the R horizon.
11. Andisols are soils derived from volcanic sources.

12. Most of the plants on earth depend on the soil for their existence.

13. The most common soil organisms are the nematodes.

14. Microorganisms are organisms that are visible to the human eye.

Fill in the Blank

1. The _____ is the progression of food energy from one species of organism to the next.

2. Minerals are the substances in the soil that are _____.

3. _____ is a common element in organic substances and is used to identify a material as being organic.

4. _____ are derived from parent material that was at one time living.

5. Soils that have a large concentration of _____ are coarse and do not hold water very well.

6. The chemical makeup of clay usually results in the particles having a _____.

7. The _____ of the soil has to do with how acid the soil is.

8. Soil forms in layers called _____.

9. The broadest soil grouping is called _____.

10. An _____ is all of the plant and animal life that lives in an area.

11. Within the _____ live billions of microorganisms of many different types.

12. _____ are plant-like organisms that contain no chlorophyll.

13. An important group of microscopic worms that inhabit the soil is _____.

14. _____ are organisms that are visible to the unaided eye.

15. The soil is full of living organisms that depend on each other in an _____.

16. Gaps or spaces in the soil are called _____ and are either filled with air or water.

17. Soil pH is adjusted by adding _____ for acid soils or _____ for alkaline soils.

Discussion

1. Explain why soil is the source of life and agriculture.

2. Name the four soil components.

3. How are alluvial soils formed?

4. Where are soil particles in a river or stream usually deposited?

5. What areas of the United States have soils formed from marine sediments?

6. What forces of nature help form soils?

7. Name the physical characteristics of the soil.

8. Why are sandy soils usually less fertile?

9. Explain the difference between acid and alkaline soils.

10. Name the types of soil peds.

11. Describe the carbon cycle.

12. Name the three basic groups of soil nematodes.

13. Explain why soil is our most precious resource.

14. Explain the importance of the soil in the food chain.

15. Describe how organic soils were formed.

16. Explain how weathering helps form soils.

17. Describe the importance of clay in soil fertility.

18. Explain why earthworms are important in the soil ecosystem and agriculture.

CHAPTER 3

CELLS: AGRICULTURE'S BUILDING BLOCKS

KEY TERMS

cellulose
semipermeable
diffusion
osmosis
homeostasis
turgid
nucleus
cytoplasm
mitosis

STUDENT OBJECTIVES

After studying this chapter, you should be able to:

✦ Describe the cell as the building block of life.

✦ Describe how one-celled organisms are important to agriculture.

✦ Explain how knowledge of the functions of cells is important to agriculture.

✦ Analyze the differences between plant and animal cells.

✦ Explain the variations in the different types of cells.

✦ Describe the structures found in cells.

✦ Give examples of how agriculture makes use of structures found in cells.

✦ Analyze the functions of the various parts of cells.

✦ Explain the process of diffusion.

✦ Explain the process of osmosis.

✦ Explain the concept of homeostasis.

✦ Explain the process of mitosis.

One of the most important developments ever made by scientists was the invention of the microscope in the 1600s. When they began using the invention, a whole new world was opened to them. No one even suspected the miracles that could be observed under the magnification power of the microscope. In 1665, one such scientist, Robert Hook, observed a thin slice of cork under the microscope and saw tiny spaces that looked like small rooms attached together. Because they looked like the tiny living quarters of monks or prisoners in a jail, he termed the structures cells. The name has remained to this day, and thousands of scientists since Hook's time have studied the structures.

Without the ability to study cells, most of the advances in agriculture would not have taken place because life processes occur on the cellular level. To develop better plants and animals, scientists must understand how cells function and reproduce. To make improvements, everything in agriculture from reproduction to nutrition must be understood on the cellular level.

All living organisms are composed of cells. In fact, some organisms are made up of only a single cell. Such organisms as amoebas, paramecia, and other protozoa are composed of single cells. These tiny creatures play an important role in nature and are important in agriculture. Protozoa can cause animal diseases that cause the loss of millions of dollars to producers of agricultural animals (Figure 3–1).

One-celled bacteria play an important role in the digestive process of certain animals. Ruminants such as sheep and cattle could not digest fibrous plants like hay and grass if it were not for the one-celled bacteria living in their digestive tracts.

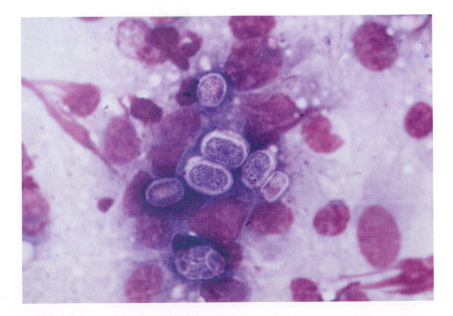

Figure 3–1 *One-celled organisms can cause animal disease.* Courtesy of James R. Duncan, The University of Georgia.

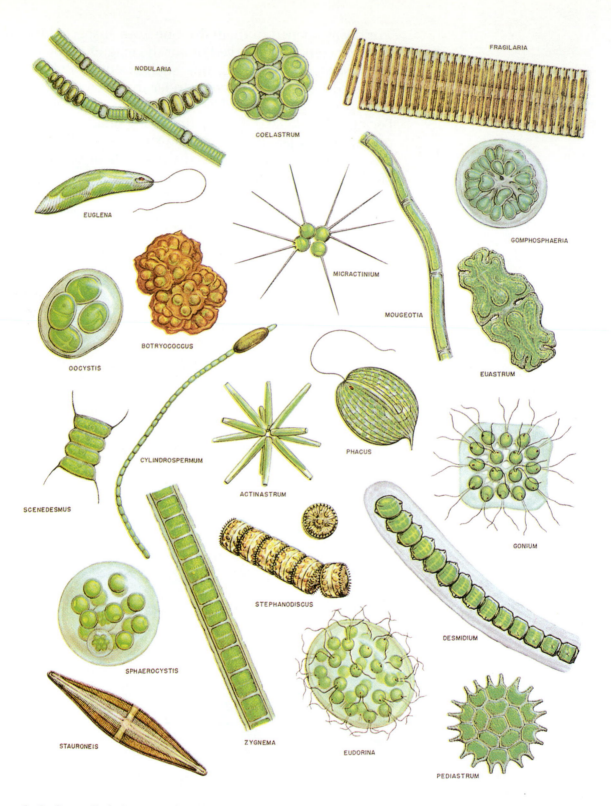

Figure 3–2 *One-celled plants provide food for aquatic animals and put oxygen into the water.* Courtesy of USDA-SCS.

Other types of bacteria are essential in placing usable nitrogen in the soil, and still other types of bacteria are responsible for the fermentation essential in making some types of food. One-celled plants called phytoplankton provide food for fish and place oxygen into the water (Figure 3–2).

Other organisms, both plants and animals, are made up of billions of cells. Cells are the building blocks of living things. In the higher animals and plants, each has many different types of cells. All plant and animal systems have specialized cells that perform distinct functions.

Cells vary greatly in size and shape (Figure 3–3). The largest known single cell is the ostrich egg. Many, such as bacteria, are so tiny that a very high-power microscope called an electron microscope is needed to see them. Cells may be round like a ball, square like a box, long and thin like a string, or shaped like a plate. Others, such as amoebas, change shape constantly. Each type of cell has a particular role to play, and the shape of the cell is related to that role.

All of agriculture is built around cells. Producers make their livelihood from the growth of plants and animals. This growth results from an increase in size or number of cells in

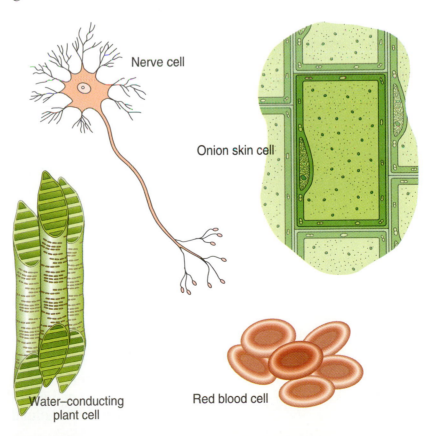

Nerve cell

Onion skin cell

Water–conducting plant cell

Red blood cell

Figure 3–3 *Cells may be found in a variety of shapes.*

the plant or animal. Producers also depend on the reproduction of plants and animals. Reproduction begins with cells. The male and female reproductive organs of both plants and animals produce cells that unite to form new organisms. Reproductive technicians work with cells when they take sperm from males for artificial reproduction and eggs from females for embryo transfer (Figure 3–4). An entire industry is built around producing eggs from poultry. All eggs are single cells.

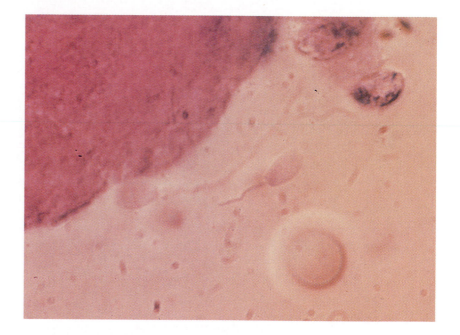

Figure 3–4 Reproductive technicians work with one-celled sperm and eggs. *Courtesy of Richard Fayrer-Hosken, The University of Georgia.*

Without adequate knowledge of cells, very few advances in agriculture would be possible. All processes of all the systems of plants and animals begin in the cells. Neither plants nor animals could live without the actions that take place within cells.

CELL STRUCTURE

Cell Walls

Several differences exist between plant and animal cells. One of the most important differences is that plant cells have a structure called a cell wall that serves as the outer barrier of the cell (Figure 3–5). Animal cells do not have this framework.

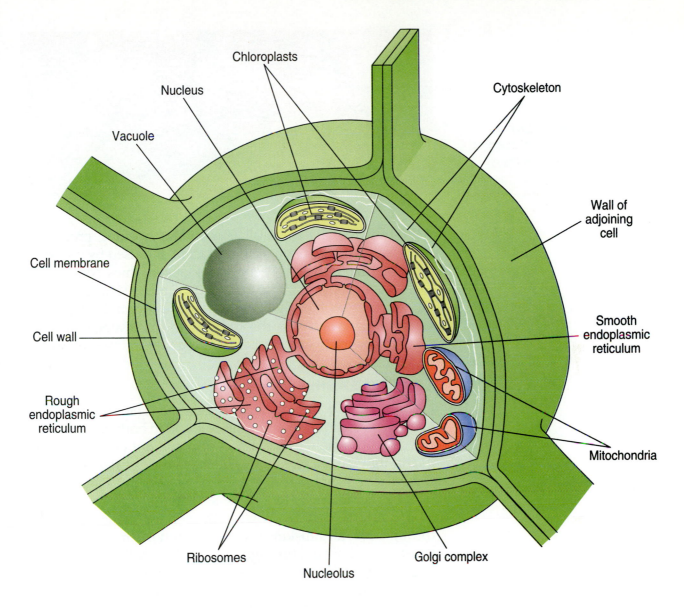

Figure 3–5 *Plant cells have cell walls that give them their shape.*

This structure gives the plant cell its shape and provides protection for the plant. The cell wall is composed of a substance called **cellulose** that provides rigidity for the walls of the cell and provides some support for the entire plant (Figure 3–6). Cellulose is very important in agriculture because such products as lumber and paper are manufactured from the cellulose found in the walls of hard plant cells such as those found in the branches and stems. In the softer plant tissues such as leaves and fruit, a substance called pectin helps give the cell

Figure 3–6 *Products such as lumber are composed of cellulose found in plant walls.* Courtesy of Georgia Pacific.

wall strength. The cell wall aids in the ripening process of fruits, and its components are extracted and used by people to process fruits into jelly. Pectin gives jelly its thick consistency. Cell walls of softer plant parts have two layers. Tissues from the harder parts, such as the stem, may have several layers. These layers all have openings through which pass water, gases, and plant nutrients. Plant cells are held together by the cell walls.

Cell Membranes

Inside the cell wall is the cell membrane. This structure, also known as the plasma membrane, is found in both plant and animal cells. All material that passes into and out of the cell must go through the cell membrane. In plants the transfer takes place through the membrane at the openings in the cell wall.

The membrane is said to be selective or **semipermeable**, which means that it allows only certain material to pass through. Not all substances pass through the membrane. The substances that pass through are usually small molecules and ions (charged molecules). Because some materials pass and others do not, the membrane is said to be selectively permeable. This membrane allows material such as water and other nutrients needed for the life processes to pass through into the cell. It also gets rid of the waste materials from these processes that would otherwise accumulate and harm the cell. The materials pass through the membrane in a process called **diffusion** (Figure 3–7). In this process, molecules in solution pass through the membrane from a region of higher concentration to a region of lower concentration. For example, in an animal's cell, fewer molecules of oxygen are inside the cell than outside the cell. Also there are usually more carbon dioxide molecules inside than outside. As the cell uses up oxygen molecules, more oxygen passes through the membrane to equalize the number outside and inside the cell.

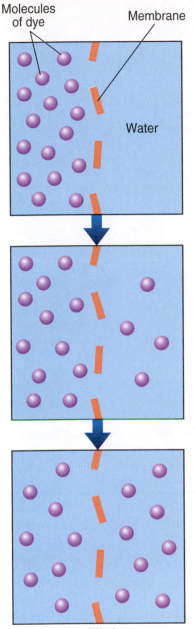

Equilibrium

Figure 3–7 *In a process called diffusion, materials pass through the cell (or plasma) membrane.*

Likewise, carbon dioxide molecules move out of the cell to an area that is less concentrated with carbon dioxide molecules. Through diffusion, the cell constantly takes in needed molecules, such as oxygen, and expels unwanted molecules such as carbon dioxide.

Water also passes through the semipermeable cell membrane. This process is called **osmosis** (Figure 3–8). As in diffusion, the water moves from a region of high concentration to a region of low concentration. The more material water has dissolved in it, the less it is concentrated. If the cell has relatively little water inside, the solution tends to "draw" water from outside into the cell through the cell membrane.

Through diffusion and osmosis, the materials move from one part of the cell to another and in and out of the cell. In all organisms this is extremely important because these processes allow the cell to remain constant even though conditions in the environment change. The ability of an organism to remain stable when conditions around it change is called **homeostasis**. For example, cells must retain the proper amount of water. The buildup of water in the cells creates an internal pressure called turgor that helps the cell retain its shape. When the cell is filled with the proper amount of water, the cells are filled out and taut. Cells in this condition are said to be **turgid**. In times of drought when plants do not have all the water they need, the cells lose their turgor and are limp and wilted (Figure 3–9). As soon as water is made available to the plants, the cells fill with water, the pressure builds, turgor is restored, and the plants are returned to their upright healthy appearance.

The Nucleus

Almost all cells have a relatively large structure called the **nucleus**. Cells that have a nucleus are called eukaryotic cells, and cells that do not have a true nucleus are called prokaryotic cells. For example, a bacterium is a prokaryotic cell, and most animal and plant body cells are eukaryotic (Figure 3–10).

The nucleus is made primarily of nucleic acids, protein, and enzymes and serves as the control center for all of the activities of the cell. Most eukaryotic cells have only one nucleus, but some cells have more than one nucleus. For example, certain types of animal muscle cells have many nuclei. The nucleus is surrounded by a double layer membrane that regulates the movement of materials into and out

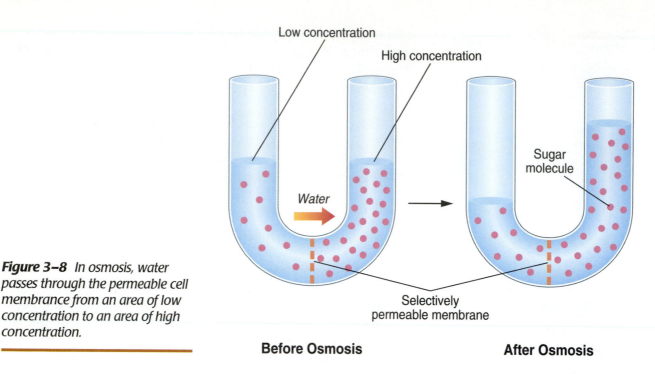

Low concentration

High concentration

Water

Sugar molecule

Selectively permeable membrane

Before Osmosis   **After Osmosis**

Figure 3–8 *In osmosis, water passes through the permeable cell membrane from an area of low concentration to an area of high concentration.*

A   B

Figure 3–9 *When plants are wilted, the cells in the plant have lost their turgor. Place A lacks water. Plant B has plenty of water.* Courtesy of James Strawser, The University of Georgia.

of the nucleus in much the same way that the cell membrane functions.

One of the most important roles of the nucleus is that it contains the genetic code that gives the organism the characteristics it possesses. This genetic code is contained in a substance called deoxyribonucleic acid (DNA). The molecules of DNA are arranged in threadlike strands called chromosomes. Segments of the chromosomes are called genes and are responsible for transferring the genetic code. The process of gene transfer is discussed more completely in Chapters 4 and 5.

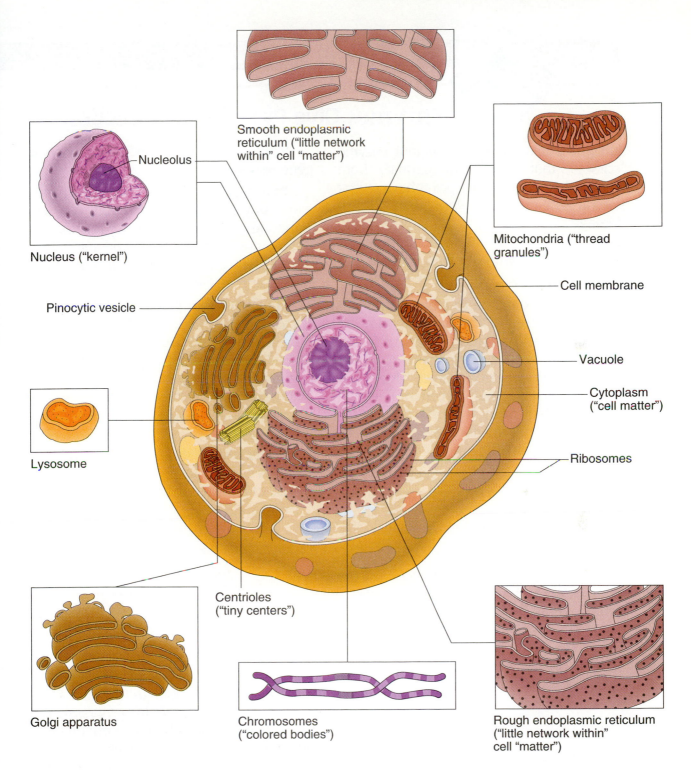

Smooth endoplasmic reticulum ("little network within" cell "matter")

Nucleolus

Nucleus ("kernel")

Pinocytic vesicle

Lysosome

Mitochondria ("thread granules")

Cell membrane

Vacuole

Cytoplasm ("cell matter")

Ribosomes

Golgi apparatus

Centrioles ("tiny centers")

Chromosomes ("colored bodies")

Rough endoplasmic reticulum ("little network within" cell "matter")

Figure 3–10 *Plant and animal cells have similarities and differences. Compare the plant cell in Figure 3–6 with this animal cell.*

BIO ✦ BRIEF

Enzymes Give Plants UV Protection

Sunlight would kill plants, without enzyme "scissors" that undo gene damage from ultraviolet (UV) rays. In fact, plants have several natural gene menders tailored to the kind of damage done, according to findings of Dr. Edwin L. Fiscus, an Agricultural Research Service scientist working with a researcher at the University of California at Davis.

Ultraviolet damage to crops is rare. But knowing the repair mechanisms may be important if UV radiation increases in the future as a result of the thinning of the Earth's protective ozone layer. The scientists used Arabidopsis, a common white-flowered plant with a small number of genes, which allows for easy tracking of genetic differences.

DNA is a series of chemical bases—A-G-C-T (for adenine, guanine, cystosine, and thiamine)—that form the alphabet of life. If they get damaged, the code is illegible; too much unreadable code, and the plant dies.

Plants may respond in several ways to gene damage.

"When your car breaks down," says ARS plant physiologist Fiscus, "you can call someone who does general repairs. But other times, a specialist may be able to perform a particular type of repair much more rapidly and efficiently."

"It's like that for plant cell damage," says Fiscus, who works in the ARS Air Quality-Plant Growth and Development Research Unit." To fix damaged DNA, there are both general repair enzymes and at least two highly specialized kinds."

Fiscus and geneticist Anne Britt at UC-Davis confirmed what others suspected: Two specialized enzymes in plants are essential for UV repair. They are both from a class of enzymes called photolyases.

The generalized repair enzyme system, says Britt, is probably designed for a wide variety of relatively rare types of damage. It

Cytoplasm

Cells are filled with a thick, clear fluid that surrounds the nucleus. This fluid, known as the **cytoplasm**, contains all of the material needed by the cell to conduct life processes. The fluid in the cytoplasm aids in moving these essential materials. This fluid is constantly moving and suspends other parts of the cell. Some one-celled organisms, such as the amoeba, move by means of the flowing of the cytoplasm within the cell. In certain types of cells, the cytoplasm contains pigments that give an organism its color.

works by excising the damaged bases, or sequences, and rebuilding them—a process that tends to be slow and inefficient.

More common kinds of damage, such as when UV light causes Ts and Cs to crosslink improperly to each other, are also repaired by specialized photolyases, which eliminate this inappropriate bond between the bases. Photolyase repair is specific, rapid, efficient, and—like excision repair—relatively error-free.

Another interesting thing about these enzymes, Britt says, is that they are activated by light, so the very cause of the UV damage is also what triggers its repair. The scientists proved photolyase enzymes are essential for plants' survival in natural light by using special mutant plants developed by Britt which can't produce the enzymes.

Fiscus, whose research station is on the campus of North Carolina State University, devised special growth chambers that delivered precise doses of various ratios of UV light and regular sunlight. The mutant plants were highly sensitive to UV light, compared to normal plants.

Source: Agricultural Research Magazine.

Scientists are discovering the mechanisms in plant cells that prevent ultraviolet damage from the sun. Courtesy of USDA-ARS.

Organelles

Within the cytoplasm of cells are small structures that serve different roles. In much the same fashion as the organs of a body support an animal, these structures, called organelles, support the cell. Among the most important of the organelles are the mitochondria, which are shaped like a peanut. They break down food nutrients and supply the cell with energy. Cells that use more energy contain more mitochondria than cells that are less active. For example, muscle cells contain more mitochondria than bone cells because bone cells require far less energy than muscle cells.

Vacuoles are organelles that serve as storage compartments for the cell. They consist of a membrane that encloses water and other material. They store the nutrients and enzymes needed by the cells. Also, vacuoles provide a storage space for the waste materials given off by the cell.

Some cells contain organelles called microtubules. They are shaped like thin, hollow tubes, composed of protein, which act as the "bones" of the cell. These structures, found in animal cells, not only support the cell and give the cells their shape but also assist the movement of chromosomes during cell division.

Microfilaments are fine fiber-like structures composed of protein. These organelles help the cell to move by waving back and forth.

Cells contain thousands of tiny structures called ribosomes. These organelles are the sites where protein molecules are assembled. Proteins are needed by all cells for growth and other important functions. Enzymes that regulate the chemical processes in the cell are composed of protein molecules.

The Golgi apparatus is an organelle shaped like a group of flat sacs bundled together. They remove water from proteins and prepare them for export from the cell.

The endoplasmatic reticulum is a large webbing or network of double membranes positioned throughout the cell. These organelles transport material within the cell.

Lysosomes are the digestive units of the cell. They digest proteins, carbohydrates, and other molecules. Any foreign materials such as bacteria that enter the cell are digested in the lysosomes. As other cell parts become worn out and nonfunctional, the lysosomes' digestive enzymes break them down. The products of the digestive actions are passed into the cytoplasm and out of the cell through the cell membrane.

Plant cells have organelles called plastids that are not present in animal cells. Three types of plastids exist. Chloroplasts use the energy of the sun to make carbohydrates. These organelles contain the chlorophyll that gives plants their green color.

Leucoplasts are plastids that provide storage for the cell. They may contain starches, proteins, or lipids (substances containing fatty acids). These organelles are abundant in the seeds of plants and contain nutrients used by the emerging plant, and they provide nutrition for animals that eat the seeds.

Chromoplasts are plastids that manufacture pigments that give fruits their color. The pigments also give leaves their brilliant colors in the fall.

Figure 3–11 *When an animal matures, it stops growing, and cell division is used to heal wounds and replace worn-out cells.*
Courtesy of James Strawser, The University of Georgia.

Mitosis

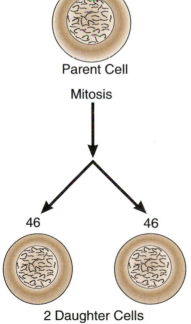

46 chromosomes

Parent Cell

Mitosis

46   46

2 Daughter Cells
(each has same number of 46 chromosome "threads")

Figure 3–12 *The nucleus of a cell divides in a process called mitosis.*

CELL REPRODUCTION

All of the growth that takes place in living organisms results from cells increasing in size or numbers. Cells have a very limited size to which they grow, so most growth results from cells reproducing or multiplying. Also, when injuries occur to either plants or animals, cells reproduce to heal the wound.

When a cell grows, it reaches a maximum size. When this size is reached, the cell divides into two cells. These cells in turn grow until they reach their maximum size and each divides into new cells. The original cell is called the parent cell, and the new cells are called the daughter cells. When a plant or animal matures, it stops growing and the process of cell division is used to heal wounds and to replace worn-out cells (Figure 3–11).

Eukaryotic cells (cells that have nuclei) divide by a process called **mitosis** (Figure 3–12). All of the genetic code for passing on traits of an organism is located in the nucleus of the cell. In the process of mitosis, the genetic coding is duplicated and transferred to the new cells. Although the process of mitosis is continual, scientists have divided the events into different phases.

The Interphase

The period when the cell is not actively dividing is called the interphase. This phase is not really a part of mitosis but is a time when the cell synthesizes materials and moves them in and out of the cell. It is during this time that the cell grows. As the cell reaches its maximum size, the DNA replicates and forms two complete sets of chromosomes. The thread-like molecules of DNA that make up the chromosomes are called chromatin and are spread throughout the nucleus. Animal cells have strands of genetic material outside the nucleus called centrioles. Most plant cells do not contain centrioles. At the end of the interphase, the cell is at the correct size, the chromosomes are duplicated, and the cell is ready to divide.

Prophase

The first phase of mitosis is called the prophase. During this phase the chromatin appears in the form of distinct shortened rod-like structures. The chromosomes are formed of two strands called chromatids that are attached at the center by a structure known as a centromere. As this formation takes place, the nuclear membrane begins to dissolve, and the entire nucleus begins to disperse. In place of the nucleus a new structure called the spindle is formed. The spindle is shaped somewhat like a football and composed of micro-tubules. In animal cells, the centrioles move to opposite sides of the cell.

The Metaphase

During the next phase, called the metaphase, the chromatids move toward the center of the spindle. The center of the spindle is called the equator. When they reach the center, the centromere of the chromatids connect themselves to the fibers of the spindle.

The Anaphase

The anaphase is the third stage of the process of mitosis. During this time the pairs of chromatids separate into an equal number of chromosomes, and the centromeres dupli-cate. When separation occurs, the chromosomes move to opposite ends of the cell.

The Telophase

The final phase of mitosis is called the telophase. The chro-mosomes continue to migrate to the opposite sides of the cell (called poles). When they reach the poles, the remains of the spindle begin to disappear, and new membranes are formed around the chromosomes. This forms two new nuclei.

To complete the cell division, a process known as cytoki-nesis occurs, which divides the cytoplasm in the cell. Because mitosis is involved with the division of the nucleus of the cell, cytokinesis is a separate process from mitosis. In animal cells a crease called a cleavage furrow begins to form in the center of the cell. This crease continues to deepen until the cell membrane divides along with the cytoplasm. One nucleus goes with each divided cell wall and cytoplasm, and the process of forming two new cells from the old cell is completed.

The walls of plant cells do not form a cleavage furrow like animal cells. Instead, a structure called a cell plate forms in the middle of the spindle and grows outward until the cell is divided into the two daughter cells.

At the completion of mitosis, the new daughter cells are genetically identical to each other and to the parent cell that divided to form them. After formation, the daughter cells then go into interphase, and the whole process of mitosis starts over. Through this continuous process, an organism grows and maintains its structure through the replacement of worn out and injured cells.

SUMMARY

All living matter is composed of cells. The processes of life occur within these cells. Because agriculture deals with the processes of life, particularly growth and reproduction, knowledge about living cells is essential. By researching the ways cells live, function, and reproduce, scientists can develop better ways of producing food and fiber.

CHAPTER 3 REVIEW

Student Learning Activities

1. Go to the library and find an article on research being done with cells. (Hint: look for articles about cancer, digestion in cattle, or reproduction.) Report to the class on the research. Tell how the research could benefit agriculture.

2. Design your own research study that deals with cells. Be sure to include your reason for the study, your hypothesis, your procedures, and how the research could be used.

Define the Following Terms

1. cellulose
2. semipermeable
3. diffusion
4. osmosis
5. homeostasis
6. turgid
7. nucleus
8. cytoplasm
9. mitosis

True/False

1. Cells are the building blocks of living things.
2. Lysosomes are fine fiber-like structures composed of protein.
3. All living organisms are composed of cells.
4. Plants and animals can live without the actions that take place within cells.
5. All cell walls of plants have a single layer.
6. The cytoplasm contains all of the materials needed by the cell to conduct its processes.
7. The chromoplasts manufacture pigments to give fruits and leaves their colors.
8. Upon completion of mitosis, the two daughter cells are identical to each other but not to the parent cell.
9. DNA replication is the first step of mitosis.
10. When cells are filled with the proper amount of water, filled out and taut, they are said to be rigid.

Fill in the Blank

1. The genetic code of a cell is in a substance called _____.
2. The mitochondria is a cell structure called an _____.
3. _____ are organelles that serve as storage compartments for the cell.
4. The _____ function to break down food nutrients and supply the cell with energy.
5. Leucoplasts contain _____, _____, or _____.
6. The process of cell division for growth or cell replacement is called _____.
7. The four phases of mitosis are _____, _____, _____, and _____.

8. The _____ is a structure that forms in plant cells to divide a cell into two daughter cells.

9. _____ taken from plants gives jelly its thick consistency.

Discussion

1. What is one of the most important differences between plant and animal cells?
2. Name some products manufactured from cellulose found in plant cells.
3. What helps give cell walls strength in the softer plant tissues?
4. Describe a selectively permeable membrane.
5. Describe the process of osmosis.
6. How do cells maintain turgor pressure?
7. What is the major function of the nucleus?
8. Why are amoebas, paramecia, and other protozoa important in agriculture?
9. Which organelle has sites where protein is manufactured?
10. Name the three types of plastids found in plant cells.
11. How does growth take place in living organisms?
12. Describe what takes place in the metaphase phase of mitosis.
13. Explain why agriculture is built around cells.
14. Explain why it is important for cells to have selectively permeable membranes.
15. Explain why plants wilt.
16. Choose two cell organelles and explain their functions in the cell.
17. Explain why some bacteria are important in agriculture.

THE SCIENCE OF GENETICS

KEY TERMS

dominant
recessive
alleles
homozygous
heterozygous
genetics
genotype
phenotype
DNA
hybrid
heterosis

STUDENT OBJECTIVES

After studying this chapter, you should be able to:

✦ Explain how Mendel developed his theories of genetics.

✦ Explain the Mendelian Laws of genetics.

✦ Analyze a Punnett square.

✦ Discuss the makeup of chromosomes in a cell.

✦ Discuss the process of DNA transfer.

✦ Specify how genetic principles are used in plant breeding.

✦ Explain how hybrid plant varieties are developed.

✦ Discuss how animal breeding has benefited the producer.

✦ Determine how the principles of genetics are used in animal breeding programs.

✦ Discuss how computers are used in the selection and breeding of animals.

When people first started planting seeds, raising crops, and domesticating animals, they noticed that seeds usually produced plants that looked like the plants they came from. They also noticed that the newborn animals looked like the parents. As a result of this observation, they selected seeds to plant from the plants that they felt were superior. Animals that best suited their purposes were saved for breeding. Although these early people saw the results of what later came to be known as selective breeding, no one understood how these characteristics were passed on to the next generation. In fact, only within the last century have theories been developed as to how these traits are transferred from parent to offspring. After an accepted theory of how the process works was developed, gigantic strides were made in the types of plants and animals that were produced.

The first person to develop a workable theory of the transfer of traits was an Austrian monk named Gregor Mendel (1822–1884). He was an educated man who had studied science and mathematics and had a natural curiosity as to how things occurred. As a result of watching plants grow in the monastery garden, he became curious as to how characteristics of living things were passed from generation to generation. He especially noticed garden peas because of the differences in the appearances of the vines and fruit. He decided they would provide the basis of a study to determine how traits are passed on.

Mendel noticed that some of the seeds were smooth and round and others were wrinkled and not perfectly round. He also noticed differences in the color of the seed and that the color variations occurred in seeds with both round and wrinkled seed coats. There were variations in the length of the stems of the plants, and the plants that had short stems produced seed that in turn produced new plants with short stems. The same was true for the plants with long stems. One characteristic of the garden pea that makes it ideal for the study of inherited characteristics is that the flower is enclosed by the petal, preventing the pollen from other plants from getting in the flower. This type of flower contains both the male stamen and the female pistil and is self-pollinating.

To begin his experiments, Mendel developed purebred strains of the pea plants. This meant that the characteristics he chose in the plants would be carried through to the next generation each time new seeds were produced. For instance, a purebred pea plant that had round seeds would only produce seeds that were round.

Using the purebred plants, he cross-pollinated the plants by opening the petals, removing the pollen-producing stamen, and placing pollen from a plant with different flower traits. For instance, he developed plants that always produced red flowers and some that produced only white flowers. Using the flowers of the red-flowering plant, he opened the petals and removed the stamen before the plant could self-pollinate (Figure 4–1). He removed pollen from a white-flowered plant and placed this pollen in the flower of the red plant. The seed

Figure 4–1 *Mendel removed the stamen from the flower before pollination could occur. Courtesy of James Strawser, The University of Georgia.*

harvested from the cross-pollinated plants produced new plants that had all red flowers. Mendel reasoned that because no white flowers were among the plants grown from this seed, certain factors must cause the red flowers to be predominant over the white flowers. These he called **dominant** factors. From this concept he developed his Law of Dominance, which says that overriding or dominant factors make certain **recessive** traits disappear. Later scientists called Mendel's traits dominant or recessive **alleles**. An allele is an alternate or different form of a gene. As will be explained later, genes contain the material that controls an organism's traits. Red and white color are alleles for the gene that controls the color of the pea plant.

To gain a better understanding of inheritance in pea plants, Mendel allowed the crossed plants from the seeds of the plants from his first experiment to self-pollinate. Some of the resulting plants had red flowers, and some of the plants had white flowers. He called the plants from the first crossing the first filial generation or the F1 generation. The plants grown from seeds of the F1 were referred to as the F2 (second filial) generation. Mendel was puzzled by the riddle of how plants with red flowers could produce plants with white flowers. He further observed that in the F2 generation, white flowers appeared about 25 percent of the time, and red flowers appeared about 75 percent of the time. He reasoned that before the pollen merged with the egg from the female structure of the flower, the factors for red flowers and the factors for white flowers had become separated or segregated. From this idea, Mendel developed the Law of Segregation. This law says that the alleles responsible for the traits from each parent are separated and then combined with factors from the other parent at fertilization. In other words, each parent provides one of the two genes for each particular trait.

Mendel performed other experiments using pea plants with different traits. He noticed that plants with red flowers and plants with white flowers could both have round or wrinkled seed, short or long stems, or pods of different colors. All together he identified and experimented with seven different traits of garden peas. His experiments with the traits led to the development of the Law of Independent Assortment. This principle states that factors (genes) for certain characteristics are passed from parents to the next generation separate from the other factors or genes that transmit other traits.

Later, a mathematician named R. C. Punnett developed a diagram called a Punnett square that illustrates the possible combinations for a particular trait. Figure 4–2 gives an example using the flower colors from Mendel's experiments. The allele responsible for the dominate red color is represented by the capital letter R and the allele responsible for the recessive white color is represented by the lowercase letter r. Along the

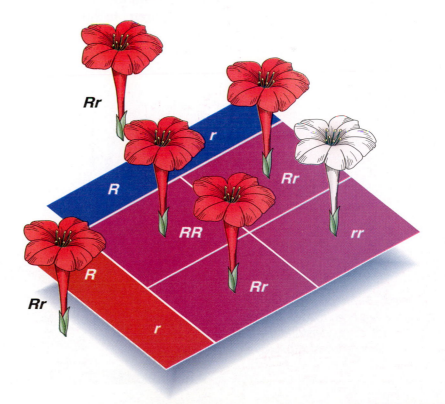

Figure 4–2 *Using a Punnett square with the colors Mendel used, it can be determined how likely it is the dominant characteristic (red color) will appear. This scheme represents two heterozygous red flowers.*

top of the square is the allele from one parent, and along the side is the allele from the other parent. Where the boxes intersect represents the possible combinations of the alleles that form the gene for that characteristic. For example, if the dominant R is paired with a recessive r, the flower will be red. If two recessive r genes are paired together (rr), the flower will be white. Genes that contain two dominant alleles (RR) or two recessive alleles (rr) are said to be **homozygous**. Genes that possess one dominant allele and one recessive allele are said to be **heterozygous**.

Combined traits can be expressed by using other letters to represent the other traits. For example, if the seed has a dominant smooth seed coat it would be written as "S," and the recessive rough seed coat could be written as "s." A plant that contains genes for homozygous red and homozygous smooth seed coat would be written as "RRSS." A homozygos red flower combined with a heterozygous smooth seed coat would be written as "RRSs." As you can see, several different combinations are possible for just seed coat texture and flower color.

Since Mendel, several generations of scientists have studied the process, but only relatively recently have scientists begun to understand just how these traits are passed from one generation to the next. The study of how traits are passed from parent to offspring is called **genetics**. Even with all of the research that has been conducted and all the knowledge accumulated, scientists still do not fully understand how traits are passed (Figure 4–3). As more research is completed,

Figure 4–3 *Although a lot of progress has been made, scientists still do not fully understand the process of gene transfer. Courtesy of USDA-ARS.*

it becomes apparent just how complicated the process of inheritance is. Scientists do, however, have enough knowledge to develop theories of how the transfer occurs. They know that plants and animals transfer traits in much the same way. Characteristics are passed from parent to offspring through a process known as gene transfer. Keep in mind that the environment in which an individual organism grows and lives in also influences the way it looks and behaves.

The genetic composition of an individual is called the **genotype**. How the allele expresses itself is called the **phenotype**. For example, two black calves may have the same phenotype (the black color) but different genotypes. The genotype for one calve may be heterozygous dominant for the black color (Bb) and the other may be homozygous dominant for the black color (BB).

GENE TRANSFER

All of the material responsible for the transfer of traits is grouped together in structures called chromosomes. Recall from Chapter 3 that chromosomes are present in the nucleus of each cell of eukaryotic plants and animals. Chromosomes are composed of long strands of protein and nucleic acid molecules that can be arranged in an almost unlimited number of ways. These molecules are called deoxyribonucleic acid or **DNA**. Within these molecules is the genetic code that determines all of the characteristics of an organism. Different segments of the chromosomes control the different traits that are expressed in the organism. Each chromosome segment that controls a trait is called a gene. One segment or gene on the long chromosome may control whether an animal's coat is white or black; one gene may control how large the animal will be when it is grown; and another may determine the shape of the animal's ears. A combination of genes may control a characteristic. For example, the potential weight of an animal at maturity may be controlled by the size of bone, the amount of muscling, or the overall length of the animal. All of these characteristics are controlled by different genes, but they all are a part of the characteristic of potential mature weight. The environment will also play a part in the eventual weight of the animal. The number and types of parasites, the amount and type of feed, and the climate may all have an effect on the animal's weight (Figure 4–4).

Figure 4–4 *Both genetics and the environment determine how an animal will look at maturity. Genetics will determine color pattern, and the environment will greatly affect body weight.* Courtesy of James Strawser, The University of Georgia.

Each cell of an organism's body contains a gene for each characteristic. Each gene contains two alleles, one contributed by each parent. Remember that an allele is an alternate form of the same gene, and one allele is responsible for passing on a characteristic to an organism. As pointed out in Mendel's Law of Dominance, one of the alleles usually expresses itself, and the other allele remains hidden.

In a process called meiosis, the chromosomes are divided so that each sperm and each egg contains only one allele for each characteristic. In other words, each parent contributes half of the genetic material for the new organism. Recall from Chapter 3 that cells reproduce through the process of mitosis. Meiosis occurs very much like mitosis, but the process is carried a step further, and the chromosomes of the daughter cells separate into a grouping of individual alleles. In the male, meiosis results in the development of four sperm cells from the original parent cell. In the female, meiosis results in the development of one egg. When the egg and the sperm unite at fertilization or conception, the newly formed cell contains two chromosomes composed of a pair of alleles (in the form of a gene) for each trait, each parent having contributed a chromosome containing an allele for each of the characteristics. The sex cells (sperm and egg) each have only one set of alleles and are referred to as having a haploid number of chromosomes. All body cells carry chromosomes in pairs. In body cells, the total number of chromosomes is referred to as the diploid number of chromosomes. This number is designated as $2n$ where n is the number of chromosomes. Different organisms have different numbers of pairs of chromosomes. For example, cattle have 30 pairs, horses have 32 pairs, and a

Red Delicious apple has 34 pairs of chromosomes. Some plants have more than two sets of chromosomes in their cells. These are referred to as polyploid numbers of chromosomes.

The sex of an animal is determined by chromosomal arrangement. Each parent contributes a chromosome that pairs with a chromosome from the other parent to determine the sex of the new animal. The sex-determining chromosomes are of two types. Some are shaped like the letter X, and others are shaped like the letter Y. The arrangement of chromosomes for a diploid vertebrate male (an animal with a backbone) includes an X and a Y diploid chromosome.

Sex Determination in Animals

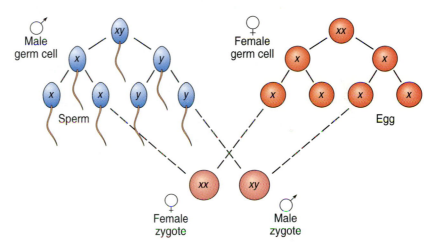

Figure 4–5 *The sex of an individual is determined by the combination of the X and Y chromosomes.* Courtesy of USDA-ARS.

Diploid vertebrate females have a pair of X chromosomes. When meiosis occurs in the male, the XY chromosome pair is separated into X and Y chromosomes, and the female XX chromosome pair is separated into single X chromosomes. At fertilization, if two XX chromosomes are paired together, the new animal will be a female. If a Y and an X chromosome are paired together, the new animal will be a male (Figure 4–5).

The code for how the organism is to be formed (all of its characteristics) is contained in the DNA that makes up the genes. The molecules forming the DNA are shaped like a coiled spiral called a helix that resembles a corkscrew. If this corkscrew-shaped helix were straightened out, it would resemble a ladder with rungs where the segments fit together. As the cell prepares to enter mitosis (during part of the interphase), the chromosome pairs with the DNA separate. During meiotic cell division (the formation of the sex cells or gametes), the chromosome pairs are also separated, with the DNA remaining intact.

BIO BRIEF

There's Treasure in the Soybean Genome Map

The soybean has three fewer pairs of chromosomes and about 1.7 billion fewer pieces of DNA units than humans. But Perry B. Cregan has shown that the same scientific tools can be used to draw the map of both the soybean and human genomes.

Cregan is a plant geneticist with the Agricultural Research Service (ARS) in Beltsville, Maryland. The United Soybean Board has given him and colleagues elsewhere in the country more than $1 million to construct a map of the soybean genome using a technique from the human genome project called simple sequence repeats (SSRs).

This approach enables Cregan to look at the smallest pieces of DNA in each of the soybean's 20 pairs of chromosomes that contain a total of 1.3 billion bases, or units. Each base is represented by a single letter of the DNA alphabet: A-G-C-T for adenine, guanine, cystosine, and thymine.

Cregan and colleagues use SSRs to find unique variations in these pieces that can serve as markers for nearby genes. With earlier technology from the human genome project, they mapped the broad outline of all 20 pairs of the soybean's chromosomes.

So far, they've found more than 700 SSR markers from within these 20 chromosome pairs. Now they're filling in the details—one unit at a time—with a new type of DNA

marker, single nucleotide polymorphism (SNP). There are an average of 65 million bases per chromosome, says Cregan. "Ideally, we'd like to have an SNP marker every 100,000 to 250,000 bases." SNP has greatly increased the rate of DNA marker discovery.

Cregan is actually working with three other teams of scientists, each of which has a different genetic map because each uses different soybean varieties, and a fourth map if in the making. Cregan's ARS colleague, Randy C. Shoemaker, is leading a team of researchers that has developed a map at Iowa State University. Researchers at the University of Nebraska at Lincoln and the University of Utah at Salt Lake City have the other two maps.

Why would the United Soybean Board spend a million dollars to draw a map of the soybean genome? One reason is a projected additional 220 million bushels of soybeans a year for American farmers to sell on the domestic and international markets. That's the number of bushels destroyed each year by just one soybean pest, the soybean cyst nematode.

Cregan's goal and that of a number of other soybean researchers is to identify genes that will provide resistance to the soybean cyst nematode so that the resistance can be bred into commercial soybean

Scientists use computers to examine DNA sequence

2 or 3 bushels an acre over America's typical yearly plantings of 75 million acres.

Cregan and his colleagues are hoping to find DNA fragments from the wild ancestor of soybean that could induce such a yield increase in commercial soybeans. He and the Nebraska team have developed more than 300 lines of soybeans that have one-eighth of their genomes replaced by DNA fragments from the wild ancestor of soybeans.

"If you could see the scrawny vine a wild soybean is, with seed pods that hold seeds not much bigger than poppy seeds, you'd realize how unlikely it seems that genetic material from the wild soybean could be used to improve the beefy soybeans on the market today," says Cregan. "Without DNA markers, it would be impossible to find just the right wild soybean DNA fragments. But this SSR genome map allows us to do a unit-by-unit search for genetic benefits. It is what gives our breeding efforts a reasonable probability of success."

And, the maps will make breeding easier and quicker. No longer will breeders have to grow plants and expose them to insects or plant diseases to find the resistant ones, says Cregan. "They can use the map markers to identify the resistant plants and only continue breeding the plants that have the genes they want."

varieties. Cregan has completed the development of mapping markers for the first two of the four genes he thinks are responsible. He's continuing work with colleagues at the University of Minnesota to find markers for the remaining two resistance genes.

Potential gains are huge for American farmers. The extra 220 million bushels is just for starters. Throw in another few hundred million bushels for other pests the soybean might also resist, if a complete genome map were available to plant breeders; and add yet another 200 million bushels or so that could be harvested, if genes were found to raise soybean yields

Source: Agricultural Research Magazine.

Picture the helix as a long ladder that has been twisted (Figure 4–6). At each point on the helix where the two halves connect are the nitrogen-containing bases adenine (A), thymine (T), guanine (G), and cytosine (C). These attach to each other at the center of the rung. These nitrogen bases are shaped so that they pair with only one other base. Adenine (A) can pair only with thymine (T), and cytosine (C) can pair only with guanine (G).

Prior to cell division, the DNA copies itself in a process called replication. During replication, the strands of DNA separate, and each half is assembled so that two strands, exactly alike, are formed.

The messages encoded in the DNA are transferred to the rest of the cell by means of a messenger substance known as ribonucleic acids (RNA). RNA uses the model of the DNA molecule to transfer the pattern for how the animal is to be constructed. This pattern determines how the molecules of DNA are arranged on the gene. As the embryo begins to develop, the cells differentiate. Some of the cells form muscle, some hair, some skin, some internal organs, and so on. In a plant, the cells form into leaves, stems, and roots. The process by

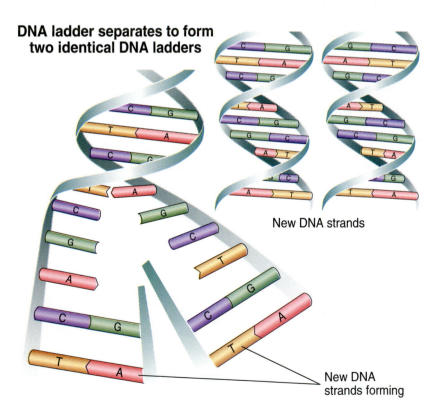

DNA ladder separates to form two identical DNA ladders

New DNA strands

New DNA strands forming

Figure 4–6 *The DNA helix resembles a twisted ladder.*

which differentiation occurs is not yet fully understood. At conception, the chromosomes from each parent unite to form the genetic code for the fertilized egg. DNA molecules can be arranged in the gene in an almost infinite number of ways. The arrangement of the molecules and how the molecules are paired at conception determine the makeup of the new animal.

PLANT BREEDING

Plant breeding is the sytematic process of matching genetic factors from parent plants to produce offspring that are superior to the parent plants. Some form of plant breeding has gone on since before recorded history. All of the plants that are now grown throughout the world were once wild plants that humans domesticated. Hundreds of different species of plants are grown by producers (Figure 4–7). These plants are known as cultivars. Within these species, plant breeders have developed thousands and thousands of varieties to produce better results under a variety of conditions. The first plant breeding was no more than locating the best of the plants and using seed from those plants to plant the new crop. After Mendel devised the principles of genetics, great strides were made in breeding plants. These advancements came about when plant breeders began using systematic methods to produce the plants they wanted.

Figure 4–7 *Hundreds of species of cultivated plants have been developed from wild species.*
Courtesy of James Strawser, The University of Georgia.

Figure 4–8 *Plant breeders have definite goals in mind when developing new varieties.* Courtesy of USDA-ARS.

Perhaps the greatest advancements in the way crops are grown came about as a result of the breeding of superior plants. How well a plant grows and produces is affected by such factors as how well it takes up water and nutrients, the size of its fruit, how efficiently it uses photosynthesis, and how many seeds the plant produces. All of these factors are greatly influenced by the breeding of superior plants. For instance, how well a plant takes in water and withstands a period of drought is affected by the size and depth of the roots. Plants with longer roots reach farther into the ground than plants with shorter roots. Plant breeders can select plants that have longer roots to use in a breeding program aimed at developing plants that can withstand drought.

Plant breeders always have a definite goal in the type of plant they want to produce (Figure 4–8). These goals usually center around the type of crop that is needed or the conditions under which the crop is grown. For example, a problem that existed for years in the vegetable industry was the high cost of hand harvesting. The obvious answer was to develop machines to do the harvesting. The main obstacle was that crops such as green beans and tomatoes matured all season long instead of the whole crop maturing at one time. The solution was to develop varieties that would mature at the same time and could be harvested in one operation (Figure 4–9). Another problem that was solved through plant breeding was the development of tomatoes that are firm enough to withstand the rigors of mechanical harvesting without tearing or bruising.

Adapting plants to differing environments has long been a goal of plant breeders. Plants that once only grew well in one area have been bred to grow in different climates. For example, different varieties of soybeans are grown in the

Figure 4–9 *Plants bred to be picked by a mechanical harvester need to mature at about the same time.* Courtesy of James Strawser, The University of Georgia.

Figure 4–10 *Soybean varieties were developed to have a growing season to match the climate where they are grown.* Courtesy of James Strawser, The University of Georgia.

South than in the northern states. The North has a shorter growing season, and the plants must mature and bear their seed in a shorter time. These differing varieties came about as a result of a scientific plant breeding program that systematically selected and bred plants that met the climatic demands (Figure 4–10).

Plant breeders learned that crossing two purebred varieties of a plant could result in a plant superior to either of the parent plants. This cross is referred to as a **hybrid**. A hybrid variety has an effect called hybrid vigor or heterosis that makes the plant take on characteristics superior to the parent plants. It may grow faster, resist disease better, or produce more fruit or seed than the parents. Plant breeders have used this phenomenon for years to produce more efficient plants.

Of all the achievements that have improved the world's agricultural output, the development of hybrid varieties of corn ranks near the top (Figure 4–11). The development of hybrid corn began in the United States around the turn of the 20th century with a botanist named George Shull. The corn variety he developed was released in 1909 and achieved greater yields than the other corn varieties that were being used. In the 1920s the use of the hybrid varieties began to make quite an impact. The development of the varieties, the use of the newly formed Cooperative Extension Service, and the teaching of Vocational Agriculture in high schools allowed the dissemination of information about the new hybrids.

Today new hybrid varieties are constantly being developed for almost all of the plants grown for agricultural use, and corn is no exception. The new varieties are developed as a result of research conducted at universities all across the country. In addition, several large commercial firms research and develop seed for sale to producers. Annually the sale of hybrid seed accounts for a sizable amount of the agricultural

Figure 4–11 *The development of hybrid corn greatly increased yields.* Courtesy of James Strawser, The University of Georgia.

Figure 4–12 *The development of a new crop variety may take as much as 10 years.* Courtesy of James Strawser, The University of Georgia.

industry. New varieties are released almost every year, and the price of the seed reflects the years of development that went into the production of the seed.

In order for a new variety to be of use it must not only reach the goals set by the plant breeder but must also breed true. This means that the seed must consistently produce plants having the characteristics that the breeder expects. This process is very time-consuming and requires several steps over several generations of plants to achieve the desired results. For example, a new variety of hybrid corn usually takes about 10 years and involves many scientists, such as geneticists, entomologists, pathologists, and statisticians (Figure 4–12).

As with the experiments of Mendel, the process begins with the development of purebred parent lines. This is done to make sure the genes that are passed along contain the dominant genes for the traits that the breeder wants. A plant breeder begins by selecting what is termed an inbred line. This means that the plant has self-pollinated over several generations and consistently produces the same traits. Not only must the scientists choose plants that hold promise for the goals they set, but they must select for other traits as well. For example, a plant breeder may want to develop a variety of corn that matures early enough to be grown in an area with a short growing season. The scientist would select a parent variety that matures early and from those plants select a small percentage of the plants that mature just a little earlier than the others. The scientist must make sure that the early maturing plants have sturdy stalks, resist insects and disease, stand up to mechanical harvesting, and produce a good yield.

To ensure that the plants possess all the desired traits, the scientist must grow several generations of the parent line. Unlike the garden peas that Mendel used, corn naturally cross-pollinates. This means that pollen from a different plant may fertilize the female gamete. The corn plant has a flower-like tassel on the top of the plant that produces the pollen (Figure 4–13). The female portion of the plant is located further down the stalk and is recognizable by the silk-like hairs that emerge from the structure. This structure is what develops into the ear of corn. In nature the wind blows pollen from one plant to the next, and it is difficult to determine just where the pollen came from. To correct this problem, plant breeders use bags to cover both the tassel and the shoots where the ear emerges to prevent unwanted pollen from getting into the ear structure

Cross Pollination

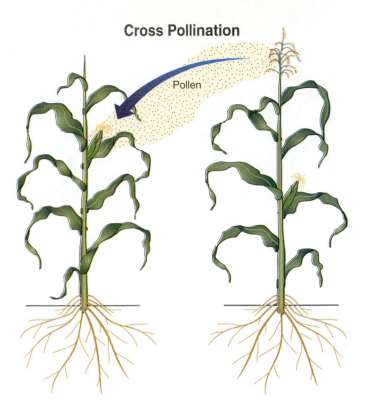

Pollen

Figure 4–13 The tassels of the corn plant produce pollen that fertilizes the female gamete inside the silks.

Figure 4–14 Bags are placed over the silk to control pollination of the plants. *Courtesy of Frank Flanders, The University of Georgia.*

(Figure 4–14). The bags that cover the tassels are used to collect the pollen for fertilization. The pollen is then dusted into the silks of the same plant and fertilization of the female gamete is achieved. By using pollen from the same plant, the genetic purity of the plant is preserved. When the seeds from this operation are harvested and planted, only the most desirable of the new plants are chosen for seed for the next generation. In fact, only about one percent of the plants are good enough to use as parents for the next generation (Figure 4–15).

When two parent inbred lines are ready to cross, the plant breeder chooses one for the female parent and one for the male variety. The tassels and silk shoots are covered in the same procedure used in producing the inbred lines, but the pollen from a plant chosen for the male parent is used. The bag is removed from the silk shoots of the plant chosen for the female parent, and the pollen is dusted in. The seed of this cross is tested numerous times to make certain that the desired results are consistent and that the new variety performs as expected. Most new varieties that are developed never are released to producers. Only those that perform the best are used in production programs.

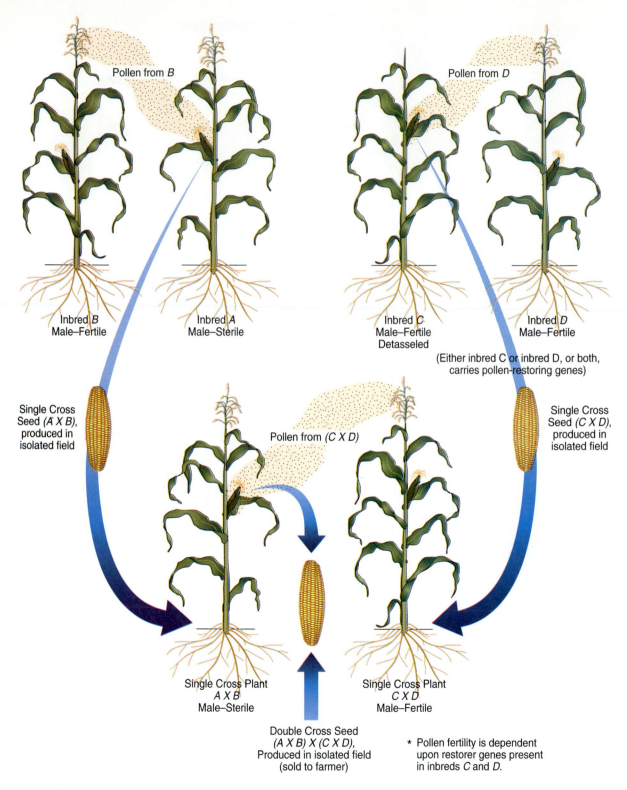

Pollen from *B*

Pollen from *D*

Inbred *B*
Male–Fertile

Inbred *A*
Male–Sterile

Inbred *C*
Male–Fertile
Detasseled

Inbred *D*
Male–Fertile

(Either inbred C or inbred D, or both,
carries pollen-restoring genes)

Single Cross
Seed *(A X B)*,
produced in
isolated field

Single Cross
Seed *(C X D)*,
produced in
isolated field

Pollen from *(C X D)*

Single Cross Plant
A X B
Male–Sterile

Single Cross Plant
C X D
Male–Fertile

Double Cross Seed
(A X B) X (C X D),
Produced in isolated field
(sold to farmer)

* Pollen fertility is dependent
upon restorer genes present
in inbreds *C* and *D*.

Figure 4–15 *Breeding a new hybrid variety takes several steps.*

ANIMAL BREEDING

The animal industry has benefited from selective breeding almost as dramatically as the plant industry. As pointed out in Chapter 1, today less than half the number of dairy cows exist in this country than there were in 1950, yet more milk is produced now. Since World War II, the size of an eight-week-old chicken has doubled, and that increase is done with less than half the amount of feed! Such increases in efficiency are largely due to advances made through selective breeding.

Most meat animals grown for slaughter are hybrid or crossbred animals. Just as in plants, hybrid animals possess the phenomenon of **heterosis** or hybrid vigor (Figure 4–16). They gain weight faster, use less feed, and are more resistant to diseases. Purebred breeds of animals are used to provide the parent stock for producing the animals that are raised for slaughter. Often breeds are developed as sire (father) breeds or dam (mother) breeds. A good example is the swine indus-

Figure 4–16 *Animal breeders also make use of the concept of hybrid vigor.*

try. Breeds of pigs such as the Landrace and the Yorkshire are developed for their mothering abilities. They bear large numbers of piglets in their litters, produce a lot of milk to feed them, and take good care of their offspring. Other breeds, such as the Durocs and the Hampshire, grow rapidly, have good feed efficiency, and produce well-muscled, good-quality carcasses. Using the Duroc or Hampshire as the sire and the Landrace or Yorkshire as the dam takes advantage of the good qualities of the different breeds.

Selective breeding has also produced new breeds of animals. A classic example is the development of the Santa

Gertrudis breed of cattle. The selective breeding program that developed that breed used two different species of cattle. Most of the breeds of cattle in this country are scientifically classified as *Bos taurus*. The zebu or Brahman breeds are classified as *Bos indicus* (Figure 4–17). About the turn of the century, cattle

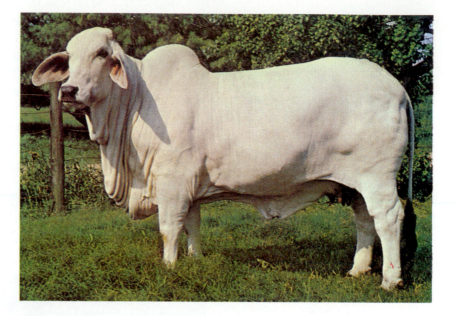

Figure 4–17 *Brahman cattle are classified as* Bos indicus, *a different species than most cattle raised in the United States.* Courtesy of American Brahma Breeders Association.

breeders on a ranch in Texas wanted to combine the docile nature and good carcass qualities of the shorthorn breed (*Bos taurus*) with the vigor and insect, disease, and heat resistance of the Brahman breed (*Bos indicus*). The process took many years of selectively breeding cattle of both breeds. The challenge was to produce animals with the desired characteristics that would breed true. When the goal of breeding true for the desired characteristics was achieved, the breeders could say that a new breed had been developed. The goal was achieved, and today the Santa Gertrudis breed is grown all across the country (Figure 4–18). Other breeds have been developed in similar fashion. The Brangus, Simbrah, Braford, and Charbray all have been developed by crossing *Bos taurus* and *Bos indicus*.

The invention of the computer has greatly enhanced animal breeders' ability to produce superior animals (Figure 4–19). Years of research has developed a measure known as heritability that can predict how much of an animal's characteristics can be passed on as a result of genetics as opposed to that characteristic being developed as a result of the environment. For example, the heritability for birthweight of pigs is only about 5 percent, whereas the heritability for the rate of

Figure 4–18 *The Santa Gertrudis was developed using the Brahman and shorthorn breeds.* Courtesy of National Santa Gertrudis Association.

Figure 4–19 *The development of the computer has greatly enhanced animal breeders' ability to produce superior animals.* Courtesy of John Wasniack, Louisiana State University.

gain from weaning to market is 40 percent. In other words, about 95 percent of a pig's birthweight is due to the environment (feed and care for the sow, for example), but only 60 percent of a pig's daily rate of gain is due to environment; the rest is genetic. Using heritability averages, breeders can concentrate on those characteristics that are most easily improved through genetics. By using computers, production data on sires and dams can be stored and analyzed. With the development of artificial insemination and embryo transplant, one animal is capable of producing large numbers of offspring. Using data from these offspring, measures such as estimated progeny differences and estimated breeding values can be assigned to a potential sire or dam. These measures can be used to select for characteristics that a producer is looking for in a sire or dam. Companies that specialize in providing semen and embryos from superior animals can use computers to match dams with sires to produce the desired characteristics in offspring.

New challenges are constantly facing animal breeders. One of the modern challenges is to produce animals that will yield a carcass with less fat in the meat. Lean animals can be produced by not feeding them as much grain and other high-energy feeds, but the meat from these animals is not as acceptable to consumers. Although fat adds flavor to the meat, at the same time it adds the cholesterol that causes consumers to be concerned. A lot of research is presently being conducted into the development of animals that will be leaner and at the same time produce meat that will be acceptable to consumers (Figure 4–20).

Figure 4–20 Research has produced animals that are leaner and still have enough fat to produce meat that will please consumers. Courtesy of American Brangus Association.

SUMMARY

Until relatively recently, researchers and breeders relied on the genes passed from generation to generation to make changes and improvements in the characteristics of plants and animals. In the next chapter, we will see how scientists are beginning to alter the genetic makeup of plants and animals to bring about desired change.

CHAPTER 4 REVIEW

Student Learning Activities

1. Imagine that a red hornless bull is crossed with a black-horned cow. Construct Punnett squares for the possible combinations of traits of their offspring if the hornless trait is dominant and the black color is dominant. Determine how many different genotypes for the offspring are possible. Also determine how many different phenotypes are possible. Remember that the genotype of the parents may be either homozygous or heterozygous.

2. Visit a farm or garden supply store that sells a wide variety of seeds or look at a seed catalog. Determine how many of the seeds are listed as being hybrid. How does the price compare with seeds that are not hybrids?

Define the Following Terms

1. dominant
2. recessive
3. alleles
4. homozygous
5. heterozygous
6. genetics
7. phenotype
8. genotype
9. DNA
10. hybrid
11. heterosis

True/False

1. Gregor Mendel developed the first workable theory of the transfer of inherited traits.
2. Each chromosome segment that controls a trait is called an allele.
3. An allele is responsible for passing on characteristics of an organism.
4. Adenine, thymine, guanine, and cystine are nitrogen bases.
5. RNA forms a blueprint for how an animal or plant is to be constructed.
6. Plant breeding has been useful only in recent history.
7. All new varieties of plants developed are released to producers.
8. Characteristics developed as a result of environment are known as the organism's genotype.
9. The phenotype of an organism is due in part to the genetic makeup.
10. Recessive genes will never express themselves.

Fill in the Blank

1. Punnett developed the _____ that can be used to show possible combinations for a particular trait.
2. Genes that posses one dominant allele and one recessive allele are called _____.
3. The letters DNA are short for _____.
4. Cells that have paired chromosomes are referred to as having a _____ number of chromosomes.
5. The molecules forming DNA are in a coiled spiral called a _____.
6. The letters RNA are short for _____.
7. About _____ of a pig's birthweight is due to genetics.
8. The study of how traits are passed from parent to offspring is called _____.

Short Answer

1. What is meant by purebred strains of plants?
2. What is the Law of Dominance?
3. How many pairs of chromosomes do cattle and horses have?
4. Name some of the factors that affect how well a plant grows and produces.
5. How do plant breeders get a hybrid variety?
6. Explain what is meant by the term "breed true."
7. Who developed hybrid corn in the United States?
8. How much of the genetic coding of an organism is contributed by each parent?
9. What is the purpose of DNA?
10. What is the purpose of RNA?

Discussion

1. Describe the impact of the computer on the science of genetics.
2. Briefly explain Mendel's theory of the transfer of traits.
3. Explain what is meant by an inbred line.
4. Describe heterosis.
5. Why do plant breeders want to adapt plants to differing environments?

GENETIC ENGINEERING

KEY TERMS

mutations

genetic engineering

E. coli

fermentation

bovine somatotropin

tissue culture

transgenetic

Animal Patent Act

APHIS

STUDENT OBJECTIVES

After studying this chapter, you should be able to:

◆ Explain how genetic mutations are used to develop new breeds of animals and new varieties of plants.

◆ Define genetic engineering.

◆ Explain the process of gene mapping.

◆ Define recombinant DNA.

◆ Analyze the methods used in gene splicing.

◆ Cite examples of how genetic engineering is currently being used.

◆ Predict some future uses of genetic engineering.

◆ Assess societal concerns about the use of genetic engineering.

◆ Summarize the laws affecting patents of genetically altered organisms.

◆ Evaluate the safeguards used in research using genetic engineering.

As pointed out in the preceding chapter, people have changed the characteristics of plants and animals for thousands of years through selective breeding. They have produced seedless watermelons, briarless blackberry bushes, docile cattle, and rapidly growing pigs that have large litters (Figure 5–1).

Another important way humans have changed the characteristics of organisms is through the exploitation of **mutations**. Mutations are variations from the normal genetic makeup of an organism. Ordinarily, the genetic transfer from parent to

Figure 5–1 *Improvements like the seedless watermelon have been accomplished through selective breeding.* Courtesy of Darby Grandbury, The University of Georgia.

offspring takes place, and the new organism takes on the characteristics of the parents. Occasionally a gene mutates or changes in such a way that new characteristics emerge. In nature these are referred to as genetic accidents or mutations. Many times when this occurs, the organism dies before it develops, but in rare instances the mutated organism lives and thrives. New varieties of plants and new breeds of animals have been developed when breeders have found mutations and developed them.

An often-used example is the development of the Polled Hereford breed of cattle (Figure 5–2). Hereford cattle are naturally horned. In the early 1900s, a rancher named Warren Gammon noticed that occasionally a polled (hornless) animal would be born in his herd of Herefords. He contacted other

Figure 5–2 *The Polled Hereford breed was developed as a result of a naturally occurring mutation.* Courtesy of American Polled Hereford Association.

breeders and bought calves from them that were born without horns. By breeding the cattle with no horns, offspring were obtained that had no horns. From these animals the Polled Hereford breed was developed.

Because breeders can take advantage of naturally occurring mutations to improve plants and animals, for many years they have dreamed of being able to create mutations artificially. Just think of the possibilities! New plants and animals could be developed that grow more quickly, have better disease and insect resistance, and higher nutritional value—the list is endless.

The design of new varieties of plants and animals has now become a reality through **genetic engineering**—the manipulation of the genes within a cell or organism to change the genetic makeup of the organism. Scientists are currently using several methods of gene manipulation to change the structure of organisms. Most of these include the removal and insertion of genetic material into organisms.

GENE MAPPING

One of the most important processes in gene manipulation is finding the location of genes on the chromosomes (Figure 5–3). This is known as gene mapping and involves locating on the strands of DNA the genes that control certain traits. Remember from the last chapter that DNA consists of long, twisted, double-stranded molecules assembled from units called nucleotides. Each nucleotide is a sugar (deoxribose), a phosphate group, and a nitrogen base that is either adenine (A), thymine (T), guanine (G), or cytosine (C). How these nitrogen bases are arranged on the DNA determines the genetic code. These substances are represented by the letters A, T, G, and C. Think of these letters as the genetic alphabet that are arranged into sequences of words. The average animal cell contains strands of DNA that if stretched out would measure over a yard long. Just think of how many combinations of the letters and words that can be arranged on these strands!

Mapping the genes on the strands of DNA involves the use of molecules that act as probes. These probes attach themselves to certain parts of the DNA where the nucleotides join each other. Remember from the last chapter that only certain nucleotides attract each other. Adenine (A) can pair only with thymine (T), and cytosine (C) can pair only with guanine (G) as

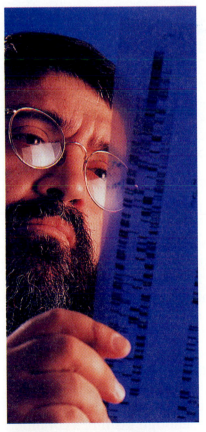

Figure 5–3 The first step in genetic engineering is to find where specific genes are located on the chromosomes. This scientist is looking at a genetic map. Courtesy of USDA-ARS.

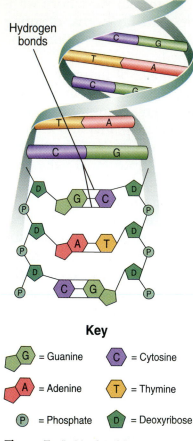

Hydrogen bonds

Key

G = Guanine   C = Cytosine

A = Adenine   T = Thymine

P = Phosphate   D = Deoxyribose

Figure 5–4 *Nucleotides are arranged on the DNA helix and are represented by the letters A, T, G, and C.*

shown in Figure 5–4. Using this fact the probe looks for combinations of where these substances join in certain sequences. For example, a genetic probe for white coat color will find and attach itself to the gene for white color and not the gene for black color. After these probes locate the nucleotides, the sequences of A, T, C, and G can be listed in a map. The genetic coding of sequences in the cell of an animal might cover 50,000 pages. It should be apparent that genetic coding is a very complicated process.

Although the coding is extremely complex, the code is the same in all organisms, even those as diverse as a one-celled paramecium and a bull that weighs a ton. The nucleotides pair only with certain other nucleotides. It is this phenomenon that makes genetic engineering both possible and mind-boggling in possibilities.

GENE SPLICING

After the DNA sequence has been located, scientists can use substances called restriction enzymes to separate the DNA at a particular location on the gene. When the pieces of DNA are removed, other DNA can be spliced in or recombined with the remaining DNA (Figure 5–5). The result is known as recombinant DNA. This new form of DNA will then reproduce with the new characteristics of the DNA that was introduced from the outside.

The first genetic splicing was done using bacteria (Figure 5–6). These one-celled organisms have circular pieces of DNA called plasmids that float freely in the cell's fluid. By selecting the proper enzyme, scientists cut out part of the plasmid DNA and insert DNA from another organism. The DNA replicates, and the new bacteria produced from the spliced DNA holds the desired characteristics.

One of the first uses of this new technology was the manufacture of human insulin. This substance is produced by the body and helps to regulate the metabolism of carbohydrates. If the body does not manufacture enough insulin, a condition known as diabetes results. Previously, insulin used to treat diabetes was harvested from the pancreases of cattle that had been slaughtered. Because only a very tiny amount could be obtained from an animal, the process was expensive. Also, cattle insulin was not exactly the same as human insulin, and some people were allergic to it.

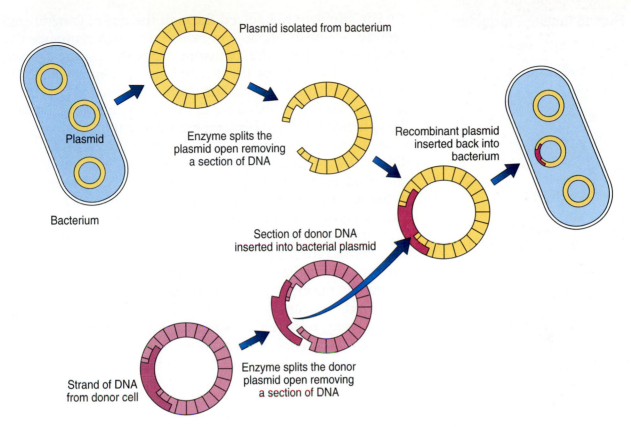

Figure 5–5 *After the sequencing has been located, pieces of DNA can be removed.*

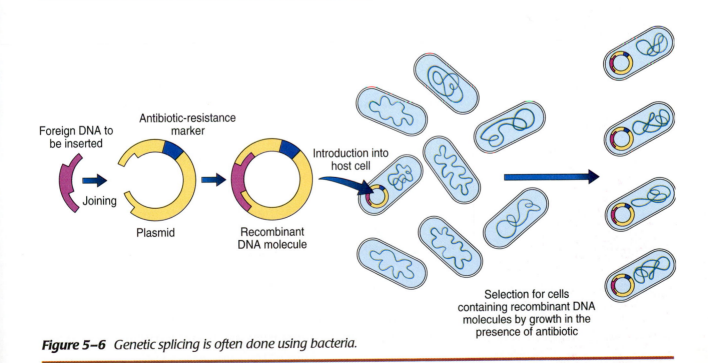

Figure 5–6 *Genetic splicing is often done using bacteria.*

Human Insulin Production

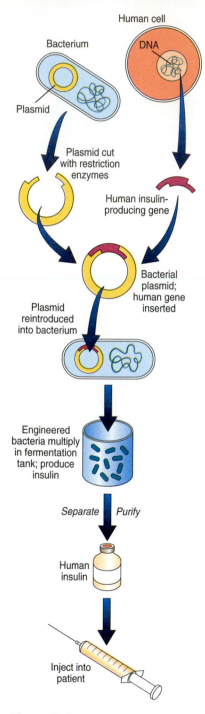

Figure 5–7 *Human insulin is produced through genetic engineering using* E. Coli *bacteria.*

This problem was solved through the use of genetic engineering. Scientists are able to isolate the DNA sequence that regulates the production of insulin. They then splice this segment into the DNA of the *E. coli* bacteria (Figure 5–7). The bacteria carrying the DNA for insulin production reproduce and pass the capability along to the next generation. Because the genetically altered bacteria are weaker than ordinary bacteria, they have to be grown under carefully controlled conditions. In a process called **fermentation**, the bacteria are produced in enormous quantities. They are given the proper nutrients, and environmental conditions are held at levels ideal for growth and reproduction of the bacteria.

When the proper number of bacteria are reproduced, they are removed from the fermentation tanks and taken apart to retrieve the insulin produced by the bacteria. The insulin is separated, purified, and the remains of the bacteria are destroyed. By keeping a supply of the genetically altered bacteria, a ready, relatively inexpensive supply of insulin is available for those people who need it.

Bacteria have become the manufacturing centers for many substances that have made the lives of humans better and more productive. Vaccines for animals and humans are produced in much the same way as insulin. Also, hormones that control growth and other bodily functions are produced using genetic engineering.

A relatively recent agricultural innovation using genetic engineering is the production of **bovine somatotropin** (BST). This substance is a hormone composed of protein that is produced by an animal's pituitary gland (Figure 5–8). BST helps control the production of milk by assisting in the regulation of nutrients into the production of milk or fat. Supplementary BST causes the cow to produce less fat and more milk. This fact has been known for many years, but the cost of producing BST has prevented commercial use. By splicing genetic material into *E. coli* bacteria, the hormone can now be produced at relatively low cost.

The development of the process of regenerating plants from a single cell (called **tissue culture**) has opened up new opportunities for genetic engineering (Figure 5–9). Through the introduction of new DNA into the single cell, a completely new plant of different genetic makeup can be grown. One method of introducing genetic material is a particle gun that literally blasts segments of DNA into plant cells (Figure 5–10). The DNA is placed on microscopic metal particles and loaded

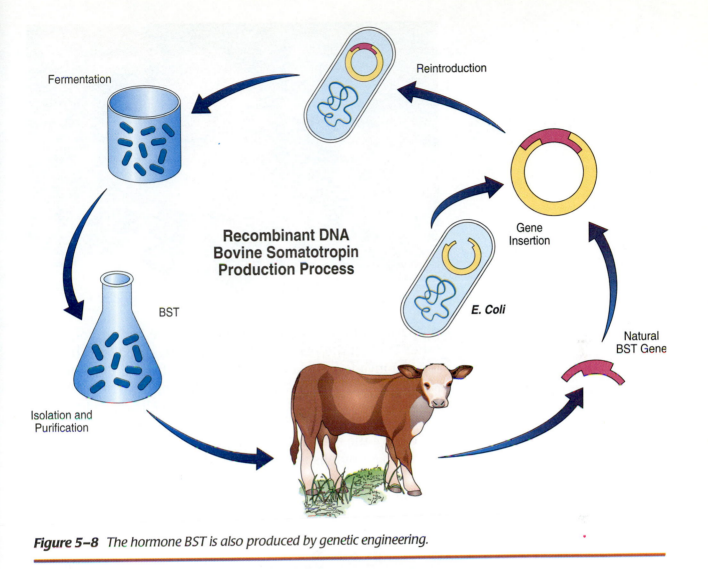

Figure 5–8 *The hormone BST is also produced by genetic engineering.*

into a particle gun containing a charge of gunpowder. The gun is fired into a plastic dish in a vacuum chamber. Inside the dish are millions of plant cells. Some of the particles of DNA enter the nuclei of the cells and combine with the cells' original genes. A new plant is produced from the cells using tissue culture. This process has been successful in developing genetically altered corn.

Another method of inserting DNA is the use of plasmids (small rings of DNA) from bacteria. The most often used bacteria, *Agrobacterium tumefaciens,* causes tumors called crown galls in plants. A plasmid causes the tumor when the bacteria invades the plant and inserts a part of the bacteria's DNA into the chromosomes of the plant. Scientists eliminate the

Figure 5–9 *The development of tissue culture has opened up new opportunities for genetic engineering.* Courtesy of USDA-ARS.

properties that allow the plasmid to cause the tumor, place other DNA in the plasmid of the bacteria, and allow the bacteria to invade the plant cells. The DNA from the genetically altered plasmid integrates with the DNA of the plant cells and becomes part of the chromosomes. The cells reproduce and create cells and plants with new characteristics.

One of the most interesting genetic engineering feats that has been accomplished occurred when scientists spliced DNA from fireflies into a tobacco plant. Genes that control the illumination of the fireflies were inserted into the tobacco plant. Under the proper conditions, the genetically altered plant

Figure 5–10 *DNA can be inserted using a particle gun that shoots DNA into the nucleus.* Courtesy of Illinois Research, University of Illinois.

would glow in the dark! The material that cause the fireflies to glow in the dark has uses in medicine. The big problem in its use is the difficulty in obtaining enough material to be practical. If the material can be generated through the growing plants, it would become readily available and inexpensive.

By far the most widely applied use of genetic engineering is in crop production. Currently dozens of different genetically engineered crops ranging from tomatoes and squash to corn and soybeans are being produced all over the world. Genetically engineered crops account for more than 86 million acres worldwide, and the amount is growing rapidly each year (Figure 5–11). Projections are that early in this century, more than 300 million acres of genetically engineered crops will be planted. Many scientists think that in the future, genetically engineered crops will be standard with conventional crops being only a minor part of production.

Figure 5–11 *Genetically engineered crops account for more than 86 million acres of crops. This amount will triple during the early part of this century. Courtesy of USDA/Dave Warren.*

The most common type of genetic engineering in crops is the insertion of genes that produce tolerance to disease, insects, or herbicides. A good example is the use of a gene from a soil bacteria *Bacillus thuringiensis* (Bt) that secretes a toxin effective in controlling many insects (Figure 5–12). This principle is that the plant secretes an insect controlling toxin that is nontoxic to mammals and most other animals. This technology is used in a variety of crops from corn to cotton. The largest concern is that insects may in time become tolerant to the Bt crops and render the genetically engineered plants ineffective in controlling insects.

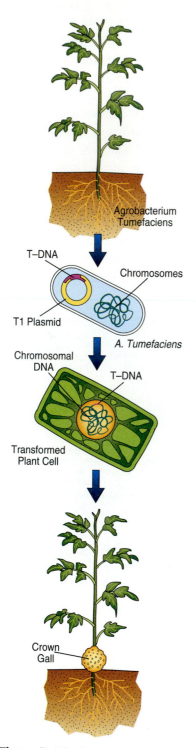

Figure 5–12 Bacteria can be used to place genetic material in plant cells.

Scientists have been able to develop plants that are resistant to herbicides—chemicals that are used to kill weeds or unwanted plants. When the weeds are growing among a crop, a herbicide must be used that will kill the weeds and not harm the crop. The use of genetic engineering can enable the inserting of genes into a crop plant, which will make the plant tolerant of herbicides. One success story is that of safflower. This crop is grown for the seed from which cooking oil is pressed. A major problem in production is that the safflower plants are extremely susceptible to herbicides. This means that chemicals that are effective in killing weeds also tend to kill the safflower plants. Through the use of genetic engineering, scientists at Montana State University developed a safflower plant that is herbicide resistant. Other crops such as soybeans also were developed to be tolerant to herbicides like Roundup®, which generally kills most plants it contacts. This greatly expands the ability of producers to control weeds in their crops.

Scientists at the University of Georgia have developed peanut varieties that are resistant to viral diseases. These diseases have always been difficult to control. By inserting a gene from the virus into peanut cells and then producing plants from the cells, the resulting plant will contain the virus gene in every cell! When the attacking virus contacts a cell, it gets the sense that the plant cell is already infected and withdraws. Because only a small part of the virus gene is used, the plant does not contract the disease as a result of the introduced virus DNA. This development has the potential for significantly reducing the cost of controlling disease in peanuts (Figure 5–13).

The use of genetic engineering has opened up many exciting possibilities. Just think of cotton that comes in a variety of colors or grain crops that produce their own nitrogen like legumes. The nutritive value of grains and other plants can be significantly increased through the manipulation of the genetic code. For example, the ability of a corn plant to store more protein in the seeds would improve the value of corn as a livestock feed. This is already a reality. Researchers at Louisiana State University introduced a gene from green beans into a rice plant. The result was a plant that produced rice with a higher protein content. Genetically altered plants that withstand cold or tolerate dry conditions could open up vast areas of the Earth that are not presently very productive.

In the area of animal agriculture, genetic engineering also has tremendous potential. Besides the production of hormones, rapid improvements in the growth and feed efficiency of animals could be achieved. For example, some scientists predict that in the 21st century, producers may be able to grow a 4-pound broiler chicken in just 25 days! This is in contrast to the 45 days it now takes. This will come about through the introduction of genetically engineered disease resistance and increased feed efficiency.

Another exciting possibility is the development of genetically developed microorganisms that live in the digestive systems (rumen) of cattle. Ordinary microbes in the rumen digest large amounts of cellulose from plants. This allows cattle to use roughages such as grass and hay. If microbes could be developed that could digest lignin, which is the binder in wood fibers, then cattle could eat woody products. Just think how cheaply cattle could be fed if they could digest and use sawdust! One possibility would be to extract DNA from termites and add it to the genetic structure of the rumen microbes. Termites digest lignin and thrive on wood.

Transgenetic animals are animals produced through transferring genetic material between different types of animals. For example, the gene that controls human growth has been spliced into the genetic material of a mouse. The mouse grew twice as large as its brothers and sisters. When the mouse reproduced, the offspring also grew twice as large as normal. The transfer of genetic material in animals occurs shortly after the egg is fertilized and before the fertilized cell begins to multiply. This is done by using an instrument called a micro-manipulator that maneuvers an extremely tiny glass needle (Figure 5–14). While looking through a microscope, the scientist places the DNA fragments in the cell at the proper location.

Figure 5–13 *Genetic material of a virus can be injected into peanut plants to make the plants immune to the virus. These peanuts have been genetically altered.* Courtesy of James Strawser, The University of Georgia.

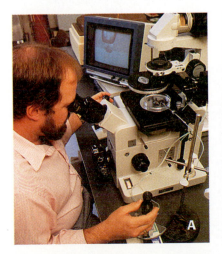
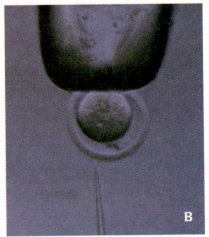

Figure 5–14 *Scientists can place genetic material into animal cells using a micro-manipulator. (A) Micro-manipulator. (B) As seen through the microscope.* Courtesy of Illinois Research, University of Illinois.

BIO BRIEF

Cah Marker Gene Aids Plant Transformation

Scientists aiming to genetically engineer crop plants are always on the lookout for genes they can use as markers in their experiments. Now, a gene known as cah may prove a useful addition to today's limited array of choices. That's according to Agricultural Research Service geneticist J. Troy Weeks in Lincoln, Nebraska. Weeks and other biotechnologists pair a marker with an experimental gene intended to give a plant some prized trait, such as increased resistance to its worst insect or disease enemies.

The marker clearly flags, for researchers, plant cells that have the new, useful gene. Other plant tissue then can be discarded so scientists can focus their efforts on the potentially useful tissue. They nurture it into laboratory plantlets and later into greenhouse plants.

The marker gene that Weeks and colleagues are testing enables plants to tolerate a chemical called cyanamide. This versatile compound has some seemingly contradictory uses. It is an environmentally friendly fertilizer that, in specific situations, can also act as a herbicide or fungicide.

Weeks' laboratory use of cyanamide exploits the first two of these uses. His cah marker gene enables callus—clumps of plant tissue—grown in petri dishes to convert cyanamide into urea fertilizer. Normally, plants cannot do this.

In petri dishes, cah-equipped cells soon appear as greenish sections against the yellow-white callus tissue. In later stages, this callus may develop healthy shoots and roots. Cells not containing the cah gene may appear brownish. The few short roots that may develop from this tissue usually won't survive. But the cah-containing cells, nourished by the cyanamide fertilizer, will thrive.

In addition to these readily apparent differences in callus, scientists can check for the cah gene with a fast, simple, and inexpensive assay. Unlike tests used to detect some other marker genes, the assay for cah does not require hazardous chemicals.

Where It's From, What It Does

The cah gene comes from a fungus found naturally in soils. Weeks saw the gene's biotech potential because he knew two things: The fungi—because of this gene—play an essential role in converting cyanamide into a useful fertilizer for plants, and without the fungi, cyanamide may kill the plants instead of feeding them.

When used as a fertilizer, cyanamide must be applied before plants emerge. Soil microorganisms break it down into urea that plants can use. The microorganisms that do this job include the fungus Myrothecium verrucaria. The cah gene was borrowed from the fungus. The gene cues the fungus—or in this case, the genetically engineered plant cells containing the cah gene—to create an enzyme called cyanamide hydratase. The enzyme enables the fungus or cah-equipped plant cell to add the water molecule necessary to convert the fertilizer form of cyanamide, called calcium cyanamide, into urea. Normally, if the calcium cyanamide fertilizer were mistakenly applied to plants, they might turn yellow and die—or at least produce lower yields. In the laboratory, Weeks moved the cah gene into wheat cells using a bioblaster, or gene gun. The gun shoots microscopic gold particles, coated with experimental genes, into plant tissue in petri dishes. Then he grew the cells on a gelatinous bed of nutrients spiked with cyanamide.

Cells with the cah gene working inside could then convert cyanamide into urea fertilizer and grow shoots and roots that were the start of new plants. So far, Weeks has produced more than 100 healthy wheat plants with the cah gene inside. He is seeking a patent for his work.

—By Marcia Wood,
Agricultural Research Service Information Staff.

Source: Agricultural Research Magazine.

A bioblaster, or gene gun, is used to place genes into plant tissue. Courtesy of USDA-ARS.

SOCIETAL CONCERNS

Genetic engineering has the potential to be revolutionary in the production of food and fiber. Scientists estimate that more than six billion people are in the world. All of these people have to be fed and clothed. With genetically altered plants and animals, the efficiency of producing food can be increased, and areas that are currently of little agricultural value can be made productive.

Although great potential is presented by genetic engineering, there is also widespread concern over the use of genetically altered organisms. One issue is ownership of genetic materials. Companies that spend millions of dollars developing organisms through genetic engineering feel that they need some way to protect their interests. Other people object because they feel that no one has the right to have exclusive ownership of genetic material, especially human material. In 1986 Congress passed the **Animal Patent Act** that allows the holding of patents for genetically engineered animals. Plants are similarly covered under the Plant Variety Protection Act. Since that time more than 200 patent applications for genetically altered organisms have been filed.

Some people are concerned about the consumption of foods that were produced from genetically engineered plants and animals. They rationalize that by altering the genetic makeup of an organism, the entire organism is changed, and the effects of the alteration are yet unknown. The fear is that the genetically engineered plant or animal could produce substances harmful to humans. Defenders of genetic engineering counter with the argument that the placing of genes is much more precise than that of selective breeding and that harmful effects can be and are controlled. No genetically altered product is released without extensive testing.

People have also been concerned about the introduction of genetically altered organisms into the environment. They cite examples of plants and animals that were introduced with good intentions and today have escaped and cause problems. A good example is kudzu, a plant that was introduced around the turn of the 20th century into the southern United States (Figure 5–15). Kudzu was brought in from Japan because it grew rapidly, was of high nutritive content that made good grazing for cattle, and it helped in the control of erosion. Soon after its introduction, it escaped into the wild

Figure 5–15 *Kudzu was introduced to the United States to help solve problems. It created other problems.* Courtesy of James Strawser, The University of Georgia.

Figure 5–16 *At least 13 species of fish have been developed through gene manipulation.* Courtesy of James Strawser, The University of Georgia.

and rapidly grew out of control. Today large areas are covered by the vines. It takes over most plants in its path and kills trees by covering them and blocking out sunlight. Many people can visualize a similar thing happening with new organisms developed through gene manipulation.

An example of this dilemma is the production of genetically altered fish. Currently at least 13 species of fish have been developed through gene manipulation (Figure 5–16). One area of development is that of growth hormone genes that are injected into newly fertilized eggs. This gene usually comes from another species of fish that grows faster than the recipient fish. The result is that the newly developed fish grow from 20 to 100 percent faster than the nonaltered species. The question is how these genetically altered fish will affect the environment. Will they consume more than their share of available food? Will their spawning habits change? What will be the result of breeding with normal fish that inhabit the waters? Because the answers to these questions are not known, these genetically altered fish have not been released into natural waterways.

Perhaps the greatest concern is over genetically altered microorganisms. The fear is that after microorganisms are released into the environment, there is no way of removing them. People note that the most commonly used organism in gene alteration is the *E. coli* bacteria that normally live in the human intestinal tract. Because the bacteria multiply so rapidly and the human body is a host for these bacteria, many people feel that we do not know enough about genetic alteration to allow manufactured strains of bacteria into the environment.

In the early 1980s, a strain of bacteria was developed that helped protect plants from frost damage. When word got out that scientists were going to field test the bacteria, many people objected. They were afraid that the altered bacteria would spread throughout the soil and replace the normal soil microbes. Also, there was concern that the scientists did not know all of the characteristics of the bacteria, and they might cause irreparable damage to the environment. The issue was settled after four years of legal challenges, and the scientists were allowed to spray plants with the bacteria. Subsequent tests have revealed no evidence of problems.

Proponents of the use of genetic engineering point out that although some imported plants and animals have escaped and caused problems, in hundreds of cases, organisms have been brought into the country and have had a highly positive impact. In fact, most of the agriculturally produced plants and animals have been brought in from other countries. Proponents feel that the same holds true for organisms that are created through genetic alteration. Although no absolute guarantees can be given that there will be no problem, if the proper precautions are taken, the risks can be held to a minimum.

REGULATION OF GENETIC ENGINEERING

Scientists conduct research on genetic engineering in closely controlled laboratory situations (Figure 5–17). The buildings and equipment are designed to prevent the escape of newly engineered microorganisms into the environment. Research in these facilities may take many years before the determination is made that the newly designed organism will be tested outside the laboratory.

The United States Department of Agriculture (USDA) has developed strict guidelines for testing genetically altered

GENETIC ENGINEERING ✦ 115

Figure 5–17 *Scientists conduct research on genetic engineering under closely controlled conditions.* Courtesy of James Strawser, The University of Georgia.

organisms if the new organisms can come in contact with the environment. The management and enforcement of these regulations come under the control of the Animal and Plant Health Inspection Service (**APHIS**), which is a component of the USDA. In addition, the USDA has developed guidelines that all USDA research agencies use prior to approving a field test of any genetically altered organism (Figure 5–18).

Agriculturists sometimes become upset at activist groups who attempt to stop innovations such as genetic engineering. Research scientists point out that these groups are often misinformed and give out misleading information. However, these groups do provide a useful service. Their protests are another way of regulating the process of experimentation. Heightened public awareness helps to ensure that those doing research remain diligent and do not become careless. It also helps everyone remain conscious of the fact that problems could result from carelessly conducted research on genetically altered organisms.

The public is concerned with the use of organisms they think might cause problems. However, with governmental regulation and the strict code of ethics used by most research scientists, the benefits of genetic engineering far outweigh the potential harm.

Jurisdiction of Federal Agencies Over Research on Recombinant DNA Technology and Associated Technologies (Office of Science and Technology Policy, 1986)

Subject	Responsible Agency or Agencies[a]
Contained research, no release in the environment	
Federally funded	Funding agency[b]
Nonfederally funded	NIH or S&E voluntary review, APHIS[c]
Foods, food additives, human drugs, medical devices, biologies, animal drugs	
Federally funded	FDA[d], NIH guidelines and review
Nonfederally funded	FDA[d], NIH voluntary review
Plants, animals, and animal biologies	
Federally funded	Funding agency[b,d], APHIS[c]
Nonfederally funded	APHIS[d], S&E voluntary review
Pesticidal microorganisms	
Genetically engineered	
Intergeneric	EPA[d], APHIS[c], S&E voluntary review
Pathogenic intrageneric	EPA[d], APHIS[c], S&E voluntary review
Intrageneric nonpathogens	EPA[d], S&E voluntary review
Nonengineered	
Nonindigenous pathogens	EPA[d], APHIS
Indigenous pathogens	EPA[d,e], APHIS
Nonindigenous nonpathogens	EPA[d]
Other releases of microorganisms in the environment	
Genetically engineered	
Intergeneric organisms	
Federally funded	Funding agency[d,b], APHIS[c], EPA[f]
Commerically funded	EPA, APHIS, S&E voluntary review
Intrageneric organisms	
Pathogenic source organism	
Federally funding	Funding Agency[d,b], APHIS[c], EPA[f]
Commercially Funded	APHIS[c], EPA[g]
Intrageneric combination, no pathogenic source organisms	EPA[h]
Nonengineered	EPA[d,h], APHIS[c]

[a]NIH = National Institutes of Health, B&E = Science and Education Administration (U.S. Department of Agriculture), APHIS = Animal and Plant Health Inspection Service (U.S. Department of Agriculture), FDA = Food and Drug Administration, EPA = Environmental Protection Agency.
[b]Review and approval of research protocols conducted by the National Institutes of Health, the Science and Education Administration, or the National Science Foundation.
[c]The Animal and Plant Health Inspection Service issues permits for the importation and domestic shipment of certain plants and animals, plant pests, and animal pathogens, and for the shipment or release in the environment of regulated articles.
[d]Lead Agency.
[e]The Environmental Protection Agency has jurisdiction over research on a plot larger than 10 acres.
[f]The Environmental Protection Agency reviews federally funded environmental research only when it is for commercial purposes.
[g]Lead agency if nonagricultural use.
[h]Under the jurisdiction of the Environmental Protection Agency, but the agency will require only an informational report.

Figure 5–18 *Several different agencies regulate the use of genetically altered organisms.*

SUMMARY

Genetic engineering has the potential to revolutionize agriculture. The endless possibilities of developing improvements in plants and animals can help solve world hunger. As scientists develop genetically modified organisms, great care must be taken to ensure that the new organisms are not harmful to humans or the environment. With the proper safeguards, genetic engineering will play an important role in the future of agriculture.

CHAPTER 5 REVIEW

Student Learning Activities

1. Take one side or the other on the issue of releasing genetically altered organisms into the environment. Go to the library and collect information on genetic engineering and public concerns. Organize your arguments either for or against and debate the topic with classmates who take the opposing view.

2. Think of a problem you consider to be of great importance to agriculture. What are some ways that genetic engineering might help the situation. What organisms would you use? What traits would you alter? Why? What would be some potential hazards of releasing the organism? Report to the class.

Define the Following Terms

1. mutations
2. genetic engineering
3. *E. coli*
4. fermentation
5. bovine somatotropin
6. tissue culture
7. transgenetic
8. Animal Patent Act
9. APHIS

True/False

1. Selective breeding to change the characteristics of plants and animals is a relatively new practice.

2. Many mutated organisms die before they develop.

3. The use of mutations to make improvements in plants and animals creates only limited possibilities.

4. Bovine somatotropin is developed from the pancreases of cattle.

5. In order to introduce new genetic material, DNA can be blasted into a cell by a particle gun.

6. Through genetic engineering, scientists have been able to develop plants that are resistant to herbicides.

7. Most of our society has a very positive outlook on genetically altered organisms.

8. Kudzu is an example of an introduced organism that has been beneficial to the environment.

9. The United States Department of Agriculture has left the regulation of genetically altered organisms mainly to the authority of state agencies.

10. Governmental regulations and the ethics of most research scientists greatly limit the potential danger posed by genetically altered organisms.

Fill in the Blank

1. Polled Hereford cattle were developed through the use of naturally occurring _____ .

2. Most methods of gene manipulation involve the _____ and _____ of genetic material into organisms.

3. The four nucleotides found in DNA are _____, _____, _____, and _____ .

4. The process of mapping genes in the strands of DNA involves the use of molecular _____.

5. The first genetic splicing was done using one-celled organisms that are called _____.

6. The manufacture of human _____ was one of the first uses of genetic splicing.

7. Bacteria can be produced in large quantities through a process called _____.

8. _____ _____ is the process of regenerating plants from a single cell.

9. The _____ is an example of one plant that has been successfully engineered for herbicide resistance.

10. The transfer of genetic material into animals is performed with the use of an instrument called a _____.

Discussion

1. Name three advancements in agriculture that have resulted from selective breeding.

2. What is the effect of supplemental BST in cows?

3. Give two examples of how genetic engineering has improved agricultural crops.

4. Why must genetically altered bacteria be grown in a carefully controlled environment?

5. Describe how new genetic information is introduced into an organism through gene splicing.

6. What are some societal concerns over genetically altered organisms?

7. What precautions are being taken to minimize the risks that might be imposed by genetically altered organisms?

PLANT SYSTEMS

KEY TERMS

stoma
chlorophyll
photosynthesis
vascular bundles
xylem
phloem
respiration
node
tap root

STUDENT OBJECTIVES

After studying this chapter, you should be able to:

◆ Explain why plants are essential for life.

◆ Explain the functions of the various layers of a leaf.

◆ Discuss the process of photosynthesis.

◆ Discuss the process of respiration.

◆ Distinguish between xylem and phloem.

◆ Discuss methods used to identify plants by their leaves.

◆ Distinguish between woody and herbaceous stems.

◆ Explain the functions of the layers of a plant stem.

◆ Distinguish between the types of root systems.

◆ Explain why plants are the basis of agriculture.

◆ List various crops grown for a specific plant part.

◆ Explain why root systems are essential in soil conservation.

Plants are the basis for all of agriculture. Because they provide food for humans and for the livestock we raise, all of our food comes directly or indirectly from plants (Figure 6–1). In fact, all of life is dependent on plants. Plants convert energy from the sun into energy for animals in the form of food. All of our grains, fruits, and vegetables come directly from plants. All of our food of animal origin—meats, milk, cheese, and other products—comes from animals that have been fed plants. Even the fish we consume were raised on feed of plant origin.

Figure 6–1 *All of our food that comes from animal sources originated as food from plants.* Courtesy of James Strawser, The University of Georgia.

Plants manufacture and store food nutrients for use by the plant. Some plants store these nutrients in the leaves, others in the stems, and some in the roots. Both humans and animals consume plant parts that contain stored nutrients. The part of the plant that is consumed depends on where the nutrients are stored. For example, carrots store nutrients in the roots. These large, orange roots taste good and are eaten by animals and humans. The stems of celery and the leaves of spinach are eaten.

In addition, plants stabilize the soil by preventing erosion. Plants keep topsoil from being moved by wind and rain. Plants growing on hills and mountains keep the soil in place and provide cover and food for wildlife (Figure 6–2).

Figure 6–2 *Plants hold soil in place on hillsides.* Courtesy of Progressive Farmer.

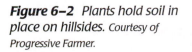

Another vital function of plants is replenishing the Earth's oxygen supply by giving off oxygen and using carbon dioxide. Animals breathe in air, use oxygen, and exhale carbon dioxide. Their bodies need oxygen to survive. Plants do the opposite—they take in air, use carbon dioxide, and give off oxygen. Through this cycle, the correct balance of oxygen and carbon dioxide is maintained in the atmosphere (Figure 6–3). Much concern has been expressed in the past few years over the destruction of the world's forests. Because so much of the world's oxygen supply is given off by plants, the large-scale destruction of plants could lead to problems with the air we breathe.

Plants also provide beauty for our world. Just think of the beautiful scenery you have seen. Most of that beauty is due to plants. The beautiful green colors of spring and the radiant colors of autumn are all produced by plants. A multimillion dollar agricultural industry has grown around the production of plants that beautify our environment (Figure 6–4). Plants beautify our homes by providing landscaping outside and potted plants inside.

Animals have digestive, nervous, reproductive, and other systems. These systems allow animals to live, grow, and

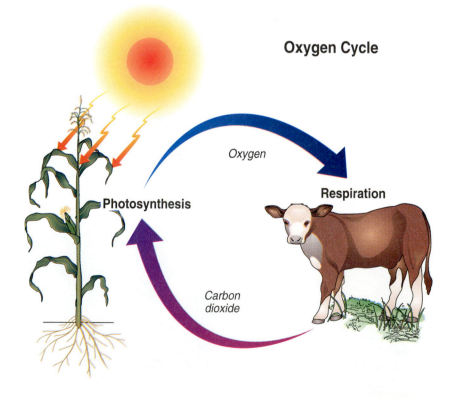

Oxygen Cycle

Oxygen

Photosynthesis

Respiration

Carbon dioxide

Figure 6–3 *The oxygen cycle helps keep the correct balance of oxygen and carbon dioxide in the air.*

Figure 6–4 *A multimillion dollar industry has developed around producing plants that beautify our homes.* Courtesy of James Strawser, The University of Georgia.

reproduce. Plants also have systems that serve the same functions. These systems include leaves, stems, roots, and flowers. Flowers are the reproductive system of most plants and will be discussed in Chapter 7.

LEAVES

Much of the lush green scenery we see about us is made up of leaves. Plant leaves come in an almost endless variety of shapes and sizes (Figure 6–5). Leaves manufacture and store food for the plant. The energy of the sun is used to convert raw materials such as water, carbon dioxide, calcium, phosphorus, potassium, and nitrogen into food for the plant. This food is in the form of sugars and carbohydrates as well as fats and proteins. Because this food is stored in leaves, animals eat the leaves. Over the years, agriculturists have discovered and developed plants with leaves that are high in nutritive value and that livestock will eat. Cattle, sheep, and horses use the leaves of grass and other plants as forage (Figure 6–6). Producers cut and dry these forages into hay that is stored and used as livestock feed. Humans also eat the leaves of plants. As with plants used to feed animals, scientists have developed plants that make good food for humans. Cabbage, mustard, lettuce, and spinach are all examples of plant leaves eaten by humans. Leaves are a primary place for food storage in plants. Many of the nutrients needed by animals and humans are obtained from eating plant leaves.

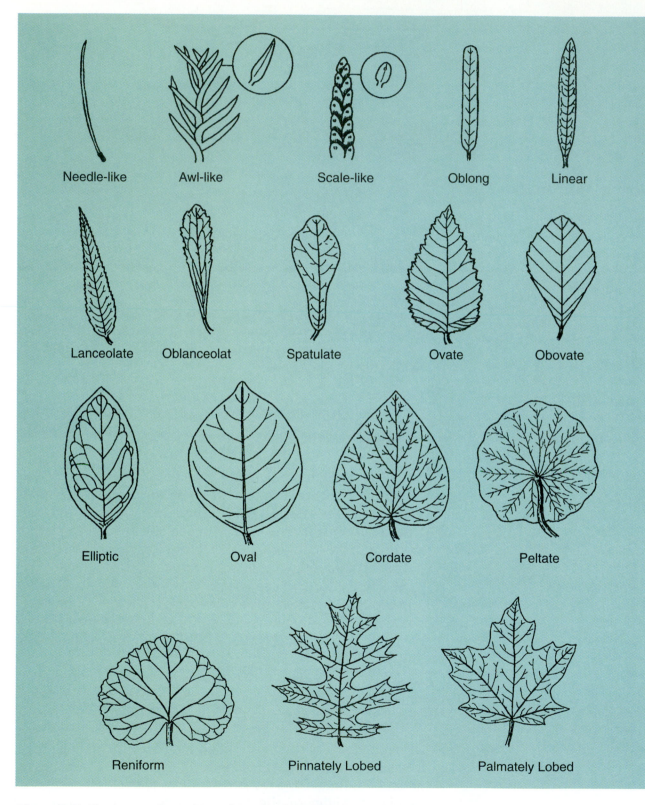

Figure 6–5 *Plants come in a wide variety of leaf shapes.*

Figure 6–6 *Cattle make use of leaves that are cut and dried into hay. Courtesy of Progressive Farmer.*

The Structure of Leaves

Leaves are made up of several different layers (Figure 6–7). The outermost layer consists of a coating of waxy material known as the cuticle. This coating helps the leaf retain water by slowing down the rate of evaporation of the water inside the leaf. It also helps prevent disease organisms from entering the plant.

Beneath the cuticle is a layer of cells called the epidermis. This layer protects the leaf and covers the entire surface of the leaf on the top and on the underside. Within the epidermis are openings called **stoma** that allow air into the leaf and water vapor and oxygen to move out (Figure 6–8). The opening of the stoma is controlled by cells on both sides called guard cells. These cells open and close due to the turgor of the guard cells. Turgor is the firmness of the cells due to the amount of water in the cells. When the concentration of water in the cells is high, the guard cells are open, and water

A Section of a Dicot Leaf

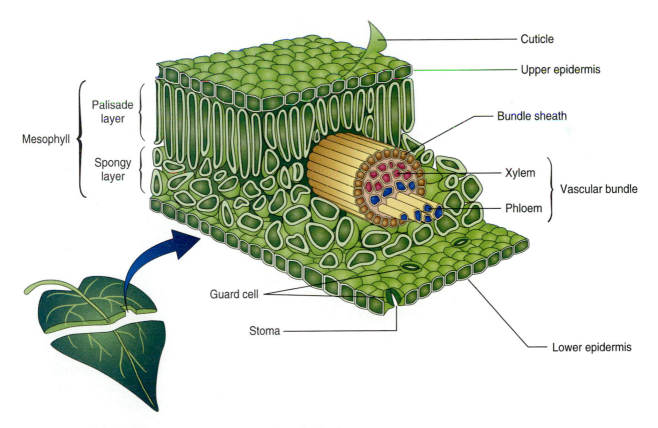

Figure 6–7 *This diagram shows the cross section of a leaf.*

Lower Epidermis of a Leaf

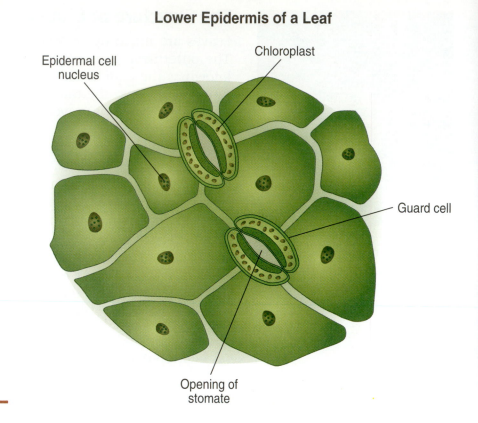

Epidermal cell
nucleus

Chloroplast

Guard cell

Opening of
stomate

Figure 6–8 *Stoma control the entrance of air and the exit of water vapor and oxygen.*

vapor and oxygen escape. When the concentration of water in the guard cells is low, the cells relax, and the stoma closes. Usually the guard cells are open in the daylight and closed in the darkness. About 90 percent of the water taken in by the plant is passed out through the stoma, and all of the carbon dioxide used by the plant is taken in through the stoma. The closing of the stoma prevents the loss of too much water.

Between the epidermis layers is the mesophyll layer. This layer is divided into two sublayers called the palisade mesophyll and the spongy mesophyll. The palisade mesophyll consists of elongated cells that contain structures called chloroplasts. These structures contain a green substance called chlorophyll and are the sites of photosynthesis. The molecules of **chlorophyll** capture light energy from the sun and use it for the chemical reaction of **photosynthesis**, the process by which the leaf makes food for the plant. "Photo" means light, and "synthesis" means to put together. The spongy mesophyll contains chloroplasts but not as many as the palisade layer. Instead, the spongy layer contains air spaces that are in contact with the atmosphere through the stoma.

Within the spongy mesophyll are the **vascular bundles**. These bundles are ridge-like veins that run through the leaves and contain the xylem and the phloem. The **xylem** are tubes that bring water from the roots to the leaves of the plant. The **phloem** are tubes that carry the products of photosynthesis to the other parts of the plant.

In photosynthesis, the broad surface area of the leaf absorbs sunlight to be used as energy. Carbon dioxide is taken from the air, and the other nutrients (including water) are taken from the soil and transported to the leaves in the water through the xylem. The chloroplasts create a chemical process that converts these raw materials into usable food for the plant.

The chemical reaction takes carbon dioxide and water and converts these materials to sugar and oxygen. The diagram for the reaction is as follows:

$$6CO_2 + 6H_2O \rightarrow C_6H_{12}O_6 + 6O_2\uparrow$$

Because this reaction requires sunlight, the importance of sun for plants to grow is apparent. Agricultural crops need a varying amount of sunlight depending on the type of plant. Some row crops such as soybeans, cotton, and corn require a lot of sunlight, and some ornamental plants such as azaleas and African violets require relatively little sunlight.

To maintain its systems and to grow, plants use some of the food stored from the photosynthesis process (Figure 6–9). This breaking down and utilization of plant foods is called **respiration**. Respiration is the reverse of photosynthesis. In respiration, sugar and oxygen are broken down, releasing carbon dioxide, water, and energy. The reaction is diagrammed as follows:

$$C_6H_{12}O_6 + 6O_2 \rightarrow 6CO_2 + 6H_2O + energy$$

About 25 percent of the sugar manufactured by photosynthesis is used by the plant for its own energy. As this energy is released in respiration, fats, proteins, and other products are also broken down for plant use. Although respiration takes place 24 hours a day, photosynthesis takes place only during the daylight hours.

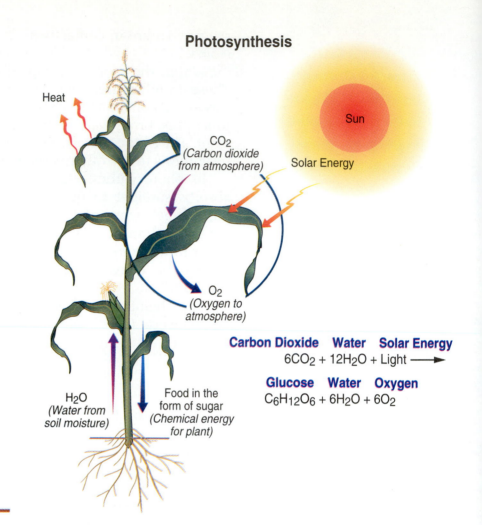

Photosynthesis

Heat

CO_2
(Carbon dioxide from atmosphere)

Sun

Solar Energy

O_2
(Oxygen to atmosphere)

Carbon Dioxide   Water   Solar Energy
$6CO_2 + 12H_2O + Light \longrightarrow$

Glucose   Water   Oxygen
$C_6H_{12}O_6 + 6H_2O + 6O_2$

H_2O
(Water from soil moisture)

Food in the form of sugar
(Chemical energy for plant)

Figure 6–9 Diagram of the process of photosynthesis.

Leaf Types

Most leaves have two basic parts, the blade and the petiole. The blade is the broad flat portion of the leaf, and the petiole is the stalk that connects the blade to the stem. Many plants (called deciduous plants) lose their leaves in the fall season. Other plants (called evergreen) lose their leaves throughout the year as the leaves age (Figure 6–10). The individual leaves age and fall off all year and not just in the autumn. The tree appears green all year round.

On the petiole, is a layer of specialized cells, called the abscission layer (Figure 6–11). It is at this point that the leaf separates from the stem and falls off. Scientists have discovered that this process is caused by a substance called auxin. During the spring and summer months, the leaf is growing and producing its own auxin. When the cooler months arrive, the growth slows down, and a greater amount of auxin is in

Figure 6–10 Evergreen trees appear green throughout the year while deciduous trees shed their leaves. (A) Evergreen (B) Deciduous tree in winter. *Courtesy of James Strawser, The University of Georgia.*

the stem than in the leaf. When this occurs, the leaf drops off. In evergreen plants, the leaf may age, or drought conditions may lessen the production of auxin and cause the leaf to fall off.

Leaf Structure of a Deciduous Plant

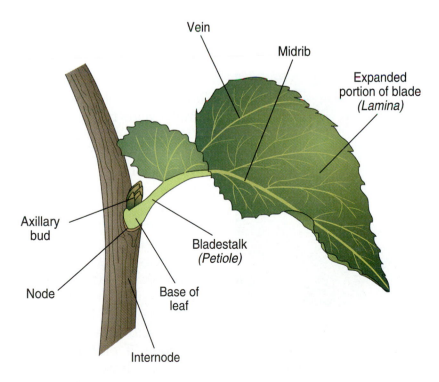

Figure 6–11 The abscission layer is located on the petiole.

In some leaves, there is no petiole, and the leaf is attached directly to the stem. These leaves are called sessile leaves. Most grasses have sessile leaves (Figure 6–12). Plants with sessile leaves usually are annuals that die each year and come back from seed.

Features of a Grass Leaf

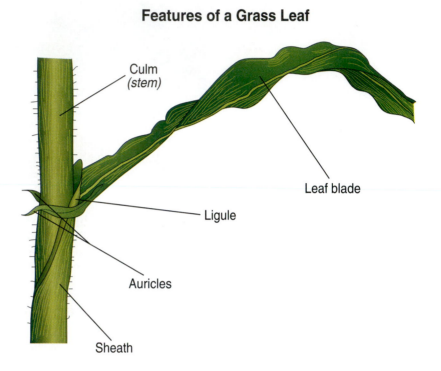

Figure 6–12 *Most grasses have sessile leaves.*

Figure 6–13 *Corn leaves have parallel veins. Courtesy of Michael Dzaman.*

Leaves have veins that run throughout the blades. In different types of leaves the veins are arranged differently. Parallel veins run the length of the leaf and are approximately the same distance apart. Corn leaves have parallel veins (Figure 6–13). Other types of leaves have veins that branch out and form a network. These are called net veins. If the veins branch from a main vein that runs the length of the leaf, the leaf is pinnate. An example of this vein design is the leaf of a red oak tree (Figure 6–14). If the leaf has several main veins originating at the base of the leaf, it is palmately veined. Maple leaves are palmately veined (Figure 6–15).

Leaves are also classified according to their arrangement on the stem. Generally, one or more leaves are attached at a node. The **node** is the part of the stem from which the leaf sprouts and grows. The space between nodes is called the internode. If only one leaf is attached at the node, it is a simple leaf. If two or more leaflets are attached at a node, it is

Silver Maple

Red Oak

Figure 6–14 *A red oak has pinnate leaf veins.*

Figure 6–15 *A maple has palmate veins.*

a compound leaf. Leaves are arranged in several ways on the stem. The three most common arrangements are whorled, alternate, and opposite (Figure 6–16). In a whorled arrangement, three or more leaves originate from the same node. In an alternate arrangement, only one leaf originates from each node, and the leaves are not directly across from each other. In an opposite arrangement, leaves originate from separate nodes directly across from each other.

Leaves are also identified by shape. Figure 6–17 shows a diagram of the shapes of leaf apexes (tip) and leaf bases (bottom).

Leaf margins are also different. "Entire" means that the edge is smooth. "Parted" means that the edge has notches. Figure 6–18 illustrates the types of leaf margins.

The venation, arrangement, shape, and margin of leaves are all used to identify plants. The identification of plants is vital to agriculturists. Knowing the identity of a weed pest or the names of ornamental plants used in a landscaping scheme is essential in carrying out jobs associated with agriculture.

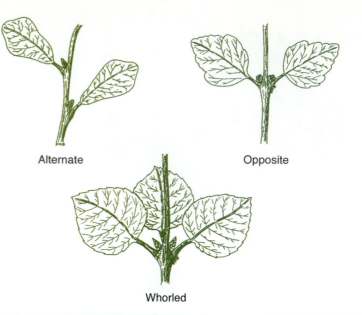

Figure 6–16 *The three most common leaf arrangements are whorled, alternate, and opposite.*

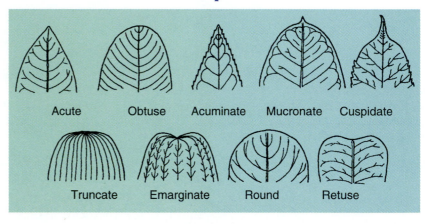

Tips

Figure 6–17 *Leaves come in an assortment of apex and bottom shapes.*

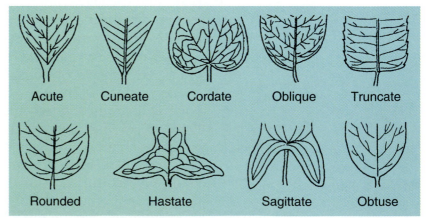

Bases

Margins

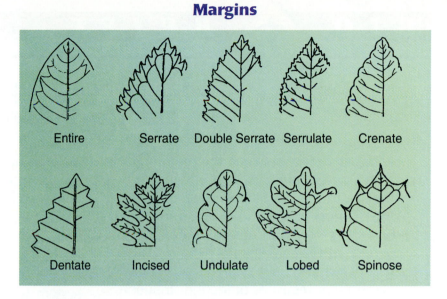

Entire Serrate Double Serrate Serrulate Crenate

Dentate Incised Undulate Lobed Spinose

Figure 6–18 *The shapes of the leaf margins are used to identify the plant.*

STEMS

Stems serve several major purposes—they support the plant, conduct water and nutrients from the roots to the branches and leaves, store food for use by the plant, and carry food manufactured by the leaves to the roots.

Plant stems have many agricultural uses. The stems of trees are used for lumber, paper, power poles, and many other uses (Figure 6–19). Fibers from the stems of plants like sisal

Figure 6–19 *Stems of trees are used for lumber, paper, and a variety of other uses. Courtesy of James Strawser, The University of Georgia.*

BIO BRIEF

Understanding Sugar Transport in Plants

A signal system between the brain and the stomach tells animals to stop eating when their stomachs are full—or, conversely, it lets them know when they need more food. Scientists for the Agricultural Research Service at Urbana, Illinois, have discovered that plants also have a specialized signaling system that regulates nutrient distribution.

ARS plant physiologist Daniel R. Bush says plants can regulate distribution of nutrients via a signal system that controls the flow of sugar from photosynthetic leaves to nonphotosynthetic tissues, such as roots and seeds. These tissues are called sinks, because they must import sugars and amino acids to support plant growth and development.

Plants convert light energy from the sun into biochemical energy that is used to synthesize the sugars and amino acids through the complex photosynthetic process. Up to 80 percent of the products of photosynthesis are transported to sink issues in the plant's vascular system.

"We have known about the mechanics by which the plant's vascular system transports organic nutrients for years. But we didn't have any information about how this system is regulated," says Bush. "What we have discovered is that sucrose, the major form of transported sugar, is also a signal molecule that the plant responds to by increasing or decreasing nutrient flow to sink tissues."

When sucrose departs from plant leaves, it flows in the elongated phloem cells that lie end to end, forming a living conduit in the plant's vascular system. A specialized sucrose transport protein, first described by scientists in the ARS Photosynthesis Research Unit at Urbana, actively loads the sugar into the phloem.

are used to make rope. The stems of the flax plant contain fibers used in making linen cloth.

There are two types of plant stems. Herbaceous stems are green and relatively soft (Figure 6–20). The support they give plants comes mainly from the pressure of water in the stem tissue. For this reason plants with herbaceous stems wilt and droop when they do not have enough water. Plants with herbaceous stems usually are annuals, which live only one year. Most row crops, forages, and vegetables have herbaceous stems. Woody stems are hard and do not wilt when water is low (Figure 6–21). Trees and most shrubs have woody stems.

The concentration of sucrose inside the phloem cells is up to 100 times greater than that outside. "This attracts water into the cells and increases the hydrostatic pressure of the leaf phloem," says Bush.

"When sucrose is released from the phloem cells in sink tissues, water also leaves the cells, creating a hydraulic pressure difference between the leaf and the sink phloem. That pressure difference causes a mass flow of solution through the phloem. In many ways, this is very similar to the pressure-driven flow of blood pumped through the human body," says Bush.

By learning more about how plants allocate photosynthetic products, scientists may be able to modify plant growth to increase yields, alter nutritional value, or overcome environmental challenges such as elevated carbon dioxide levels.

Source: Agricultural Research Magazine

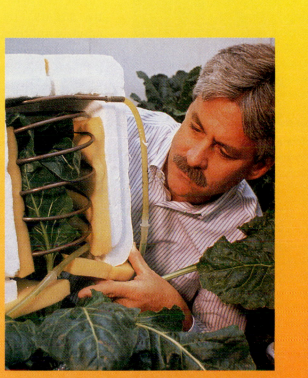

Scientists have discovered that plants have a specialized signalising device that regulates nutrient distribution.

Woody stems are composed of many layers (Figure 6–22). The outermost layer is the bark, which protects the plant. Bark is divided into inner and outer sections. The outer section consists of an epidermis that protects the plant, a cork that contains a waxy waterproof substance that keeps the inner tissues from drying out, and a cork cambium that contains merisystematic cells. Merisystematic cells divide and cause growth. These cells divide constantly, and more material is built up on the outer side of the stem. When the outer cells die, they form the cork layer. The inner bark consists of the cortex, an area for food storage, and the phloem. The phloem has tubes that conduct nutrients from the leaves to the roots and other parts of the plant.

The nutrients move through the phloem in a fluid called sap. The type and quantity of sap varies with the plant. The sap of some plants contains a high amount of sugar and can be harvested and used for food (Figure 6–23). For example, the sap of sugar maple trees is harvested and boiled to further concentrate the sugars. The resulting fluid is known as maple syrup. Sugar cane is another example. The canes are harvested and pressed to extract the sap. The sap is either processed into dry granulated sugar or concentrated into molasses (sorghum syrup).

Beneath the bark is a single layer of cells called the cambium. The cambium layer separates the phloem and the xylem. This layer is made up of growing, living cells that add cells to both the phloem and the xylem.

The wood in a woody stem is made up of tubes that bring water and nutrients from the roots. The xylem forms rings within the stem of the plant. Each year a new ring is added, and the old ring dies. In trees, this ring is called the annual ring and is used to determine the age of a tree. Water and nutrients are transported in the outermost ring only. The rings that are no longer used for this transporting are used to store food and as support for the plant.

Figure 6–20 *Herbaceous stems are green and relatively soft.*
Courtesy of James Strawser, The University of Georgia.

Figure 6–21 *The stems of woody plants are hard and do not wilt.*
Courtesy of James Strawser, The University of Georgia.

Figure 6–22 *Woody stems are composed of many layers.*

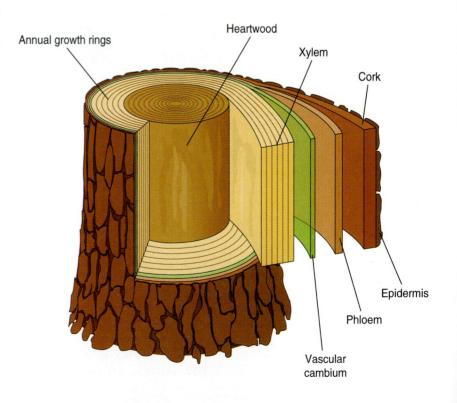

Figure 6–23 *The sap of the sugar maple is used for food.* Courtesy of Vermont Bureau of Tourism.

At the center of woody stems is the pith. This is made up of large cells with many air spaces between them. The pith is a storage area for the plant. As the stem matures, the pith usually dies.

Modified Stems

The stems of most plants grow vertically above the ground. However, the stems of some plants grow horizontal to the ground and may even grow under the ground. These stems are usually involved with the reproductive process and will be discussed in Chapter 7.

ROOTS

As much as half the weight of the entire plant may be in the root system (Figure 6–24). This system often reaches far into the ground and may extend out longer than the branches of the plant. Roots serve several purposes. They take water and nutrients from the soil, store food manufactured by the plant, and anchor the plant.

There are two main types of root systems. **Tap root** systems have a strong central root that grows downward into the soil. Often the main root grows to be quite large and stores a lot of plant nutrients. The carrot is a good example of a tap root system (Figure 6–25). Most dicots have a tap root system. Dicots are plants with seeds having two cotyledons or seed halves, such as beans. Fibrous root systems have no large central root but have many branches of fine roots. This complex

Figure 6–24 *As much as half the weight of a plant may be in the root system. Courtesy of USDA-ARS.*

network of roots holds soil in place and prevents erosion. Grasses and most other monocots have fibrous root systems. Monocots are plants with seeds that have only one cotyledon or seed half, such as corn or wheat.

Some plants have specialized root systems. In plants such as some vines, roots may sprout from a stem or leaf. These are called adventitious roots. Other roots may not grow in the soil

Tap and Fibrous Roots

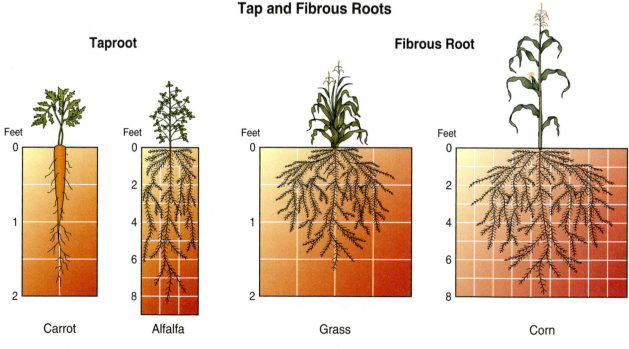

Figure 6–25 *There are two types of root systems: tap and fibrous.*

Figure 6–26 *Aerial roots are suspended in the air. Orchids have aerial roots. Courtesy of James Strawser, The University of Georgia.*

at all. Aerial roots are are suspended in the air. The roots of orchids are aerial (Figure 6–26). Some roots are suspended in the water. Aquatic roots take nutrients from the water or a combination of water and the soil beneath the water. Water lilies are an example of plants with aquatic roots.

Nutrient Absorption

As mentioned earlier, the primary function of the root system is to take water and nutrients from the soil and send them to other parts of the plant. These materials are taken in through very fine roots called root hairs. Water and minerals enter the root hairs by osmosis and diffusion (Figure 6–27). The membranes of the root hairs have tiny pores that allow the passage of water molecules. This is called a semipermeable membrane. Water is inside and outside the membrane. On the inside the water is in the root; on the outside it is in the soil. If there is a higher concentration of water in the soil than in the root, water passes into the root through the semipermeable

Water Movement from Soil Through Root

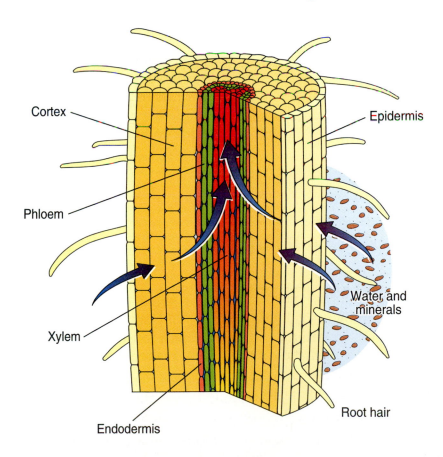

Figure 6–27 *Water and minerals move into the root hairs through osmosis.*

membrane. The movement of the water creates a pressure called osmotic pressure. This builds up and causes water to move into the vascular cylinder of the root. This section contains the phloem that transports sugar and other dissolved nutrients to the rest of the plant.

Through a symbiotic relationship, some plants provide their own supply of the vital nutrient nitrogen. (A symbiotic relationship means that two organisms live together and both benefit.) Certain plants have swellings called nodules on their roots. Within these root nodules live rhizobium bacteria. The bacteria invade the root hairs of the plant, and the plant responds by creating a nodule around the bacteria. These bacteria take nitrogen from the atmosphere and combine it with oxygen to form NO_3 or NH_3. These forms of nitrogen can be used by the plant. Because plants cannot use pure nitrogen, the symbiotic relationship with the bacteria greatly aids the plant. Plants that host nitrogen-fixing bacteria are known as legumes. These include soybeans, clover, beans, and peas.

SUMMARY

Plants grow all over the world. Without plants most life on Earth would cease to exist. Through agriculture, plants useful to humans are grown and improved to provide us with a constant source of food and fiber. Agricultural scientists will continue to look for new plants and new uses for the plants we currently grow.

CHAPTER 6 REVIEW

Student Learning Activities

1. Bring to class five examples of plant systems such as roots and leaves that are used as agricultural products. Report to the class on why and how each is used. What characteristics of the system make it useful?

2. Collect 20 different leaves from plants of agricultural importance and classify each according to shape, margin, venation, evergreen or deciduous, simple or compound. Share the collection with the class.

3. Collect four woody and four herbaceous stems. Explain the differences in the stems.

Define the Following Terms

1. stoma
2. chlorophyll
3. photosynthesis
4. vascular bundles
5. xylem
6. phloem
7. respiration
8. node
9. tap root

True/False

1. All of the food we eat comes either directly or indirectly from plants.
2. Plants take in air, use oxygen, and give off carbon dioxide.
3. The main function of leaves is to give beauty to the plant.
4. Humans eat the leaves of some plants.
5. Stoma are in the spongy layer of leaves.
6. The manufacture of food by the plant is called respiration.
7. Photosynthesis takes place only during the day.
8. Corn leaves have parallel veins.
9. The xylem carries water and nutrients up the plant from the roots, and the phloem carries nutrients from the leaves to the rest of the plant.
10. Most grasses have tap root systems.

Fill in the Blank

1. Plants convert _____ from the sun into energy for animals in the form of _____.

2. If it were not for plants, much of the Earth's _____ would be eroded by _____ and _____.

3. Plants beautify our homes by providing landscaping outside and _____ _____ inside.

4. Within the epidermis of a leaf, there are openings called _____ that allow air into the leaf and _____ and _____ to move out.

5. The palisade mesophyll consists of _____ that contain many structures called _____.

6. Photosynthesis (photo meaning _____ and synthesis meaning to _____ _____) is the process by which leaves make _____ for the plant.

7. The breaking down of plant foods is called _____ and is the reverse of the _____ process.

8. Most leaves have two basic parts, the _____ and the _____.

9. Most _____ have sessile leaves. Plants with sessile leaves usually are _____ that _____ each year and come back from seed.

10. The node is the part of the stem from which the leaf _____ and _____ . The space between the nodes is called the _____.

11. Stems serve several major purposes: they support the plant, and they conduct _____ and _____ from the roots to the _____ and _____.

12. The inner bark of plants consists of the cortex, an area for _____ _____ and the phloem.

13. The stems of some plants grow _____ to the ground and may even grow _____ the _____.

14. The primary function of the root system is to take _____ and _____ from the _____ and send them to _____ _____ of the plant.

Discussion

1. Explain the differences between photosynthesis and respiration.

2. What are the differences between woody and herbaceous stems?

3. What are some uses of plant stems?

4. Why do deciduous plants lose their leaves in the fall?

5. Why is an understanding of plant systems important to an agriculturist?

6. What purposes do leaves serve in plants?

7. What are some ways that leaves are identified or classified?

8. What uses do people make of tree sap?

9. List the purposes plant roots serve.

10. Explain how legumes are different from other plants.

CHAPTER 7

PLANT REPRODUCTION

KEY TERMS

asexually
zygote
perianth
meiosis
endosperm
monoecious
dioecious
cultivars
grafting
tissue culture
explant
in vitro culture

STUDENT OBJECTIVES

After studying this chapter, you should be able to:

- ✦ Discuss the differences between sexual and asexual reproduction in plants.

- ✦ Explain the importance of plant reproduction to producers.

- ✦ Explain how plants reproduce sexually.

- ✦ Identify and discuss the functions of the parts of a flower.

- ✦ Explain the process of gamete formation in plants.

- ✦ Discuss the pollination process.

- ✦ Describe the role of insects in the pollination process.

- ✦ Explain how plants reproduce asexually.

- ✦ Explain the principles of growing a new plant from cells.

- ✦ Discuss the different methods producers use to propagate plants asexually.

Like all living things, plants must reproduce to go on living. The entire industry of agriculture is built around the reproduction of plants. Plants that we eat, feed our livestock, and use to beautify our surroundings must multiply in order to be useful.

Plants reproduce in two different ways, sexually or **asexually**. Plants that reproduce sexually must produce two gametes, the sperm and the egg. Plants that reproduce asexually do so vegetatively. This means that a new plant can be created without the joining of an egg and a sperm. These

143

plants reproduce by using part of the parent plant to start a new plant. In agriculture, plants may propagate themselves naturally, or the producer may use one of several methods to induce the plant to propagate. Whether natural or induced, vegetative propagation results in new plants that are genetically the same as the parents. In many types of plants, reproduction can take place by both methods.

SEXUAL REPRODUCTION

In sexual reproduction, fertilization must occur. This involves the uniting of the male gamete (the sperm) with the female gamete (the egg) as in Figure 7–1. This process usually takes place in the plant's flower and results in the formation of a **zygote** or fertilized egg. The zygote then grows into the seed.

A flower develops from a bud on the stem or branch of a plant (Figure 7–2). The flower is attached to the stem or flower stalk by an enlarged structure called the receptacle. An enclosing structure called a calyx protects the developing flower bud. The calyx is usually made up of green petal-like structures called sepals that encase the flower. When the growing flower opens, the corolla emerges. The corolla is made of petals. The calyx and the corolla make up the **perianth**, which

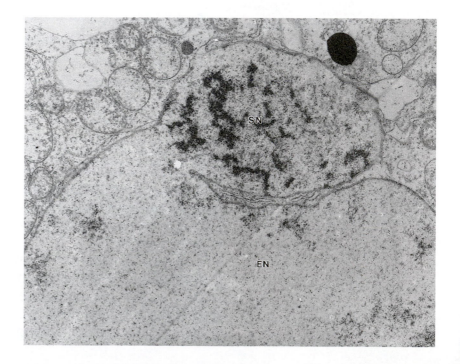

Figure 7–1 *Fertilization occurs when the two gametes unite. SN = sperm nucleus; EN = egg nucleus.* Courtesy of H. Lloyd Morgensen, Northern Arizona University.

Parts of a Flower

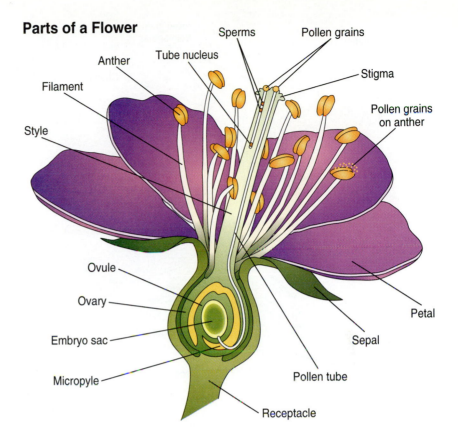

Figure 7–2 *Flowers contain the reproductive structures of plants.*

is the colorful head of the flower. A perfect flower contains both male and female parts (Figure 7–3).

The female organs within the flower are known as carpels. The carpel is made of a slender tubelike structure called the pistil. At the top of the pistil is a broadened area that receives the pollen grains. This part is called the stigma. Connected to the stigma is a tube called the style that leads to the bottom of the pistil. At the bottom of the pistil is the ovary that contains a hollow structure called the ovule. After fertilization, the ovule develops into the seed, and the ovary develops into fruit.

The process of producing the female gamete or egg begins inside the ovary with a type of cell division called **meiosis**. The mother cell divides by meiosis and forms four haploid cells. Remember from Chapter 5 that diploid means that a cell contains two complete sets of chromosomes. Haploid means that the cell contains only one complete set of chromosomes. The gametes (the egg and the sperm) contain only one set of chromosomes and are haploid cells. At fertilization, the gametes unite, and the two complete sets of chromosomes create a diploid cell called the zygote. Three of the four haploid cells from the female deteriorate, and only

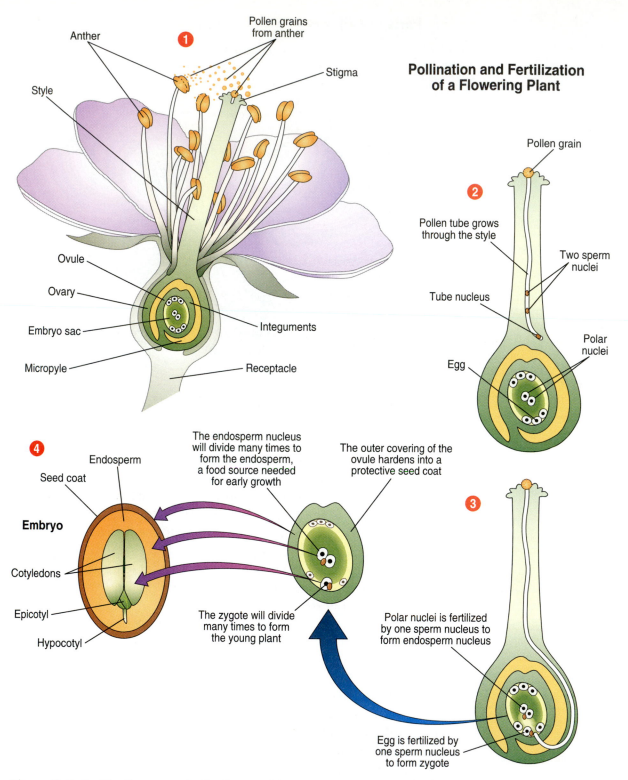

Figure 7–3 *Fertilization occurs in the flower of the plant. The result of fertilization is the seed. The endosperm is a food-storing mass of cells.*

one functional cell remains. This cell increases in size and forms a cavity called the embryo sac. Within this sac the cell divides and redivides, forming a group of haploid nuclei. Certain of these nuclei become the gametes, and some become structures known as polar nuclei.

Around the pistil are slender stalklike structures known as the stamen. These organs produce the pollen that serves as the sperm to fertilize the ovule. The stamen is attached to the bottom of the flower by a thin stalk called a filament. At the top of the filament is the anther that produces the pollen. Inside the anther are found cells called diploid mother cells. The mother cells divide by meiosis and form haploid cells that in turn develop into grains of pollen.

After the gametes (sex cells) are formed, fertilization can take place. After the pollen come in contact with the stigma of the pistil, the stigma takes in the pollen grain and holds it. The pollen grain begins to dissolve, and the cytoplasm from the wall of the pollen breaks down. From this cytoplasm, a pollen tube is formed that extends downward to the ovary. By the time the pollen tube has extended all the way to the ovary, the male gamete will have divided into two male gametes. These two gametes pass through the pollen tube to the ovary. Remember that the female gamete has divided into a female gamete and a polar nucleus. When fertilization occurs, one male gamete (sperm) unites with the female gamete (egg), and the other unites with the polar nucleus. The union of the egg and sperm creates the zygote that develops into the embryo or the seed. The union of the polar nuclei with the other sperm develops into the **endosperm**, a food-storing mass of cells surrounding the seed. For example, in a seed or grain of wheat, the part that is ground and used as flour is the endosperm. Another common example is the fluffy white part of popcorn.

Plants can be pollinated by other plants or by their own flowers. Flowers that contain sepals, petals, a stamen and a pistil (or carpel) are complete flowers. Some flowers, however, do not contain all of the parts and are called incomplete flowers. For example, the flowering tassel of a corn plant has stamens but no pistil and is known as a staminate (male) flower (Figure 7–4). The pistil is on a separate flowering part (called a pistillate flower) located on the newly emerging ear. These are the silks that extend from the end of the ear of corn. **Monoecious** plants such as corn have separate staminate and pistillate flowers on the same plant. **Dioecious** plants, such as soybeans, have staminate and pistillate flowers on separate plants.

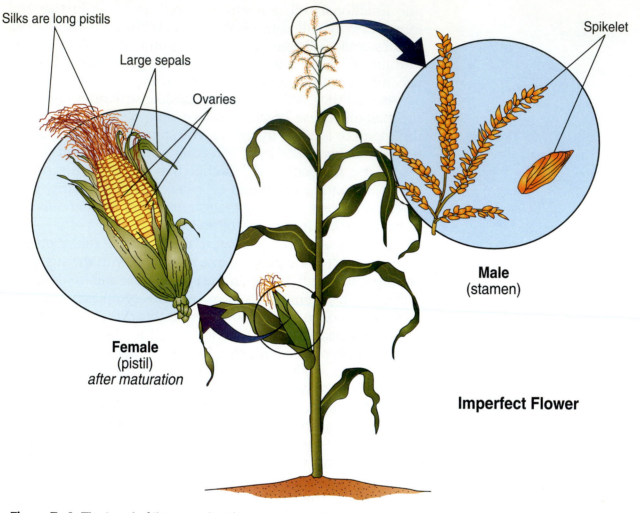

Figure 7–4 *The tassel of the corn plant is an example of a staminate (male) flower.*

If fertilization (pollination) takes place from flowers on the same plant, the plant is said to be self-pollinated. If the flower is pollinated by pollen from another plant, it is said to be cross-pollinated. Because some plants have only male or only female flower parts, they must be cross-pollinated. Most plants cross-pollinate to some degree if there are other plants of the same type in the vicinity. Some plants are able to fertilize themselves and are called self-fertile plants. Plants that cannot fertilize themselves are called self-sterile plants. Usually those plants that are self-fertile can also be fertilized from other plants. In doing this, the plants receive genes from other plants, and the genetic makeup is different than if they are self-pollinated.

Figure 7–5 Red Delicious apple trees must be pollinated by another apple tree. *Courtesy of James Strawser, The University of Georgia.*

Some plants, especially certain **cultivars**, are self-sterile. Cultivars are plants that are cultivated and grown by humans. These plants may have been bred to the point where they are sterile unless pollinated from other plants. Red Delicious apple trees are a good example (Figure 7–5). By themselves, they are sterile and need another apple tree to cross-pollinate. A Yellow Delicious tree nearby can produce pollen that is excellent in fertilizing the Red Delicious trees.

The pollen must be transported in order to enter the pistil through the anther. In some plants, like corn, this is accomplished by the wind. The wind blows across a flower and carries pollen grains to another flower. Some plants, however, are fertilized by insects (Figure 7–6).

To attract bees and other insects, plants have developed brilliantly colored flower petals. Although insects probably do not perceive the colors of the flowers the same way as we do, they are attracted by the petals. The flowers secrete a sweet fluid known as nectar that the insects use for food. As they crawl or fly into a flower to collect nectar, grains of pollen are caught on their legs and other body parts. When they go to the next flower, the grains of pollen are knocked off and remain in the flower.

Honeybees are particularly adept at pollinating. Many insects go to flowers of different kinds. Bees, on the other hand, work a particular kind of flower for a period of time. For

Figure 7–6 Insect pollinate plants as they gather nectar from blooms. *Courtesy of USDA-ARS.*

Figure 7–7 *Fruit growers bring in bees to pollinate crops. Courtesy of California Agriculture, University of California.*

example, honeybees may work apple blossoms for several days until the flowers are gone and then work a different type of flower. By doing this they go from apple blossom to apple blossom and pollinate the trees. This ensures that the blossoms are thoroughly pollinated.

Fruit growers hire beekeepers to bring in truckloads of bees in the spring as the trees are blooming (Figure 7–7). The bees live in wooden, boxlike structures called hives with a separate colony of bees in each hive. The hives are easy to handle and can be loaded on a truck with the bees still in them. The owner of the bees can then move the hives from orchard to orchard for a fee from the fruit or crop producer. In addition, the producer can harvest the honey and sell it at a profit.

ASEXUAL PROPAGATION

Many plants reproduce not only through the production of seed but also by growing a new plant from part of the old plant. This type of reproduction is called asexual or vegetative reproduction. The new plant is genetically identical to the plant that produced it. Remember that plants produced from seed obtain half of their genetic makeup from each of two parent plants and are not genetically identical to either.

The production of plants by vegetative means has advantages to producers. Because the plants are genetically identical to the parent plants, superior plants can be reproduced without losing the qualities that make them superior. For example, if a particularly beautiful rose is produced, the same color and brilliance might be difficult to produce in the next generation if the new plants are produced from seed (Figure 7–8). The reason is that through seed production, half of the genes must come from another plant. If, however, the rose can be propagated vegetatively, it will have the same genetic makeup as the parent and the same type of flower. Plant breeders go through a very extensive process of crossing plants to produce the types they want, but these plants often cannot reproduce sexually and must be produced vegetatively. Most modern varieties of fruit trees are reproduced in this manner.

Plants reproduced vegetatively can be propagated much faster than those propagated from seed. The production of seed involves the entire cycle of flower production, pollination, seed maturation, and germination. If the new plant is

Figure 7–8 *If a particularly beautiful rose is discovered, the plant is propagated asexually. Courtesy of James Strawser, The University of Georgia.*

grown from a part of the parent plant, it can be produced without going through these cycles. This means that the producer can produce higher quality plants in a shorter period of time.

Another advantage of vegetative propagation is that seedless plants can be grown. This is particularly important for certain plants that are grown for food. For example, seedless grapes are usually preferred by consumers for eating fresh. With vegetative propagation, new grapevines can be produced that will generate grapes without seeds.

Types of Asexual Propagation

Some commercially produced plants reproduce vegetatively with little help from the producer. These plants have specialized bud stems and roots that propagate themselves. Two methods to obtain new plants from specialized plant parts are division and separation. In separation, the plant parts are merely pulled apart because the plant naturally separates the parts for the production of the new plants (Figure 7–9). In

**Separation of
Bulblets from Bulb**

**Division of
Herbaceous Perennials**

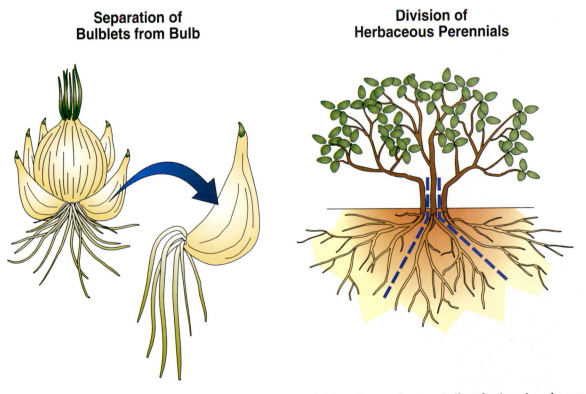

Figure 7–9 *In separation, the plant parts are pulled apart. In division, the producer cuts the plant part and grows a new plant from each section.*

Strawberry Plant

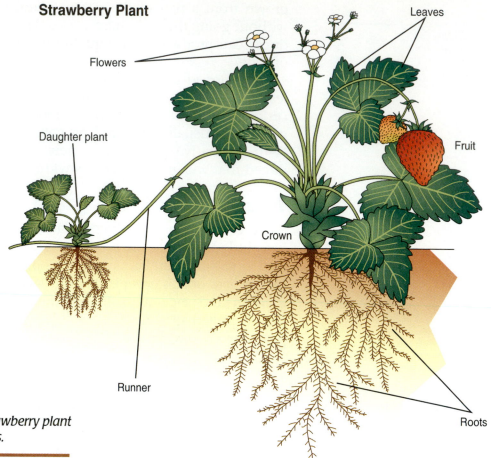

Figure 7–10 *The strawberry plant propagates by runners.*

division, the producer cuts the plant part into sections and grows a new plant for each section.

Specialized parts of the plants that are used for vegetative propagation include bulbs, corms, tubers, stolons, rhizomes, and crowns. Runners or stolons are specialized stems that grow on top of the ground and reach out horizontally (Figure 7–10). As the nodes of the runner come into contact with the ground, new roots form and a new plant begins. Growers can accelerate this process by pinning the runners to the ground. This method (called layering) will be discussed later. One common example of this process is the strawberry plant. As the runners reach out across the ground, new plants are grown at the nodes. The producer can then detach the runner from the parent plant, dig out the new plant and transplant it in a different place.

Rhizomes are also specialized underground stems that are used in propagation (Figure 7–11). They are generally larger

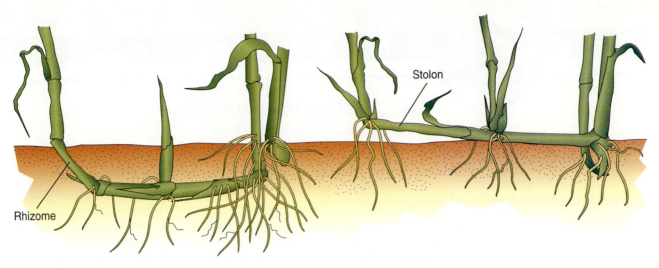

Figure 7–11 *Stolons grow above ground, and rhizomes grow underground.*

than stolons and can be broken or cut apart, and the parts are used to establish a new plant. Irises are easily propagated by rhizomes. Also, many grasses, such as Bermuda and Johnsongrass, are propagated by rhizomes.

Sometimes a stolon or rhizome may have an enlargement called a tuber on it. These tubers can be cut apart and new plants grown from them. For example, a potato is a tuber (Figure 7–12). On the surface of the tuber are several buds called eyes. Each bud is capable of sprouting a new plant. Producers plant chunks of potatoes that contain the buds. The resulting plant is an exact genetic copy of the parent.

Tuber Division

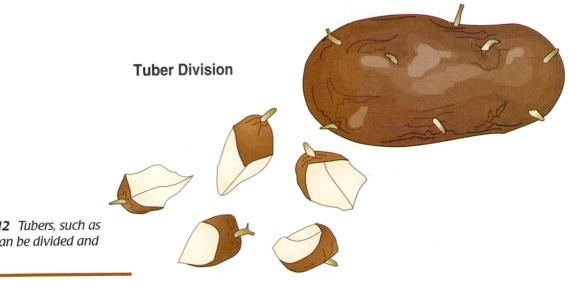

Figure 7–12 *Tubers, such as potatoes, can be divided and planted.*

BIO BRIEF

Ice Crystals Will Not Stop Winterfat Seeds

Winterfat, *Eurotia lanata*—a native of the spinach family, is a hardy, low-growing shrub found from the Yukon to Mexico and is an important food source for cattle in the western United States. With the exception of the Antarctic nematode, which can survive with ice in its cells, few organisms have the ice tolerance of the winterfat shrub. Tests on winterfat seeds show they can survive, even soaked with water, at temperatures at least as low as −22°F. Ice crystals in the seed embryo means death to most seeds, so what is this shrub's secret?

Terry Boot, a rangeland scientist with USDA's Agricultural Research Service in Cheyenne, Wyoming, is conducting research to find out how winterfat seeds survive and if this ability can be transferred to other major plant species. In Booth's earlier tests, he soaked and froze the seeds to simulate western freeze-thaw cycles, which can be so deadly to seeds. These harmful cycles occur often in the spring because seeds get wet from snowmelt during the day and freeze overnight, or seeds start sprouting during a warm spell and are hit with a sudden winter-like storm. Most seeds have a greater risk of freezing damage as the water content increases, but winterfat seeds often performed better when they had been soaked at cold temperatures before freezing. Booth also discovered that the plants have evolved so that the seed's germination requirements fit local climate variations, making it more desirable to plant locally collected seeds.

More research is needed to uncover the factors that contribute to the ice hardiness of the seed, but Booth says, "A key part of winterfat's tolerance to freezing seems to be

Many plants are propagated by means of bulbs. Two kinds of bulbs exist, tunicate and nontunicate (Figure 7–13). Tunicate bulbs have dry outer layers of membranes that are the result of last year's growth. They are made up of layers of leaflike membranes. When cut into cross sections, they appear to have concentric rings. The onion is a good example of this type of bulb (Figure 7–14). Tunicate bulbs propagate naturally by growing small bulblets around the bulb. Because these bulblets may each grow into a separate plant, growers periodically dig these bulbs and separate the bulblets to produce new plants. These bulbs may also be cut into segments. Each cut portion of the parent bulb is stored for one to two

its ability to hydrate, dehydrate, and rehydrate again without significant damage." Booth suspects that winterfat fights ice with ice. The hairy layers that cover winterfat seeds appear likely to promote ice crystal formation. If ice crystals form first in the outer layers, these crystal may suck water from the embryo, aiding a freeze-dehydration process that limits damage caused by embryo ice. It may keep the largest crystals out of the embryo. Booth points out that more time and tests are needed before reaching my conclusions.

Partial funding for the research is provided by Ducks Unlimited of Canada because winterfat provides nesting cover for ducks on Canadian prairies. Also, the U.S. Bureau of Land Management, the USDA Forest Service, and mining companies plant winterfat on degraded rangelands and strip-mined areas.

Source: Agricultural Research Magazine

Seeds of the winterfat plant, Eurotia lanata, *can be harvested in October*

weeks. After that they are planted, and they produce bulblets from the bulb cutting.

Nontunicate bulbs are also known as scaly-type bulbs because of the layers of scales on them. Each of the outer scales can be separated and planted. From each of these scales a new plant will grow. Lilies are examples of a nontunicate or scaly-type bulb.

Corms are another type of underground stem used in propagation. Corms differ from bulbs in that they are solid and have nodes and inner nodes (Figure 7–15). Propagation is through the production of new corms (called cormels) from the old corm. Growers separate the cormels out and plant

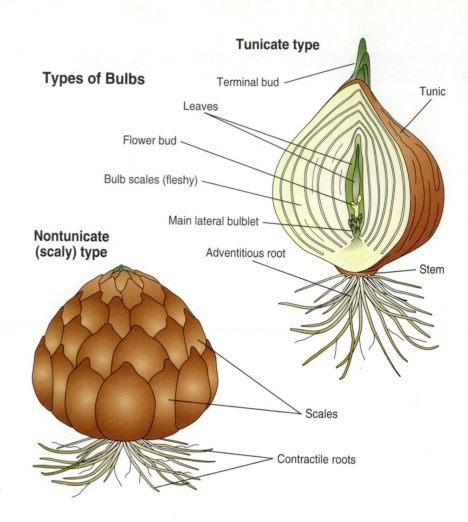

Types of Bulbs

Tunicate type

Terminal bud

Leaves

Flower bud

Bulb scales (fleshy)

Main lateral bulblet

Tunic

Adventitious root

Stem

Nontunicate (scaly) type

Scales

Contractile roots

Figure 7–13 Bulbs may be tunicate or nontunicate

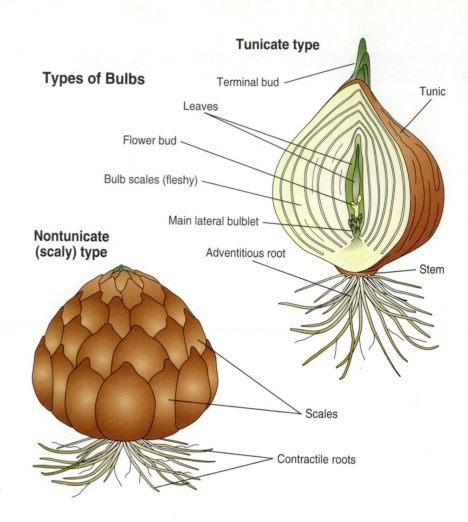

Figure 7–14 The onion is a good example of a tunicate bulb. *Courtesy of James Strawser, The University of Georgia.*

them. Also, the corms may be divided into sections, with each section producing new plants. Examples of plants that have corms are gladiolus and garlic.

Propagation by Cuttings

Many plants may be propagated by cuttings taken from the plant (Figure 7–16). Depending on the type of plant, cuttings from leaves, stems, roots, or buds may be used to grow new plants. Plants are completely different from most animals in that they can regenerate an entire plant from a severed part of the parent plant.

Cuttings offer some advantages over other methods of propagation. With a large plant, a lot of material may be used to reproduce the plant. Theoretically, any cell of the plant should be able to replicate itself because all of the genetic coding for the entire plant is held within each cell. A large tree or shrub can provide a lot of stems, leaves, or roots from which to grow new plants. Also, growers are able to control

Corm

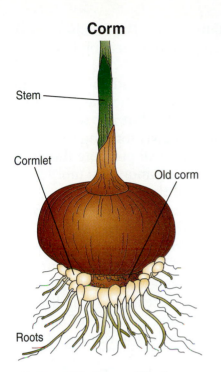

Stem

Cormlet

Old corm

Roots

Figure 7-15 Corms differ from bulbs in that they are solid and have nodes and inner nodes. Propagation is through cormlets.

the process because they do not generally have to wait on the plant to produce bulbs, tubers, or corms. However, in some plants, cuttings are best taken at certain times of the year. For example, hardwood cuttings grow best if they are taken during the dormant season.

For the plant to regenerate itself, the grower must create conditions that will allow the development of roots and shoots from the plant cutting. Because the cuttings do not have roots, the grower must create conditions that retard the growth of leaves and stems until the roots can grow and develop. These environmental conditions include the proper temperature, moisture, air movement, and light. Proper temperature is important because the photosynthesis process is controlled by temperature. Generally, the higher the temperature, the higher the rate of photosynthesis. For a plant cutting, a high rate of photosynthesis may not be desirable. The process encourages the use of stored plant energy for the production of leaves and buds. At a lower rate of photosynthesis, this energy can be used to promote the growth of roots on the cutting. Although the proper temperature may vary with the type of plant being propagated, the best temperature is usually between 70 and 80°F in the daylight hours and 60 and 70°F during the night. Soil temperature should be around 5 to 10°F higher. The higher temperature in the root area

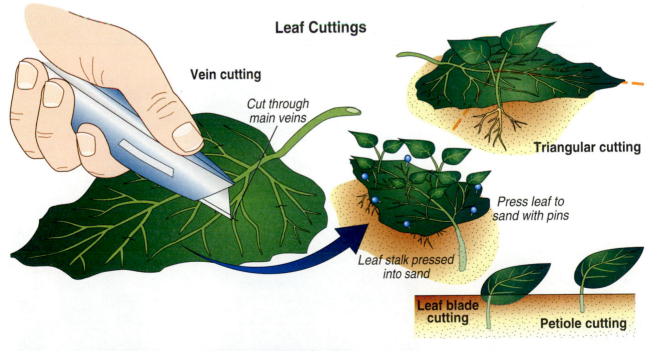

Leaf Cuttings

Vein cutting

Cut through main veins

Triangular cutting

Press leaf to sand with pins

Leaf stalk pressed into sand

Leaf blade cutting

Petiole cutting

Figure 7-16 Some plants may be propagated by leaf cuttings.

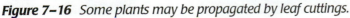

stimulates the oxidation of fatty acids. This creates a substance known as suberin, which promotes healing in the area where the cutting was severed.

Moisture is important in the media and in the air surrounding the plant. Media is the material the plant cutting is placed in to root. It must be kept moist but not wet because excessive moisture will prevent the newly forming roots from getting the proper amount of air and will promote the decay of the cutting. Relative humidity is the amount of moisture in the air. This is expressed in terms of the percent of moisture the air will hold before condensation occurs. Leaves must retain their moisture in order to produce carbohydrates and plant hormones (auxins) necessary for the production of roots and the rest of the plant systems. A high relative humidity prevents the plant cutting from losing too much moisture through transpiration. Too high of a relative humidity will promote the decay of the plant cutting. A range of 60 to 80 percent is generally considered ideal for the rooting of most plant cuttings.

Roots require oxygen to grow and thrive. Proper circulation of air through the media is important to allow oxygen to reach the part of the cutting that will grow the new roots. This means that the media must be of the proper consistency to allow the flow of air to the roots. Also, the media must not be saturated with water for extended periods.

Light is also important in producing new plants from cuttings (Figure 7–17). Remember from Chapter 6 that light is

Figure 7–17 *Light is essential in the production of new plants. Greenhouses are designed to allow for the proper amount of light. Courtesy of USDA-ARS.*

Figure 7–18 *Plant cuttings are dipped in root-growing hormones to stimulate root growth.* Courtesy of James Strawser, The University of Georgia.

the energy used by the plant for photosynthesis. The more light a plant receives, the greater the rate of photosynthesis. Although a certain amount of light is necessary for the cuttings to take root, too much light can be detrimental. As with too high a temperature, the energy is put into the growth of leaves and not roots. To properly root, plant cuttings should be shielded from intense light.

Growth Regulators

As explained in Chapter 8, the reproduction of cells is controlled to a large degree by plant hormones called auxins. When roots begin to grow on plant cuttings, the process is stimulated by growth hormones. In certain plants, these hormones are released naturally, and propagation can readily occur using cuttings. Some plants react well to externally applied regulators that are supplied by the producers. The cutting is dipped in a powder containing the root-growth stimulating hormone or soaked in a solution containing the substance (Figure 7–18).

Layering

Some plants are difficult to root from cuttings. These plants can often be propagated by a process called layering. In layering, a portion of the plant is covered with soil or other material, and the growth of roots is stimulated while still attached to the parent plant. This method is used when the other methods do not work well. There are disadvantages to layering. It costs more than many other methods because of the labor required to complete the task. Also, fewer new plants can be started by layering than with cuttings.

The process of simple layering involves cutting a notch in a branch or stem of a plant (Figure 7–19). The wounded place is covered with soil with the tip left in the air. The wound in the plant stimulates the growth of callus cells. These cells are not differentiated. This means that they are not genetically programmed to become a certain type of tissue, such as a leaf, root, or stem. These cells become programmed to differentiate into root cells; new roots begin to grow; and a new plant is begun. The portion with the new roots is cut free from the parent plant and planted elsewhere. Broadleaf evergreen trees and shrubs such as rhododendron and magnolia are propagated by this method.

In mound layering, the top portion of the parent plant is pruned back in the dormant season (Figure 7–20). In the late

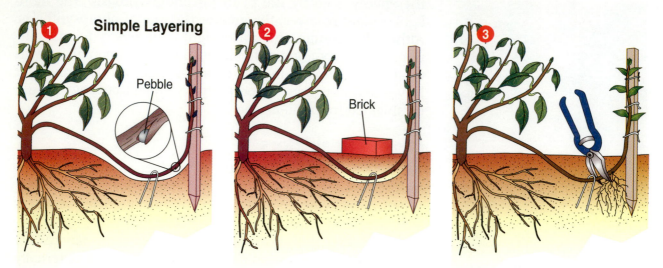

Figure 7–19 *Simple layering involves cutting a notch in a stem and covering with soil.*

spring when new growth appears, soil is mounded up around the base of the plant, and part of the new growth is covered. By the end of the next dormant season, the shoots that were covered have developed new roots. These new rootings are

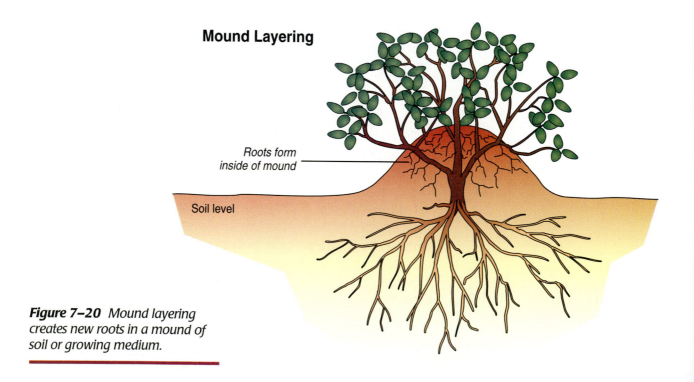

Figure 7–20 *Mound layering creates new roots in a mound of soil or growing medium.*

separated from the parent plant and planted in another place. Currants and gooseberries are propagated by mound layering.

Tip layering is a method by which only the new growth of the tip of a plant is placed in the ground and covered with soil (Figure 7–21). The new growth continues to grow downward in the soil. As the growth continues, roots form and new

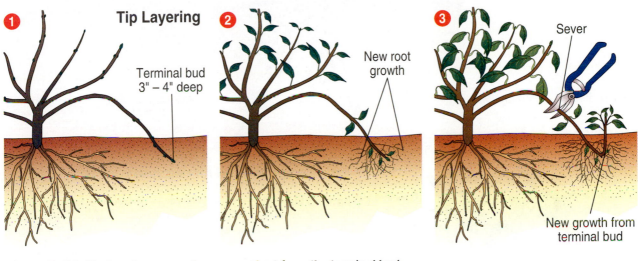

Figure 7–21 *Tip layering generates a new plant from the terminal bud.*

shoots are sent upward out of the ground. The new shoot is removed from the ground and separated from the parent plant. Tip layering is used to reproduce such plants as blackberries, raspberries, and dewberries.

Trench layering is used to propagate vines such as muscadine grapes and philodendrons (Figure 7–22). This method involves covering an entire stem in the soil. The nodes of the stem are notched before the stem is covered, and new roots form at the notched nodes. Several new plants can be started from a single limb using this method. A variation of this method is called serpentine layering. In this variation only the nodes are covered.

Layering can also be done without covering plant parts with soil. This method is called air layering because the process is done in the air instead of the soil (Figure 7–23). When propagating woody plants using this method, a healthy, fast-growing limb is selected, and a ring of bark is removed. The wound is covered with a ball of sphagnum moss. The moss is covered with polyethylene and tied to prevent the

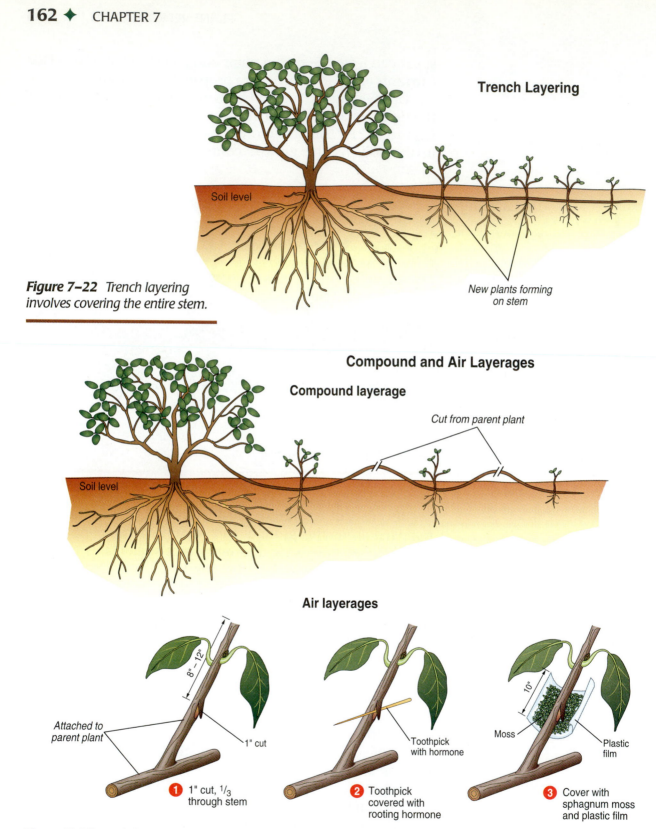

Trench Layering

Soil level

New plants forming
on stem

Figure 7–22 *Trench layering involves covering the entire stem.*

Compound and Air Layerages

Compound layerage

Cut from parent plant

Soil level

Air layerages

Attached to
parent plant

8" – 12"

1" cut

❶ 1" cut, ⅓
through stem

Toothpick
with hormone

❷ Toothpick
covered with
rooting hormone

Moss

10"

Plastic
film

❸ Cover with
sphagnum moss
and plastic film

Figure 7–23 *In air layering, new roots are generated above ground.*

escape of moisture. After several weeks, new roots form at the place where the bark was removed. The branch is severed, and the new plants are planted. Herbaceous plants are air layered in a similar manner except a notch is cut in the limb and a toothpick is placed in the notch. The wound is then treated in the same way as the woody plants. Air layering is used to propagate large houseplants such as rubber plants. Some citrus trees are also reproduced this way.

Grafting

A method of propagation that has been used for hundreds of years is **grafting**. In this technique, material from two plants is joined together to form one plant. This method is used with plants, such as almonds and apples, that are difficult to propagate by other methods. For example, a hybrid apple tree might be difficult to grow from seed because the tree from seed will be different from either of its parents (Figure 7–24). If a young sprout is combined with a portion of a desirable tree, the results can be a tree that grows and produces like the desirable tree. This method is also used to produce specialty trees such as dwarf trees and fruit trees that grow several different varieties of fruit. Trees are also grafted onto rootstock that is stronger or more suited to the area. English walnut trees are grafted onto black walnut rootstock because of the superior root system of the black walnut (Figure 7–25).

Grafting involves cutting two parts of a plant, usually a tree. The lower part is called a rootstock, and the upper part is called a scion (sometimes spelled cion). The rootstock may be larger than the scion. This gives the advantage of faster growth and earlier maturity.

The proper technique in grafting is to align the rootstock and the scion so that the cambium layers match. The wounding from the cut causes the production of callus cells. Remember that callus cells are undifferentiated. The callus cells of the scion and of the root stock intermingle, grow, and differentiate to form new cambium cells. The new cambium cells develop into the tubes that transport water and nutrients from the roots to other parts of the plant. These tubes are called the xylem and phloem.

Many types of grafting have been developed to fit specific purposes and specific plants. The procedure may involve using stems, roots, or buds. All of the methods use the same basic techniques. The cut portions of the rootstock and scion must be aligned at the cambium. They are bound tightly

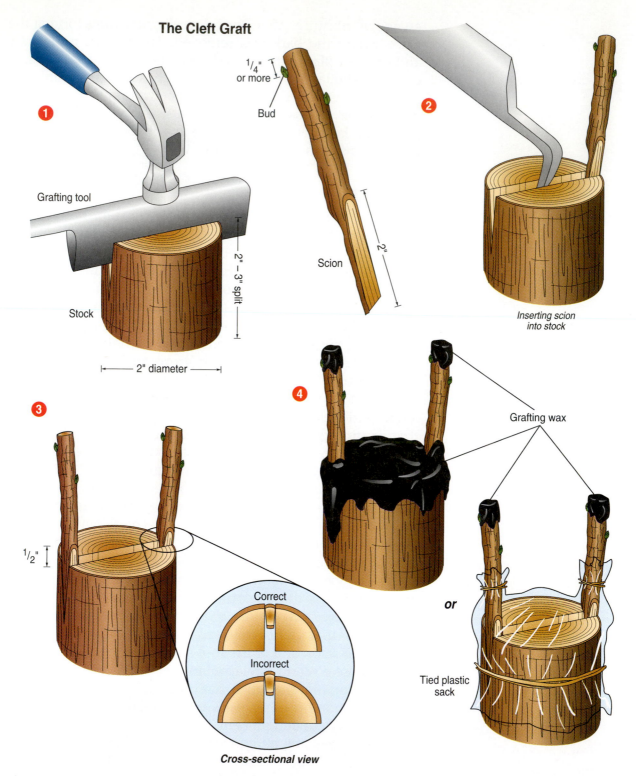

The Cleft Graft

1

Grafting tool

2" – 3" split

Stock

2" diameter

$\frac{1}{4}$" or more

Bud

Scion

2"

2

Inserting scion into stock

3

$\frac{1}{2}$"

Correct

Incorrect

Cross-sectional view

4

Grafting wax

Grafting wax

or

Tied plastic sack

Figure 7–24 *Grafting is used to propagate new trees from a hybrid fruit tree.* Courtesy of James Strawser, The University of Georgia.

Figure 7–25 *English walnut trees are grafted onto the roots of a black walnut. This provides a better root system.* Courtesy of Jack Kelly Clark, University of California IPM Project.

Figure 7–26 *Tissue culture generates new plants from cells of the parent tree.* Courtesy of USDA-ARS.

together and covered with a wax or other material to prevent the parts from drying out. Figure 7–25 shows the procedures used in most types of grafting.

Tissue Culture

One of the newer technologies in plant propagation is **tissue culture** (Figure 7–26). This procedure involves the use of a small amount of tissue from a plant to grow a new plant. There are two big advantages to this procedure. The first is that only a tiny portion of the parent plant is used, so the potential to grow a large number of genetically identical plants is greatly increased. The second advantage is that the new plants are free of disease-causing organisms.

Each plant cell contains all of the genetic coding for the entire plant. This means that theoretically, each cell should be capable of growing a new plant. Chapter 8 discusses how growth takes place through the multiplication of cells called meristem cells. The genes in each cell are programmed to differentiate. This means that some cells become stem cells, some become leaf cells, some become root cells, and so on. In meristem tissue, the cells are still undifferentiated. In the process of tissue culture, these cells are cut away from the plant before they are programmed to differentiate. By placing the cells in an environment that causes them to grow, they multiply and differentiate into roots, stems, and leaves that the plant must have to live.

Cells are stimulated to differentiate by a number of factors. These include plant hormones, location of the cells on the plant, light, nutrients, and temperature. Several other factors probably influence cell differentiation. Scientists still do not fully understand the process. For tissue culture to work, all of the factors mentioned have to be taken into consideration.

Most plant parts can be used for tissue culture, but the tissue is usually taken from the very end of a stem or root. The end contains the meristem cells. The tissue that is removed is called the **explant**. Care is taken to get the proper amount of tissue from the correct place.

The explant is thoroughly cleaned and sterilized. It is then placed in an enclosed glass container. This arrangement is called **in vitro culture**. "In vitro" literally means "in glass." Within the glass container is a solution of sterile minerals, nutrients, and hormones. Because this solution is an ideal environment for the growth of microorganisms, the explant, container, and solution must be sterile. The sterilization of all of the material can be a tremendous advantage because the new plant will be disease-free. As the new plant begins to grow, it is transferred to other containers of growth media as the different stages of growth occur.

As new methods of tissue culture are developed, this type of propagation is likely to become even more commonplace. Today, several large companies use tissue culture as the basis for their propagation. The future holds promise for many applications of this technique.

SUMMARY

In nature, plants reproduce to survive. Agriculture uses this natural phenomenon as the basis for the agriculture industry. All of agriculture is dependent on plant reproduction.

Developing increased efficiency in the ways that desirable plants reproduce means increased efficiency for agriculture. Agricultural research and development has brought about a means of efficiently reproducing plants in order to better feed and clothe the world population.

CHAPTER 7 REVIEW

Student Learning Activities

1. Collect 10 different flowers. Determine whether they are complete or incomplete flowers. Locate the pistil and the stamen. Remove the pollen and examine it under a microscope. Sketch the different shapes of the pollen.

2. Locate several plants that have stolons or rhizomes. Follow the stolons or rhizomes and count the plants that are connected together. Compare your notes to others in the class.

3. Locate and make a list of plants that propagate by seed and by stolons. Determine which would be the best way to propagate the plants commercially. Give your reasons.

Define the Following Terms

1. asexually
2. zygote
3. perianth
4. meiosis
5. endosperm
6. monoecious
7. dioecious
8. cultivars
9. grafting
10. tissue culture
11. explant
12. in vitro culture

True/False

1. In the sexual reproduction of plants, fertilization does not have to occur.
2. Asexual reproduction is also known as vegetative propagation.
3. The stamens of a flower are the male structures that produce pollen.
4. The corn plant is monoecious.
5. New varieties of plants that have been formed in nature are known as cultivars.

6. Flowers have developed a strong scent to attract bees and other insects.

7. Bees are the most efficient of all insects at pollinating flowers.

8. Sexual propagation produces a plant that is genetically identical to the parent plant.

9. A potato is a tuber.

10. A high rate of photosynthesis is not desirable when attempting to root a cutting.

11. Photosynthesis is affected only by light and not by temperature.

12. When grafting, it is important that the rootstock and the scion be identical in size.

Fill in the Blank

1. The male gamete is called a _____ , and the female gamete is called an _____.

2. The union of egg and sperm creates a _____.

3. The _____ of a flower is composed of petals that give the flower its color.

4. The top part of the pistil is broadened to receive pollen and is called the _____ .

5. Complete flowers contain _____, _____, _____, and _____.

6. _____ is the substance secreted by a cutting that promotes healing at the place where the cutting was severed.

7. _____ are plants that are cultivated and grown by humans. Many of them are sterile.

8. To produce a plant that is genetically identical to the parent, one must use a(n) _____ form of propagation.

9. Two methods of obtaining new plants from specialized plant parts are _____ and _____.

10. In _____, plant material from two separate plants is joined together to form one plant.

11. In the grafting procedure, the lower part is called the _____, and the upper part is called the _____.

Discussion

1. What is the function of the anther in the flower?

2. Explain the difference between self-pollination and cross-pollination?

3. What are two ways in which pollen can be transported?

4. Name four specialized plant parts that may be used in vegetative propagation.

5. Name the ways in which corms differ from bulbs.

6. What are the advantages of cuttings as opposed to other methods of propagation?

7. Why is light important in the production of new plants from cuttings?

8. Define layering. Explain the various methods of layering.

9. Briefly describe the grafting process.

10. What are the two advantages of tissue culture?

11. Why must materials and instruments used in tissue culture be kept sterile?

PLANT GROWTH

STUDENT OBJECTIVES

After studying this chapter, you should be able to:

✦ Discuss the importance of plants to life on earth.

✦ Summarize the importance of seeds in our lives.

✦ Distinguish between dicot and monocot seeds.

✦ Explain the processes involved in seed germination.

✦ Describe the ways that a plant grows.

✦ Explain the role of hormones in plant growth.

✦ Discuss the effects of light on plant growth.

✦ Explain how photoperiodism is used in agriculture.

✦ Explain the effects of gravity on plant growth.

✦ List the essential plant nutrients.

✦ Describe the symptoms of nutrient deficiency in plants.

All life on earth depends on the growth of plants. As pointed out in Chapter 6, plants play such an important role in the production of oxygen and in the food chain that animals could not survive without them. Almost all aspects of agriculture depend on the ability of plants to grow and reproduce. The entire food industry is based on the growing of plants to be used directly as food or to feed livestock. The efficient growth of plants provides the basis of producers' livelihood. Those who produce plants such as vegetables, fruits, and grains for human consumption depend on the plants' abilities to grow and produce sufficient quantities to market. The same is true for those who produce forages or grains to be used for animal feed. Many millions of dollars have been spent to research and develop ways to enhance the growth of plants.

Figure 8–1 *Most plants begin their growth from seed. Courtesy of James Strawser, The University of Georgia.*

Figure 8–2 *Cooking oil is made from the oil in plant seeds. Courtesy of James Strawser, The University of Georgia.*

SEED

Importance of Seed

Most plants begin their growth from seeds (Figure 8–1). Remember from Chapter 7, seeds are the result of the union of the male and female gametes. Seed contains the genetic material for producing a new plant. By producing new and better seed, advances in the production of agricultural crops are achieved.

Although some plants reproduce without the production of seeds, most agricultural crops have their beginnings as seed. The production of seed is one of the most important links in the entire agricultural industry. In fact, around 70 percent of our diet comes directly from seeds. The bread or bowl of cereal that you had for breakfast this morning was made from the seed of grain plants. Cooking oil is derived from such seeds as safflower, soybeans, or cottonseed (Figure 8–2). Snack foods, such as peanuts and popcorn, are made from seeds.

Seeds are also used for a variety of other purposes. Beverages such as coffee and cocoa come from the seeds of trees and shrubs. Oils and dryers in paints and wood finishes come from seeds. Oil from the seed of the castor bean was

Figure 8–3 *Seeds come in a variety of shapes and sizes. Courtesy of Illinois Research, University of Illinois.*

once used for medicine and is now used as an oil in hydraulic systems. Seeds are also used to manufacture cosmetics, ointments, and pharmaceuticals.

Types of Seeds

Seeds are about as diverse as the plants that produce them. Seeds range in size from the almost microscopic seeds of some species of orchids to the seeds of coconuts, which can weigh as much as 40 pounds. Seeds come in a variety of shapes (Figure 8–3). They can be almost flat like an almond seed or almost perfectly spherical like the seed of the avocado.

Seeds are generally classified as being dicotyledonous (**dicot**) or monocotyledonous (**monocot**). The most obvious difference between the two are the cotyledon or seed leaves (Figure 8–4). A bean leaf is a dicot; it can be opened into two distinctive halves. A corn seed is a monocot and cannot be broken down into two separate halves because it has only one seed leaf. The seeds of most legumes, such as beans, peas, soybeans, and alfalfa, are dicots, and most grasses, such as corn, rice, wheat, and barley, are monocots. Within the monocot seed is an inner structure called the **endosperm** that stores food energy for the young plant. Dicots have no endosperm and store energy in the cotyledons. Inside the dicot is a large embryo that develops into a young plant upon **germination**; monocots have a much smaller embryo.

Seed Germination

Germination is the process that causes a seed to begin to grow into a new plant. Although scientists still do not fully understand the process, many aspects about the beginnings of plant growth are known.

Figure 8–4 *Dicots have two seed leaves; monocots have only one.*

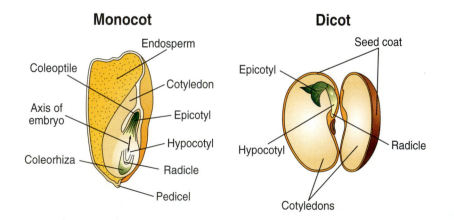

Figure 8–5 *Seeds are sometimes artificially dried to ensure proper storage.* Courtesy of James Strawser, The University of Georgia.

Water.　Mature seeds have a relatively low moisture content to prevent them from decaying. Seeds usually mature in the fall when weather conditions are driest. This allows the seeds to dry to a low enough moisture content so they will store. Producers sometimes artificially dry soybeans and grain after harvest to ensure proper storage (Figure 8–5). In order for seeds to germinate, they must take in water.

Seeds are covered with an outside coating called a **testa**. As long as the seed stays relatively dry, the testa remains intact, and the seed is dormant. When the proper moisture is present along with other favorable conditions, the testa begins to soften, swell, and finally ruptures. In fact, the entire seed takes in water and becomes larger due to the swelling caused by the water.

Oxygen.　Seeds require oxygen to germinate. Oxygen in the presence of sufficient moisture causes respiration to begin. Respiration is the process that converts the food stored by the seed into energy that is used by the plant in the germination process.

Temperature.　Proper temperature is essential to the germination of seeds. Most seeds must go through a period of cold before they will germinate. This helps prevent the seeds from sprouting as soon as they mature and are dispersed. By going through the winter and being subjected to the cold, the seeds remain dormant until the following spring when the warm temperature helps to trigger the germination process. The proper soil temperature for the germination of most seed ranges between 68 and 85°F, although the seeds of some cold-climate plants may germinate at a lower temperature.

Light.　In most seeds, light plays an important role in the germination process. Some seeds need almost total darkness to germinate, and others require some degree of light. This is why producers pay particular attention to the depth seeds are planted. If seeds are planted too deeply, they may not receive the proper amount of light and may not have enough oxygen. If they are too shallow, they may have too much light.

When a seed is developed from the parent plant, it usually goes into a period of dormancy in which the seed is alive but is not growing. When the proper conditions occur, the seed comes out of dormancy and begins to grow, and a new plant springs into life. Dormancy is necessary in order to preserve a seed for the proper time when conditions are optimum for

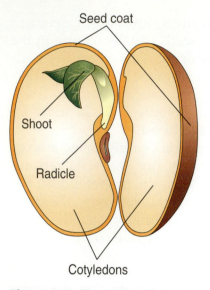

Figure 8–6 *The seed coat covers and protects the seed.*

growth. Most plants produce seed at the end of the growing season, just before cold weather sets in. If the seed sprouted (germinated) at that time, the new plant probably would be killed before it could reproduce. To prevent this from happening, several restrictors or inhibitors within the seed have to be overcome before germination can occur.

Some seeds have to undergo a period of cold weather before they will germinate; other seeds simply have to have a certain amount of time between maturation and germination. The period of cold or the interval of time allows the seed to survive the winter.

Some seeds have a hard coat that resists the entrance of water. This seed coat has to be weathered or eroded away before moisture can enter the seed and the germination process can begin (Figure 8–6). This process is known as **scarification**. Sometimes seeds are scarified by using mechanical abrasion, soaking in water for several hours, or by subjecting the seed to acid.

If seeds are buried deeply in the soil, the lack of oxygen or light can keep seeds in a state of **dormancy**. Deep plowing can unearth seeds that have been dormant for years. When they near the surface of the soil and temperature and moisture are favorable, the seeds will sprout.

Although the optimum time for viability (capability of living or germinating) appears to be about one year, seeds can live in a dormant state for many years. In fact, seeds have been discovered in the Arctic that were deposited more than 1,000 years ago. When given the proper conditions, the seeds germinated and produced new plants. However, producers know the importance of keeping new, fresh seeds that are highly viable. The tag on the bags of seed tells the year they were produced and the results of a germination test (Figure 8–7). This test tells what percentage of the seed are expected to germinate. Commercially produced seeds are usually expected to have a germination rate greater than 95 percent.

The Process of Germination

When the previously mentioned conditions have been met, the seed begins to germinate, and the growth process of the new plant starts (Figure 8–8). First the seed coat or testa softens, moisture enters the seed, and the seed begins to swell. When the testa ruptures, oxygen enters the seed and respiration begins. Respiration involves the release of enzymes that convert the insoluble starches in the seed to soluble sugars

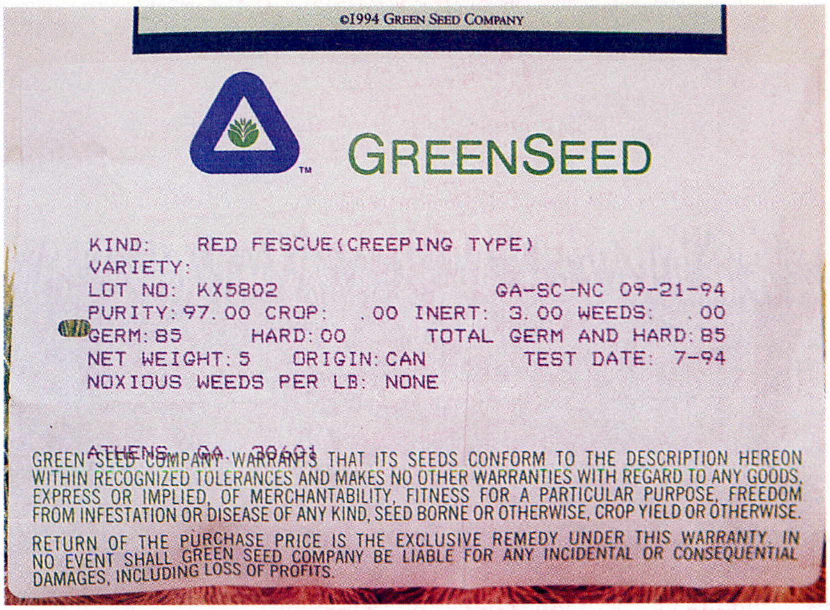

©1994 GREEN SEED COMPANY

GreenSeed

KIND: RED FESCUE(CREEPING TYPE)
VARIETY:
LOT NO: KX5802   GA-SC-NC 09-21-94
PURITY:97.00 CROP: .00 INERT: 3.00 WEEDS: .00
GERM:85   HARD:00   TOTAL GERM AND HARD:85
NET WEIGHT:5   ORIGIN:CAN   TEST DATE: 7-94
NOXIOUS WEEDS PER LB: NONE

ATHENS GA 30601

GREEN SEED COMPANY WARRANTS THAT ITS SEEDS CONFORM TO THE DESCRIPTION HEREON WITHIN RECOGNIZED TOLERANCES AND MAKES NO OTHER WARRANTIES WITH REGARD TO ANY GOODS, EXPRESS OR IMPLIED, OF MERCHANTABILITY, FITNESS FOR A PARTICULAR PURPOSE, FREEDOM FROM INFESTATION OR DISEASE OF ANY KIND, SEED BORNE OR OTHERWISE, CROP YIELD OR OTHERWISE.

RETURN OF THE PURCHASE PRICE IS THE EXCLUSIVE REMEDY UNDER THIS WARRANTY. IN NO EVENT SHALL GREEN SEED COMPANY BE LIABLE FOR ANY INCIDENTAL OR CONSEQUENTIAL DAMAGES, INCLUDING LOSS OF PROFITS.

Figure 8–7 *Tags on the bag of seed provide the producer with information about the seed.* Courtesy of James Strawser, The University of Georgia.

Figure 8–8 *When all conditions are met, seeds germinate and push up out of the soil.* Courtesy of James Strawser, The University of Georgia.

that are used for energy by the young germinating plant until photosynthesis can begin. The release of the energy in the form of soluble sugars allows for the growth of plant cells.

The cells multiply and growth takes place in two directions. In dicot seeds, a stem-like structure called the hypocotyl grows upward and emerges as an arch-shaped structure with both ends still in the ground (Figure 8–9). One end of the hypocotyl is attached to the seed leaves or cotyledons. The cotyledons contain a structure called a plumule that is located between the cotyledons. The plumule develops into the primary stem of the plant. Also on the cotyledons is a structure known as an epicotyl that develops into the first true leaves of the plant. The cotyledons serve as the first leaves by transferring stored food to the rest of the newly germinated plant.

At the end opposite the hypocotyl is the radicle, which grows downward and develops into the primary root for the plant. From this structure grow roots that go out in a horizontal direction.

The germination of a monocot seed differs in that the plumule grows straight upward, and the cotyledon remains in the ground (Figure 8–10). Also, the radicle grows down and forms a temporary root that is replaced by permanent roots that grow above the remnants of the seed. The plant's first leaves emerge from the plumule, and photosynthesis begins after the stored food from the endosperm is exhausted.

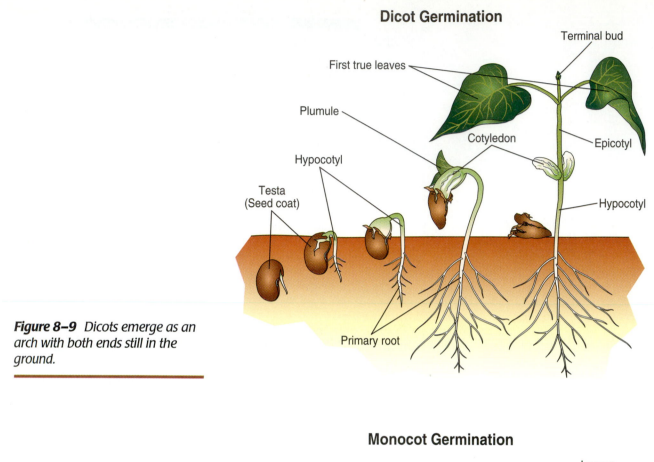

Dicot Germination

Terminal bud

First true leaves

Plumule

Cotyledon

Epicotyl

Hypocotyl

Testa (Seed coat)

Hypocotyl

Primary root

Figure 8–9 Dicots emerge as an arch with both ends still in the ground.

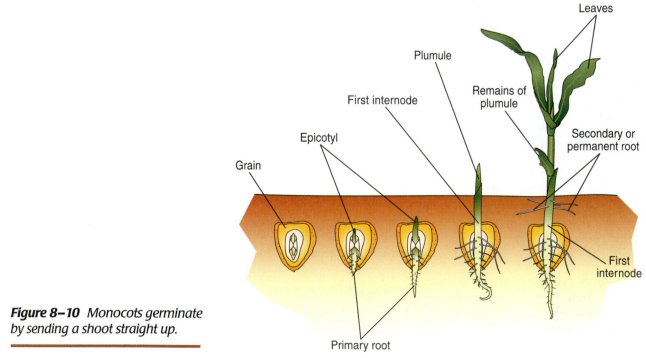

Monocot Germination

Leaves

Plumule

Remains of plumule

First internode

Epicotyl

Secondary or permanent root

Grain

First internode

Primary root

Figure 8–10 Monocots germinate by sending a shoot straight up.

GROWTH AFTER GERMINATION

After germination, plants can grow into huge structures. Just consider the size of the giant redwood trees in northern California! Large plants result from tiny seeds that germinate and produce a small seedling. Plants may also be very tiny when fully mature. Small, delicate plants are grown commercially as potted plants that add to the beauty of our homes. Plants also vary greatly in the rapidity of their growth. Plants such as oak trees may take many years to produce substantial growth while some tropical and subtropical vines grow several inches every day.

The processes that control growth are complicated. A lot of factors cause the plant to increase in size. Generally, plant growth is of three types. One is the increase in the size of the plant cells. A second is the increase in the number of plant cells, and the third is cell differentiation. Cell differentiation is the converting of cells to perform the different functions of the plant.

Growth takes place in the portions of the plant called the **meristem** tissue (Figure 8–11). This tissue is composed of cells that undergo mitosis (see Chapter 3) and divide rapidly. The meristem tissue is found at the tips of stems and branches and at the tip of roots. This tissue above (or below) the meristem tissue is divided into different regions. The first is called the elongation zone, where cells absorb water and grow longer. Above or below this zone (depending on whether the meristem tissue is on the roots or on the stems and branches) is the differentiation zone, where cells take on specialized functions. For example, some may become leaf cells, some stem cells, and others flower cells. As cells in the meristem region multiply, the cells in the other two regions change. Cells that were once in the meristem zone become cells in the elongated zone, and those cells become cells in the differentiation zone. Through this process the plant becomes larger. Plants such as trees usually put on a new layer of xylem and phloem each year. This makes the tree grow larger in diameter.

Plant Growth Hormones

Plant growth is controlled by substances called hormones. Hormones are chemical substances that are found in both plants and animals that control or influence the activities of

Root Structure

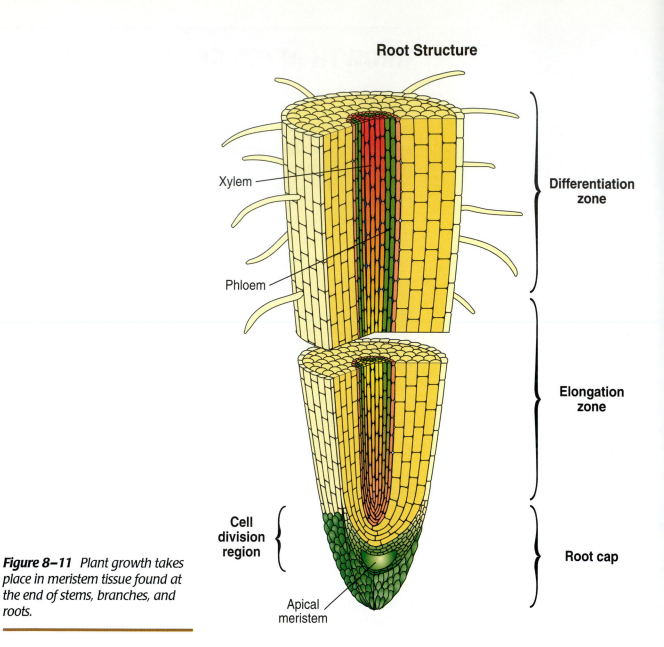

Xylem

Phloem

Differentiation zone

Elongation zone

Cell division region {

Root cap

Apical meristem

Figure 8–11 *Plant growth takes place in meristem tissue found at the end of stems, branches, and roots.*

organs or tissues. The proper balance of the hormones is essential to keeping the plant growing and functioning properly. Plant hormones function together to stimulate or inhibit plant growth. Both inhibition and stimulation of growth are important in order for the plant to grow in the proper manner and produce the proper type of mature plant. Three major groups of hormones control the growth of plants: auxins, cytokinins, and gibberellins.

Auxins. Auxins are hormones that can be found in most of the parts of a growing plant but have the highest concentration in the very end of the growing tissue. These areas are called apical meristems. At the end of the main growth shoot of the plant is a bud known as an apical bud where auxins are produced. These hormones cause cells to grow by elongation. The substance causes the plant cell walls to soften. When this happens, water is let in, and the cells become larger. When the cells become the proper size, the cell walls again harden, and the cells retain their larger size. Only a very tiny amount of auxins are needed to stimulate cell elongation, and larger amounts can inhibit cell growth. From the apical bud, the auxins are sent in such quantities to the lateral buds (buds that produce growth in the branches) that the growth of the lateral buds is slowed down. This causes the plant to grow taller with less growth to the branches. Because the apical bud controls the growth of the lateral buds, this process is called apical dominance. This phenomenon is used frequently in agriculture. If a producer wants a plant that has more limbs and is bushier, the apical bud is removed. This is a common practice in the ornamental plant industry. For the plant to produce a fuller, thicker growth with less plant height, the producer pinches the tip of the bud from the main stem of the plant (Figure 8–12).

Auxins are also involved in other plant processes. They are produced in the seeds of flowering plants. The seeds play a role in the ripening process of the fruit. As long as the seeds are developing, they produce auxins. When the seeds mature, they stop producing the auxin, and the fruit ripens and falls off.

Producers use auxins in several other ways. One way is by treating cuttings taken from plants. If the cuttings are dipped in an artificially produced auxin, root growth is enhanced on the cutting. Another way auxins are used is through the use of herbicides. A herbicide is a chemical that is used to kill unwanted plants. Remember that only very minute amounts of auxins are required to produce plant growth. When large amounts are used, the plant cells expand and grow so rapidly that they rupture, and the plant dies. A widely used herbicide called 2-4D makes use of this process. The chemical kills broadleaf plants but has little effect on grasses.

Cytokinins. Cytokinins stimulate mitosis in plant cells, creating growth by cell division and multiplication. This hormone

Figure 8–12 *Producers often pinch off the end of the stem to encourage thicker growth.* *Courtesy of James Strawser, The University of Georgia.*

BIO BRIEF

Saving Seeds for the Long Term

Agricultural Research Service plant pathologist Christina Walters is attempting to determine how long seeds can stay in storage and still germinate. Seeds stored under optimal conditions can last for hundreds of years. Scientists have discovered that seeds contain glass compounds when cooled to very low temperatures. These glass compounds have properties similar to the silica glass compounds found in window glass, except that the glass in seed forms at hundreds of degrees lower than the window glass.

These glass compounds play a role in keeping seed viable. All seed goes through an aging process, and all see eventually deteriorates no matter how they are stored. Dr. Walters is researching how all the different factors interrelate to affect the aging of seed. She says that a major goal is to identify seeds that store well and to be able to predict the rate of deterioration. She has found that water binding in the seeds has a major influence on seed storage. Glasses in the seed control the aging rates by controlling the rate of chemical reactions. Glasses make the seed cells very viscous, and this makes the molecules slow down. Glasses in seed that are too dry are too porous, and glasses in seed that are too moist are too fluid. Either condition lessens the ability of the seed to be stored well. The slower the molecules move, the slower the chemical reactions and the aging process.

Dr. Walters has discovered that if seeds are either too moist or too dry, they don't store very well. She preconditions the seed by holding them at 5°C at 25 percent relative humidity for a few weeks to achieve the

stimulates the beginning of bud growth. Some scientists think that cytokinins also aid in keeping leaves alive and growing. When the plant no longer circulates cytokinins to the leaf, the leaf stops growing, dies, and falls off.

Gibberellins. This group of hormones was discovered by a group of Japanese scientists who were studying a disease outbreak in rice fields during the 1920s. The disease caused the rice plants to grow so tall that they fell over because the stem could not support the additional weight of the extraordinary growth. They found that the disease was caused by a substance produced by the fungus *Gibberella fujikuori*. The hormone was named gibberellin after the fungus that produced it. Later, scientists found out that the substance is a naturally occurring hormone produced in many plants. It stimulates

correct moisture content; then the seeds are stored long term at −18°C. This seems to be the correct procedure for long-term seed storage.

Storing seed long term is important to protect varieties of plants that might otherwise become extinct. The National Seed Storage Laboratory in Fort Collins, Colorado stores some 300,000 different types of seed representing more than 8,000 species of plants. It is the largest gene storage bank in the world. Scientists preserve, evaluate, and distribute the vast collection of seed to breeders who use them to develop new varieties. The goal is to keep the seeds available to plant breeders all over the world. Dr. Walter's work is an important part of ensuring that the seeds are stored for future generations.

Source: Agricultural Research Magazine

Dr. Walters uses vats of liquid nitrogen to store seeds at very low temperatures for a long period of time. She studies all of the factors associated with the aging of seed.

the elongation of plant cells and also plays a part in the flowering and germination of some seeds.

The Effects of Light on Plant Growth

Plants respond to light in many ways. In Chapter 7 it was pointed out that light is necessary for photosynthesis to occur. Light is important in other processes affecting the growth of plants. It is commonly known that plants grow toward the light (Figure 8–13). If the light comes to the plant at an angle, the plant will grow in an arch toward the sun. Scientists are not sure why this occurs. The leading theory is that the presence of light causes auxins to move from the light side of the plant to the dark side. Because auxins stimulate the growth of plant cells by making them longer, the dark

Figure 8–13 *Plants grow toward the light. Courtesy of USDA-ARS/Bruce Fritz.*

side grows faster than the lighted side. This process makes the plant curve toward the light as it grows. The movement of the plant toward the light is known as phototropism. "Photo" means light and "tropism" refers to the tendency of a plant to turn toward a stimulus.

Light also has an effect on the time when plants bloom, shed their leaves, and go into dormancy. The process of plants responding to the length of daylight and darkness is called photoperiodism. For seeds to develop and mature, the flowers have to open and pollinate at the proper time. Flowering intervals are controlled to a large degree by the length of light periods. In nature this is controlled by the length of the daylight hours. In the summers, daylight hours are longer, and the plants are subjected to more sunlight than in the fall and winter months. Flower producers use this by artificially controlling the amount of light plants get each day. For example, producers of poinsettias must have flowers that bloom around Christmas. If they bloom before or after Christmas, their market will be lost. To control the time the plants will bloom, the producers place the plants on a strict schedule of light periods (Figure 8–14). Producers know that the amount of darkness is the crucial factor. If the darkness is interrupted, the flowers will not bloom at the proper time. Plants that bloom when the hours of daylight are longer are called long-day plants. Those that bloom when the daylight hours are shorter are called short-day plants. Some plants appear not to be affected by the length of daylight and are called day-neutral plants.

Figure 8–14 *Poinsettia growers control the number of hours the plants are exposed to light. Courtesy of James Strawser, The University of Georgia.*

The Effect of Gravity on Plant Growth

Another type of tropism is geotropism, or the movement of the plant toward gravity. Think about the way a plant grows. The plant actually grows in opposite directions—the stem grows upward, and the roots grow downward (Figure 8–15). The cause of this effect is the influence of gravity on plant growth. Because stems grow away from gravity they are said to be negatively geotropic. Roots grow toward gravity and are said to be positively geotropic. At one time, producers turned large seeds such as potatoes so that the eyes that produced the stem growth would be pointed upward. Research has shown that no matter how the potatoes are planted, the roots grow downward, and the stem grows upward.

PLANT NUTRITION

Just as animals need a constant intake of food to survive, plants also need nutrients. All of the plant's systems must have nutrition to function. Energy is needed to maintain the plant and to reproduce. Energy comes from the nutrients taken in by the plant and, just as animals have to have a balanced diet, plants must also have the proper balance of nutrients. These nutrients are absorbed by the plant from both the air and from the soil. Some of the nutrients are needed in large amounts. These are called macronutrients and include

Figure 8–15 *The roots and stem of a plant grow in opposite directions.*

carbon, hydrogen, oxygen, nitrogen, phosphorus, potassium, sulfur, calcium, and magnesium. Other elements are needed in relatively small amounts and are called micronutrients. These are iron, boron, zinc, manganese, copper, molybdenum, and chlorine.

Nonmineral Nutrients

Most of the nutrients required for plant growth and development are minerals. A mineral is a substance of inorganic origin. Plants must have three **nonmineral** elements. These are carbon, hydrogen, and oxygen.

Carbon. The element carbon is present in all living things whether plant or animal. This element accounts for most of the dry weight of a plant. In the 1600s, a Dutch man named Jan Baptista van Helmont conducted an experiment in which he weighed a willow tree and the soil in a container in which he planted the tree. The tree weighed 5 pounds, and the soil weighed 200 pounds. For a period of five years the tree received only rainwater and was not fertilized artificially. At the end of this period, van Helmont weighed the soil and the tree. He found that while the tree had gained more than 160 pounds, the soil had lost only 2 ounces. From this data, he concluded that all of the weight gain of the tree came from water. He attributed the loss of 2 ounces of soil weight to experimental error. Later research has shown that his conclusions were erroneous. The loss of soil weight was from the removal of minerals from the soil by the tree. The gain in weight of the tree was due to the intake of elements from the air. Most of this weight was due to the intake of carbon. Carbon is a major part of the plant compounds of starches, cellulose, fats, oils, and lignin. In fact, about 95 percent of a plant's dry weight is made up of carbon.

Carbon is taken into the plant as carbon dioxide (CO_2), an element that makes up about .03 percent of the atmosphere. Remember that a large portion of the soil volume is composed of air. Due to the action of microorganisms in the soil, the content of carbon dioxide in the soil air is higher than in the atmosphere above the ground, and some carbon dioxide is taken in through the roots. However, most of the plant's carbon dioxide is taken in through the leaves.

In photosynthesis, carbon taken from the air is converted to carbon in a form that the plant can use. Carbon is one of the major building blocks of the simple sugars. Without carbon, the manufacture of these sugars could not take place.

Some of the carbon is returned to the atmosphere through respiration. When the plant dies, the decay process returns carbon to the atmosphere where it can be used by plants again. Remember from Chapter 6 that animals breathe air, use oxygen, and give off carbon dioxide. Plants take in air, use carbon dioxide (and a small portion of oxygen), and give off oxygen. This cycle helps keep nature in balance.

Hydrogen. The element hydrogen is taken in through the air, but the major source of this nutrient is the water taken in by the plant. Like carbon, hydrogen is used in photosynthesis to create the sugars that the plant uses for energy.

Oxygen. This element makes up about 21 percent of the atmosphere and is an essential element in the photosynthesis process. Although oxygen is abundant in the air, the plant derives some of its needed oxygen from water.

PRIMARY NUTRIENTS

The macronutrients are divided into two groups, **primary nutrients** and **secondary nutrients**. The primary nutrients are nitrogen, phosphorous, and potassium. These elements, known by their chemical symbols as NPK, are the primary ingredients of commercial fertilizer. Ratios of these nutrients designate the type of fertilizer being sold (Figure 8–16). For example, the numbers 6-12-12 refer to the percentage of nitrogen, phosphorus, and potassium in the fertilizer. In other words, 100 pounds of the fertilizer would contain 6 pounds of nitrogen (N), 12 pounds of phosphorus (P), and 12 pounds of potassium (K). The remainder of the fertilizer would be secondary nutrients, micronutrients, and filler.

Nitrogen. Nitrogen is a key element in plant growth and development. It is essential in the production of chlorophyll (see Chapter 6) and gives plants a rich, green color. In fact, a symptom of a nitrogen deficient plant is a pale green or yellow color (Figure 8–17). Producers know that the addition of nitrogen to crops not only promotes rapid growth but also enhances the protein content of plants. This is because the plant uses nitrogen to build cell proteins.

Most of the nitrogen taken in by the plant comes from nitrogen in the soil that is bound up in organic matter. The decaying of this material releases nitrogen in an available

Figure 8–16 *The tag on the bag of fertilizer lists the percentages of the different nutrients in the fertilizers.* Courtesy of James Strawser, The University of Georgia.

Figure 8–17 *Plants that are deficient in nitrogen are pale green or yellow in color.* Courtesy of Potash & Phosphate Institute.

form used by the plant. Because plants can effectively use relatively large amounts of nitrogen, the natural process is inadequate for most agricultural crops. Producers add nitrogen in the form of nitrates and ammonia.

Phosphorus.   This primary nutrient plays an important role in the reproduction of seed plants. About half of the phosphorus content of a mature plant may be found in the seed and fruit. The chromosomes and genes of the plant contain this element, and it is needed to form the genetic material. In addition, phosphorus promotes rapid root growth at germination and helps the young plant to create a good root system. Plants that have a deficiency of phosphorus may have a purple tinge to the leaves (Figure 8–18).

Figure 8–18 *A phosphorus deficiency shows up on the leaves as a purple tinge.* Courtesy of Dekalb Genetics Corporation.

Because such a large portion of the plant's phosphorus is found in the seeds and fruits, the soil must be replenished each year. Most agricultural crops are grown for their seed or fruits, and this produce is harvested and removed from the field. This means that new phosphorus must be added to the soil each year to replenish the supply.

Potassium.   This element is necessary in the creation of the starches and sugars used by the plant as energy. Potassium assists in the development of resistance to disease, insects, and fluctuating weather conditions. It also aids in the process of stomate opening and closing. Without a proper amount of potassium, plants cannot grow well and have less resistance to diseases. Plants that are deficient in potassium will appear to have the tips of their leaves burned (Figure 8–19). They may also have yellow or white streaks in the veins of the leaves.

Figure 8–19 *Plant tips that appear to be burned are a symptom of potassium deficiency.* Courtesy of Potash & Phosphate Institute.

Secondary Nutrients

The secondary nutrients are those macronutrients that are not required in quantities as large as the primary nutrients but in larger quantities than the micronutrients. The secondary nutrients are calcium, magnesium, and sulfur.

Calcium.   This nutrient is needed for the formation of strong cell walls. Calcium is also instrumental in the formation of young, growing cells, particularly in the root area. It also aids in the use of the other plant elements. Calcium from plants is essential in the building of the skeletal systems of animals.

The leaves of plants that are deficient in calcium appear to be hooked, and the tips eventually die (Figure 8–20). Also, the root system will not be as well developed as in healthy plants.

Figure 8–20 *The leaves of a plant that is not getting enough calcium will curl and appear to be hooked.* Courtesy of Potash & Phosphate Institute.

Magnesium.   This element is used in photosynthesis and is a major component of chlorophyll. It is also used in the production of carbohydrates and fats and aids in the movement of other nutrients throughout the plant.

A symptom of magnesium deficiency is abnormally thin stems. The leaves become spotted with yellow while the veins remain green.

Sulfur.   Sulfur is used by the plant to form protein and increase the growth rate of roots. Insufficient amounts of sulfur may cause the plant to lose its lush green color and turn a pale shade of green. Plant growth is stunted.

Micronutrients

The micronutrients are those required by the plant at relatively minute amounts. Requirements for these elements are usually measured in parts per million. Even though they are used in tiny amounts, they are just as essential as the macronutrients. The following table shows the function of each of these nutrients and the symptoms of deficiency.

Plant Micronutrients

Nutrient	Function	Deficiency Symptom
Iron	Carrier for enzymes; production of chlorophyll	Stunted growth splotchy leaves
Chlorine	Aids in the use of enzymes	Symptoms of deficiency are unclear
Copper	Aids in respiration; helps plant use iron	Wilted young leaves; multiple buds
Zinc	Aids in reproduction; used in plant metabolism	Small, thick, spotted leaves; short internodes
Boron	Movement of sugars; water absorption by roots	Short, thick stems twisted leaves
Manganese	Aids in plant metabolism; helps in the use of nitrogen	White or yellow streaking of leaves; young leaves die
Molybdenum	Used in growth and reproduction; carries enzymes	Curly, yellow leaves; margins of leaves die

SUMMARY

As a result of techniques devised by humans, plants can grow more efficiently. Agricultural science has given us a variety of ways to make plants grow faster, bigger, and better. All of these methods use the principles of biology found in nature. Research has found ways to use these principles to increase the efficiency of growing food and fiber.

CHAPTER 8 REVIEW

Student Learning Activities

1. Keep track of all the food eaten by your family in a week. List all of the foods that come from seeds. (Do not forget such things as oils and spices.) Compare your list to those of others in your class.

2. Locate examples of plants that grow best in shade and those that grow best in open sun. What are the similarities of each group?

3. Locate plants that are suffering from nutrient deficiencies. Bring samples to class and explain the symptoms of the deficiency.

4. Research the use of gibberellins to affect plant growth. Report to the class.

Define the Following Terms

1. dicot
2. monocot
3. endosperm
4. germination
5. testa
6. scarification
7. dormancy
8. meristem
9. nonmineral nutrient
10. primary nutrients
11. secondary nutrients

True/False

1. Due to vegetative propagation only one-fourth of our diet comes from the seeds of plants.

2. Monocots have no endosperm and store their energy in the cotyledons.

3. Most seeds must go through a period of cold before they will germinate.

4. The plumule of an emerging monocot develops into the primary stem of the plant.

5. Growth activity takes place in the portions of the plant called the epicotyl.

6. Plant hormones can stimulate or inhibit plant growth.

7. The phenomenon of the apical bud controlling the growth of the lateral buds is called lateral submission.

8. Large amounts of auxins in plants result in increased growth.

9. Geotropism is the movement of the plant toward gravity.

10. A mineral is an organic source of plant nutrients.

11. Potassium is a secondary nutrient.

Fill in the Blank

1. A corn seed is a _____ because it has only one seed leaf.

2. The process that causes a seed to begin to grow into a new plant is known as _____.

3. The _____ is the outside coating of a seed.

4. With a monocot the _____ develops into the primary stem, and the _____ develops into the first true leaves of the plant.

5. _____ is the converting of cells to perform the different functions of the plant.

6. The three major groups of hormones are _____, _____, and _____.

7. The movement of a plant toward light is known as _____, and the process of plants responding to the length of daylight and darkness is called _____.

8. The three primary nutrients are _____, _____, and _____.

9. _____ is the secondary nutrient needed by the plant for the formation of strong cell walls.

Discussion

1. Name five products derived from seeds.

2. Explain the differences between dicot and monocot seeds.

3. What factors influence the germination of a seed?

4. In what ways may the germination of a seed be inhibited?

5. Explain the differences in the germination of a monocot and a dicot.

6. Describe the three types of plant growth.

7. What is one function of auxins? Cytokinins? Gibberellins?

8. How is the phenomenon of apical dominance used in the agricultural industry?

9. What are the three nonmineral elements a plant must have? How does a plant obtain these from the environment?

10. What are the symptom(s) of a nitrogen deficiency? Phosphorus deficiency? Potassium deficiency?

ANIMAL SYSTEMS

KEY TERMS

cartilage
periosteum
vertebrae
synovial fluid
myofibrils
striate
tendons
myoglobin
monogastric
ruminant
plasma
hormones

STUDENT OBJECTIVES

After studying this chapter, you should be able to:

✦ Discuss the interdependency of the systems of an animal's body.

✦ Name the major animal systems.

✦ Discuss the major parts and functions of the skeletal system.

✦ Discuss the major parts and functions of the muscular system.

✦ Discuss the major parts and functions of the digestive system.

✦ Discuss the major parts and functions of the respiratory system.

✦ Discuss the major parts and functions of the circulatory system.

✦ Discuss the major parts and functions of the nervous system.

✦ Discuss the major parts and functions of the endocrine system.

Have you ever wondered what makes a car run? After all, it is composed mostly of steel and plastic, and certainly no magic is involved. What is so mysterious about how the machine operates? The secret lies in the different systems of the car that make use of the laws of nature. Within each automobile is an electrical system, a fuel system, a chassis that supports the car, an intake and exhaust system, and a gear system. All of these systems go together to make a complete automobile. Each system serves a distinct purpose, yet each is dependent on the others. If one fails, the auto will not function properly or may not function at all. Animals are very much like cars in that they are composed of a variety of interdependent systems that perform separate functions. No one system can function entirely on its own, and each depends on the other systems for support. The animal systems include the skeletal, muscular, digestive, respiratory, endocrine, nervous, and reproductive systems. The reproductive system will be discussed in Chapter 10.

Producers of agricultural animals pay close attention to the systems of the animals they grow. In order to keep animals healthy, contented, and growing efficiently, producers have to make sure that all of the systems of their animals' bodies function properly (Figure 9–1). This is done through the selection process, the formulation of feed, a medication program, and management practices.

Figure 9–1 *Animal producers have to make sure their animals are healthy so the animals' systems will operate properly.* Courtesy of *DeRon Heldermon,* Limousin World.

THE SKELETAL SYSTEM

The skeletal system of an animal provides the frame and support for all of the other systems and organs (Figure 9–2). This system is made up of bones, connective tissue, and **cartilage**. Cartilage is composed of firm tissue that is not as hard as bone and is somewhat flexible. The skeletal systems of immature animals contain a larger proportion of cartilage than

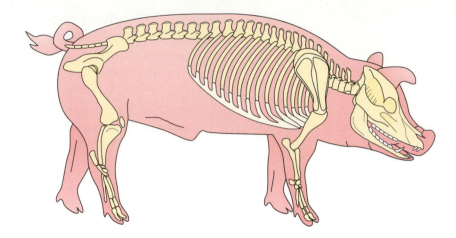

Figure 9–2 *The skeletal system provides support for the animal.*

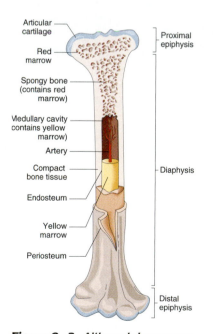

Articular cartilage
Red marrow
Spongy bone (contains red marrow)
Medullary cavity contains yellow marrow)
Artery
Compact bone tissue
Endosteum
Yellow marrow
Periosteum
Proximal epiphysis
Diaphysis
Distal epiphysis

Figure 9–3 *Although bones are mostly composed of hard mineral material, they are living tissue.*

mature animals. As bones grow, the cartilage portions of the bone harden into bone, and the bone increases in length. As an animal matures, cartilage in some parts of the skeleton continue to turn to bone. One method to determine the age of a slaughtered animal is to examine the amount of cartilage that has turned to bone in the ribs and backbone of the carcass. Other cartilage, such as that in the ears and nose, remains as cartilage throughout the animal's life.

The mature size of an animal is determined by the size and length of the bones in its body. Bones provide a place to attach the muscles and a means of movement. They also protect the internal organs of the body and are a key storage area for the body's mineral supply. The hollow portions of the bones serve as sites for blood formation.

Bones are covered with an outer layer known as the **periosteum** that cushions the hard portion of the bone and aids in the repair of a broken bone. Beneath this covering is a layer of hard mineral matter, called the compact bone, that is made of living cells, minerals, and protein (Figure 9–3). Most of the hard mineral matter is made up of calcium phosphate and calcium carbonate. This layer gives bones their strength. Inside the hard outer layer is a spongy-appearing structure called the spongy bone. The ends of bones are filled with this material, and the hollow portions of the bones are lined with it. The cavities of the spongy bone are filled with a soft substance called red marrow. This substance is responsible for the formation of blood cells. Inside the hollow portion of the bone is another type of marrow called yellow marrow. This marrow is yellow because it is mostly made up of fat storage cells. This serves as an energy storage area for the animal.

Bones grow and develop in very much the same manner as other tissues in the body. However, they are structured differently. Several types of bones are in an animal's body. These are classified according to the purpose the bone serves.

Long Bones

Long bones are, as the name implies, the longest in the animal's body. They support the body by giving it the rigidity necessary to stand and move. The large bones in the legs provide locomotion for the animal. The long bones act as levers that aid animals in moving about (Figure 9–4). A lever is a simple machine that consists of a bar that is used to pry

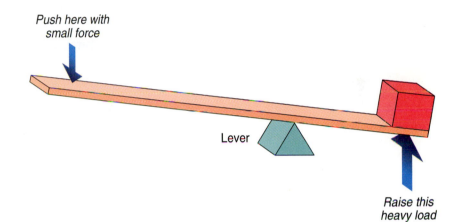

Push here with small force

Lever

Raise this heavy load

Figure 9–4 Long bones can act as levers.

against a load. The use of a lever magnifies the amount of force exerted. The longer the lever, the greater the force that is exerted. Think of a racehorse. An animal with long legs and long bones in the hip region has a distinct advantage over horses with shorter bones. The greater leverage of the longer bones allows the horse to have more power as it propels itself forward (Figure 9–5).

Figure 9–5 A greater length of the long bones, gives a horse an advantage in power and speed.

Producers often select animals based on the size and length of the long bones. Research has shown that the length of the canon bone (the bone between the ankle or pastern and the knee) of young animals is a good predictor of how tall the animal will be at maturity. Charts have been developed to help producers determine the mature size of cattle based on the hip height of the animal at a certain age. Obviously the hip height is directly proportional to the length of the long bones in the legs (Figure 9–6).

The pelvic bones, which are modified long bones, are important to producers in the selection of breeding stock. Producers want females that can give birth easily, and the size and shape of the pelvic bones can be a significant indication. For example, cattle that have long pelvic bones (the distance between the hooks and pins) usually have greater calving ease because the birth canal can open wider (Figure 9–7).

Figure 9–6 (A) In the 1950s, cattle were selected for short hip height. (B) Today, cattle are selected for a tall hop height.
Courtesy of Polled Hereford World.

Figure 9–7 *Cows with long pelvic bones usually have greater calving ease.* Couresy of Santa Gertrudis Association.

Another group of modified long bones, the ribs, are an important point of selection. The ribs form the thoracic cavity and protect internal organs such as the heart and lungs. They are long relatively flat bones that attach to the backbone on one end and to the sternum (the bone on the chest floor) at the other end. Producers know that animals with larger rib cages usually have larger hearts and lungs than animals with smaller rib cages (Figure 9–8). A widely sprung rib cage and wide floor to the thoracic cavity indicate an animal that is vigorous and that will grow efficiently.

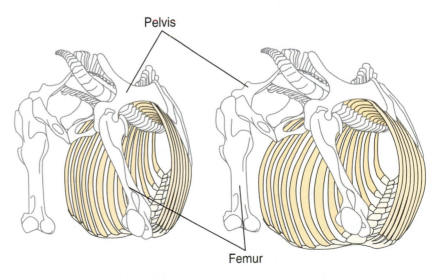

Pelvis

Femur

Narrow ribcage   **Large capacity ribcage**

Figure 9–8 *A widely sprung rib cage provides more room for the internal organs.*

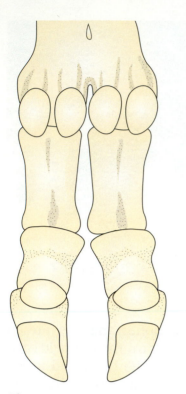

Figure 9–9 *Most short bones are found in joints and help the joint to be flexible.*

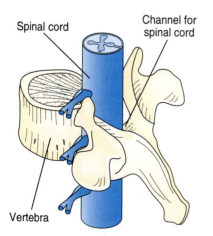

Spinal cord

Channel for spinal cord

Vertebra

Figure 9–10 *The vertebra is an example of an irregular bone that protects the spinal cord, which runs through the cavity.*

Short Bones

The short bones are smaller than the long bones and serve a different purpose. These bones may be about as large around as they are long. Most short bones are found in the joints and serve as hinges that help the joints to be flexible (Figure 9–9). Also, they help cushion shock and protect the long bones by being flexible or "giving" before the long bones are injured. Strong joints are important to the producer because the comfort and mobility of animals depends on the joints that hold the bones together. A thorough discussion of joints will be provided later in the chapter.

Irregular Bones

Irregular bones have an irregular shape. Their function is support and protection. Most agricultural animals are vertebrates. This means that they have a vertebra or backbone running the length of their bodies. Despite its name, the backbone is not a single bone but a series of irregular bones called **vertebrae**. The bones begin behind the head and continue like a chain to the end of the animal's tail. These bones provide an attachment either directly or indirectly for all of the limbs and other bones. They flex and bend to give the animal movement.

In the center of the vertebrae is a channel through which the spinal cord runs (Figure 9–10). This channel contains soft tissue and fluid that protects the cord from injury.

The vertebrae are divided into several different areas (Figure 9–11). The region of the neck from the skull to the first rib is called the cervical region. Almost all agricultural animals (mammals) have seven cervical vertebrae. These bones support and allow movement of the head and neck.

The thoracic region extends along the rib cage. Each vertebra has a rib attached to each side, although sometimes ribs at the front or rear of the rib cage have no attachments to the backbone. Cattle have 13 thoracic vertebrae, horses have 18, and pigs have either 13 or 14 depending on the breed.

The area of the spinal column from the last rib to the pelvis is called the lumbar region or the loin. If you have judged sheep, you probably have measured this area with your hand. Animals that have long lumbar vertebrae are more desirable because this is from where some of the most expensive cuts of meat come.

The sacral region of the vertebrae extends through the pelvic area. The bones in this region are usually fused

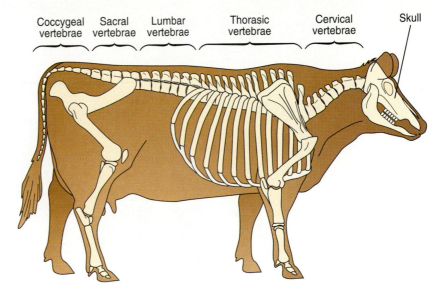

Coccygeal vertebrae | Sacral vertebrae | Lumbar vertebrae | Thorasic vertebrae | Cervical vertebrae | Skull

Figure 9–11 *The vertebrae are divided into five regions.*

together to form a rigid section called the sacrum. This structure provides an attachment for the pelvis.

The vertebrae continue to the end of the tail of the animal. This section is called the coccygeal region. In most animals the bones at the top of the tail are larger than the bones at the end. The size tapers until the ones on the end have no channel.

Flat Bones

The flat bones are relatively thin and flat and protect organs. The bones of the head are flat bones (Figure 9–12). They encase the brain and protect this delicate organ. The flat bones of the skull have openings through which the animal takes in air and nourishment. The openings or passageways are called sinuses. The ears, nose, eyes, and mouth all have openings in the skull.

Joints

All of the bones in an animal's body are connected to make up the skeletal system (Figure 9–13). They are held together by bands of tough tissue called ligaments that bind the bones at the joints. Some joints, like the bones of the skull, do not move. Others, such as the vertebrae, move slightly; others, such as the knee or shoulder, move a lot. The joints can move like a hinge (the knee), as a ball and socket (hip), or can move by a gliding motion (the vertebrae). The joints give the animal freedom of movement. The ligaments give the joints flexibility and serve as shock absorbers to protect the ends of the bones. The

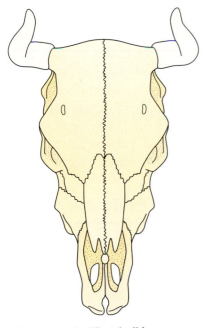

Figure 9–12 *The skull is composed of flat bones.*

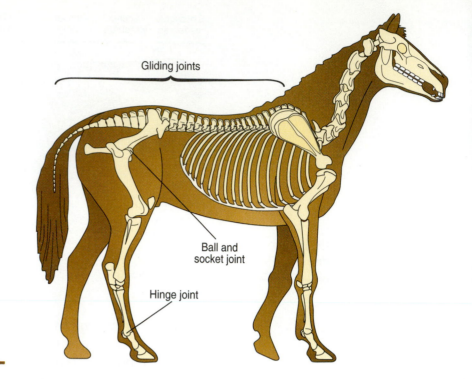

Gliding joints

Ball and
socket joint

Hinge joint

Figure 9–13 *Joints serve as hinges for the skeletal system.*

ends of the bones in the joint are covered with cartilage to aid in the absorption of shock and in the lubrication of the joint. Also within each joint is a pocket of **synovial fluid** that lubricates the joints.

Producers pay close attention to the joints of animals they select for breeding (Figure 9–14). Strong joints can help ensure that the animal moves about freely and is capable of mating. This is particularly true of animals raised on concrete floors. The legs must be set properly under the animal for an animal to move freely and efficiently. If the joints are set at an improper angle, the animal will have problems walking. These animals are not as productive as animals with proper feet, legs, and joints.

THE MUSCULAR SYSTEM

The skeletal system supports the animal and allows movement. However, the skeleton cannot move without the muscular system. Throughout the animal's body, a complex system of muscles provides the means for the animal to move about and for the proper functioning of the organs.

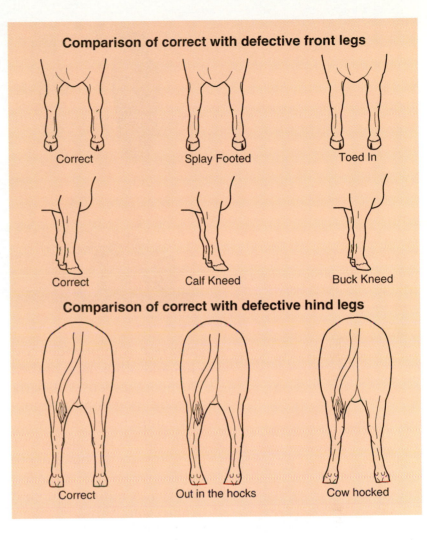

Comparison of correct with defective front legs

Correct | Splay Footed | Toed In

Correct | Calf Kneed | Buck Kneed

Comparison of correct with defective hind legs

Correct | Out in the hocks | Cow hocked

Figure 9–14 For breeding animals, producers select animals without irregularities in the joints.

Muscles are a major component of an animal's body. For example, a beef animal may have 35 to 40 percent of its total body weight in muscle. Producers of beef and other meat animals are in the business of producing muscles. It is the muscles of animals that are processed into the meat that occupies such an important part of our diet (Figure 9–15).

There are basically three types of muscles: skeletal, smooth, and cardiac.

Skeletal Muscle

Skeletal muscles make up the largest portion of the muscles. These muscles are long bundles of fibrous tissue called **myofibrils**. (*Myos* means muscle, and *fibrilla* means fibers.) Skeletal muscle cells are long and narrow and contain many nuclei. The cells are **striate** or striped in appearance. Their

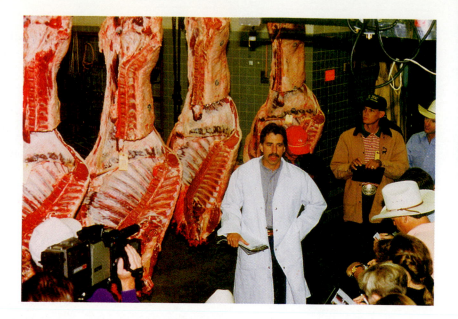

Figure 9–15 *Muscles are processed into meat.* Courtesy of Shea Gentzsch, *Limousine World.*

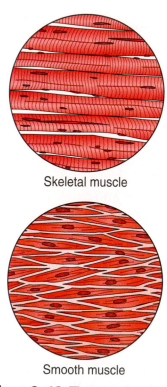

Skeletal muscle

Smooth muscle

Figure 9–16 *There are two types of muscle cells.*

function is to provide movement for the bones of the skeleton and other parts of the body. These muscles cause movement in response to particular circumstances, like moving toward food or away from a predator. They are called voluntary muscles, in contrast to involuntary muscles that maintain basic functions, like the heart and intestines.

Movement comes about by the contraction of muscles. In other words, the muscle that causes movement becomes shorter. A muscle seldom works by itself. Usually movement comes about through the coordinated effort of several muscles. As an animal walks, the movement of a leg is not controlled by a single muscle but by a group of muscles that surround the bones of the leg. These muscles work opposite each other. When one muscle contracts another relaxes and vice versa. For example, when the leg is extended forward, the muscles on the top of the leg contract, and the muscles on the bottom of the leg relax. When the leg is drawn back, the opposite occurs—the muscles on the bottom of the leg contract, and the muscles on the top relax.

The muscles move the bones through connective tissue called **tendons**. This tough tissue binds the muscle to the bones. A torn tendon means that a muscle has been torn loose from the bone.

Two types of muscles control skeletal movement—red muscle and white muscle (Figure 9–16). Although both types of muscles are throughout the skeletal system, one type usually dominates. A good example is the difference in the white and dark meat of a chicken. Red muscles dominate in the thigh regions, and white muscles dominate in the breast region.

Red muscles derive their color from the concentration of blood flowing through them and the supply of an iron rich compound called **myoglobin** that helps give blood its red color. Red muscles contain many mitochondria or organelles in which respiration takes place (see Chapter 3). These muscles contract slowly but are capable of continually contracting for relatively long periods of time.

White muscles contain fewer mitochondria, less blood flow, and a lower myoglobin content. By contrast these muscles contract faster and are stronger. They do, however, fatigue faster than red muscles.

Smooth Muscle

The smooth muscles of an animal's body control the movements of the internal organs and are found in the walls of the digestive tract, urinary tract, and other organs. The cells of these muscles form sheets of muscle tissue rather than bundles like the skeletal muscles. The cells of smooth muscles contain only one nucleus and are not striated as are the skeletal muscles.

The movement of the smooth muscles is called involuntary because the processes of the internal organs occur automatically.

Cardiac Muscle

The third type of muscle is cardiac muscle. As the name implies, these muscles control the heart. In fact, most of the heart is made of this muscle. Cardiac muscle has some characteristics of both skeletal and smooth muscle. Like the skeletal muscle, its cells are striated and are arranged in bands. Like smooth muscle, cardiac muscle has only one nuclei and is operated automatically. These muscles have amazing stamina. They must act almost continuously from before the animal is born until it dies.

THE DIGESTIVE SYSTEM

For the other systems to function, energy must be supplied. Energy comes from the food taken in or ingested by the animal. The digestive system takes the food ingested by the animal and converts it into a form that the animal can use (Figure 9–17). This conversion involves breaking down the food into components that can be absorbed and used by the cells of the animal. Basically there are two types of digestive systems in agricultural animals, **monogastric** and **ruminant** systems.

Figure 9–17 *The digestive system takes food ingested by the animal and converts it to a form the animal can use.* Courtesy of James Strawser, The University of Georgia.

Monogastric Systems

Monogastric systems are often referred to as simple stomach systems. This means that animals with this system have only one compartment in their stomachs. These include the pig, horse, dog, cat, and birds (Figure 9–18). The horse is different

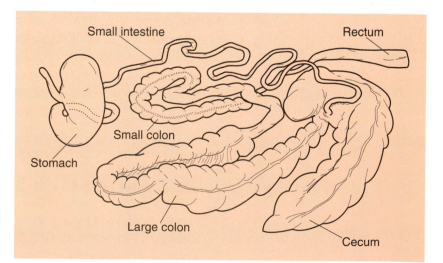

Figure 9–18 *The horse has a pouch known as the cecum that allows it to digest roughage.*

from the other monogastrics in that it has an enlargement known as a cecum that enables it to utilize high-fiber feeds by means of microbial fermentation, much as do ruminants (Figure 9–19). Simple stomach animals cannot digest large amounts of fiber like ruminants. Feed for monogastrics is referred to as a "concentrate" because of the relatively high concentration of nutrients and the low level of fiber.

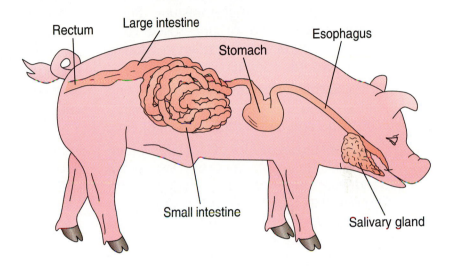

Figure 9–19 *The pig is an example of an animal with a monogastric system.*

The mouth is the organ that begins the digestive process. The tongue is used for grasping the food, mixing, and swallowing. The teeth tear and chew the feed into smaller particles that can be swallowed and further broken down. The mouth contains salivary glands, three pairs of glands that excrete saliva. Saliva contains several substances and serves several purposes: water to moisten, mucin to lubricate, bicarbonates to buffer acids in the feeds, and the enzyme amylase to initiate carbohydrate breakdown.

The hollow muscular tube that leads from the mouth to the opening of the stomach is the esophagus. It moves food from the mouth to the stomach. This is accomplished by muscles within the esophagus that contract and push the food down.

The stomach is a hollow muscle that causes further breakdown of foods by the contraction and relaxation of the muscles. This action causes the food to be pressed together, massaged, and mixed with the digestive juices secreted by the lining in the stomach. The actual process of breaking down the food is by chemical action. A strong substance known as hydrochloric acid begins to dissolve the food. Other secretions

BIO BRIEF

Better Diets for Dairy Cows

Dairy cows have known it for some time: They make more milk or get fatter when their diet includes high-moisture, finely ground corn instead of dry, rolled corn. Now, Agricultural Research Service studies have shown this scientifically.

"If you change harvesting and processing methods, you can increase corn's energy value," says animal scientist Barbara Glenn.

Earlier studies at ARS's U.S. Dairy Forage Research Center in Madison, Wisconsin, found that high-moisture, finely ground corn ferments rapidly. Feeds that are rapidly fermented in the rumen and fully digested in the intestines provide more energy for the cow to use to produce milk.

"But how much more?" asked Glenn and former colleague Vic Wilkerson at the ARS Nutrient Conservation and Metabolism Laboratory in Beltsville, Maryland. Wilkerson is now with Land O'Lakes' Western Feed Division in Portland, Oregon.

At the Beltsville lab three decades ago, ARS scientists first measured the energy value of feedstuffs for milk production—known as net energy of lactation (NEL). Today the lab is still one of a handful worldwide equipped with calorimetry chambers for net energy measurements.

"Any time we get NEL data, it's very valuable. There's very little data published because of the cost of doing the studies," says Bill Weiss, associate professor of animal science at Ohio State University. Weiss is a member of a National Research Council subcommittee that is revising the nutrient requirements of dairy cattle, including the energy values of feeds.

Feed consultants and dairy farmers rely on NRC's published values to formulate animal rations, but measured NEL values for new corn sources and types are rare; most values are estimated.

Dry corn might have been good enough in the past, but not for today's top milk producers. With many corn hybrids and storage and processing methods to pick from, says Glenn, "Corn isn't just corn anymore."

Wilkerson and Glenn measured the energy value of diets containing high-moisture corn compared with dry corn. Glenn says high-moisture corn—cut early, while still moist, and then ensiled—is popular with dairy farmers in the North Central and Northeast regions.

The researchers also compared the effect of grinding corn versus rolling it. Small ground particles are reportedly more digestible and thus able to provide more energy, she says. The different corns were mixed with alfalfa, soybean meal, and a powered mineral supplement.

Wilkerson calculated each corn's contribution to the energy value of whole diets. He wasn't surprised to find that high-moisture corn provided 14 percent more energy than dry corn, instead of the 4 percent different stated in the NRC handbook.

"Dairy farmers were getting fat cows when they substituted high-moisture corn

A large animal calorimeter is used in feeding tests that determine the energy valve of high-moisture, finely ground yellow corn. Courtesy of USDA-ARS.

milk production. It inflates the feed bill, and excess nutrients either add body fat or exit the cow as potential pollutants.

Dairy farmers do want more milk. In the ARS study, cows produced more than four pounds more milk daily with high-moisture corn than dry corn, in the alfalfa-based diet. Processing also made a difference. Finely ground corn provided five percent more energy than the big chunks of rolled corn, increasing milk production by about five pounds a day, says Wilkerson.

Weiss says his committee will consider the data in revising the energy values for dairy feedstuffs, noting that the values may be higher than the committee will agree on, "All net energy values we use now are estimated on very old numbers. This new values will help in devising rations that will make milk production more efficient."

for dry corn," he notes. "That suggested there was more energy available than what's shown in the handbook."

But farmers don't want fat cows any more than they want overly thin ones, especially when they stop making milk. "If a cow's too fat or lean, she won't breed," says Wilkerson.

Farmers also don't want cows getting more nutrients then they need for optimum

Source: Agricultural Research Magazine.

act on specific food components. For example, pepsin breaks down proteins into the amino acids, and rennin curdles the casein in milk. Another secretion, gastric lipase, causes the breakdown of fats to fatty acids and glycerol.

After the food is sufficiently broken down by muscular contractions and chemical reactions in the stomach, it moves into the next organ in the system, the small intestine. The entrance to the small intestine is controlled by a sphincter muscle that helps move food into and through the tract. This digestive organ consists of a long hollow tube that leads from the stomach to the large intestine. The small intestine is made up of several segments—the duodenum, the jejunum, and the ileum. Food passes through all of these segments and processes occur in all parts.

The duodenum is the first segment of the small intestine. This section uses secretions from the pancreas to break down proteins, starches, and fats. The intestinal walls also secrete intestinal juices that contain enzymes that continue the process of breaking down the food.

After the food leaves the duodenum, it enters the segments of the small intestine known as the jejunum and the ileum. In these are the areas where nutrient absorption takes place. Absorption is the process by which the nutrients are passed from the intestine into the bloodstream. The walls of these sections are lined with small fingerlike projections called villi that absorb the food nutrients into the bloodstream and lymph system through membranes that surround the villi (Figure 9–20). These are what is known as semipermeable

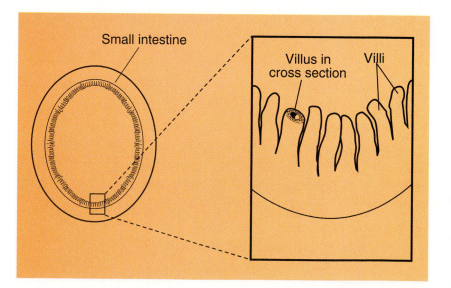

Figure 9–20 *The walls of the small intestine are lined with villi that absorb food nutrients.*

membranes. This means that the membranes allow particles to pass through in a process called diffusion.

The large intestine is the last organ of the digestive tract. The first section of this organ is a blind pouch called the cecum. Although the cecum is of little purpose in most monogastric animals, it serves a very important function in animals such as the horse. It is in this area that fibrous food, such as hay and grass, is broken down into usable nutrients.

The second segment of the large intestine, called the colon, provides a storage space for the waste from the digestive process and is the largest part of the organ. Water is removed from waste, and the process of decomposition of fibrous materials begins through microbial action.

The terminal end of the large intestine and the entire digestive system is the rectum. It serves to pass waste material through to the anus where it is finally eliminated from the body.

Ruminant Systems

A large group of animals including cattle, goats, and sheep eat large quantities of fibrous material such as grass and hay (Figure 9–21). This type of feed is called roughage. The reason these animals are able to digest all the fiber is that they have multicompartment stomachs that break these materials down into forms that are usable by the body. These animals are often called "cud chewers" because they regurgitate chunks of feed called boluses, masticate (chew), and reswallow. If you observe cows lying in the pasture you will often see them chewing on their "cud." The digestive systems of ruminants differ from monogastric systems in several ways.

The beginning of the system, the mouth, is different in ruminants because there are no upper front teeth. Instead a dental pad works in concert with the lower front teeth (incisors) in tearing off forages and other foodstuffs (Figure 9–22). The forage is then chewed between the upper and lower jaw teeth (molars). It is necessary that the mouth of ruminants produce large quantities of saliva to begin the process. This saliva is highly buffered (has a high pH) and contains phosphorus and sodium that aids the microorganisms that live in the rumen. Unlike most monogastric animals, there are no enzymes in the saliva, although some urea is released. This substance has a high level of nitrogen that is used by the bacteria in the rumen.

Figure 9–21 *Ruminants are able to digest large amounts of fibrous materials.* Courtesy of Kyle Haley, Limousin World.

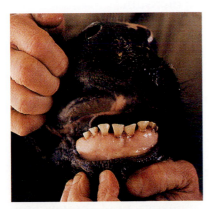

Figure 9–22 *Ruminants have a dental pad that helps tear off forage.* Courtesy of James Strawser, The University of Georgia.

Ruminants are often said to have four stomachs. This is incorrect. They have only one stomach, but it is divided into four separate and distinct compartments (Figure 9–23). The four compartments are the reticulum (honeycomb), rumen

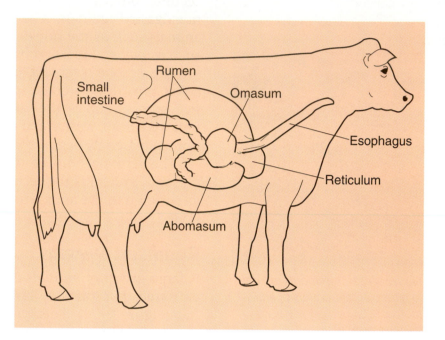

Figure 9–23 *The ruminant digestive system has several compartments that perform specific purposes.*

(paunch), omasum (many piles), and the abomasum (true or glandular stomach). In addition, the young ruminant animal has a structure called an esophageal groove or heavy muscular fold that allows milk from the mother to bypass the rumen and reticulum to go directly to the omasum.

The Reticulum

From the mouth, the esophagus leads to both the reticulum and the rumen. Ruminants sometimes pick up objects such as nails and small stones that are not digestible and that might harm the digestive system. These materials are heavier than most of the material swallowed, and they fall into the reticulum. The inside of the reticulum is lined with tissues called mucous membranes. Mucous membranes secrete a viscous, watery substance called mucus. These membranes form subcompartments that have the appearance of honeycomb (Figure 9–24). They trap and store "hardware" that does not float. This material remains in the reticulum, thus preventing dangerous objects from proceeding through the rest of the digestive tract.

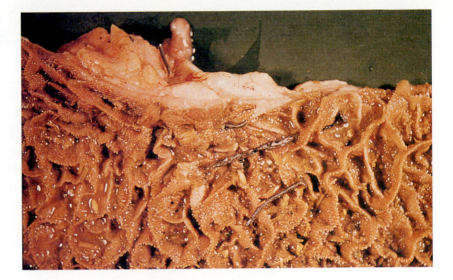

Figure 9–24 *The reticulum resembles a honeycomb and collects foreign material. Notice the nail in the center of the picture.*

The reticulum also stores, sorts, and moves feed back into the esophagus for regurgitation or into the rumen for further digestion. The process of breaking down roughages begins with a contraction of the reticulum and muscles in the esophagus to move roughage and fluid to the mouth. Excess fluid is squeezed out, and the material is reswallowed.

The food next moves to the rumen, which serves as a storage vat where food is soaked, mixed, and fermented by bacteria. These bacteria live in the rumen in a symbiotic relationship with the ruminant. This means that one organism lives in another, and both of the organisms benefit from the relationship. Bacteria thrive in the rumen environment and break down fibrous feeds for the ruminant.

The rumen is a hollow muscular paunch that occupies the left side of the abdominal cavity and contains two sacks, each lined with papillae (fingerlike projections) that aid in the absorption of nutrients. Here, carbohydrates are broken into starches and sugars. Volatile fatty acids are released as the carbohydrates are broken down, and these fatty acids are absorbed through the rumen wall to provide body energy.

Bacteria in the rumen also use nitrogen to form amino acids, and the amino acids form proteins. The bacteria can also synthesize water-soluble vitamins and vitamin K.

The Omasum

From the rumen, the food material passes through to the omasum. The omasum is a round organ located on the right side of the animal and to the right of the rumen and reticulum. The walls of the omasum contain many folds that are lined with blunt muscular papillae that grind roughage.

The Abomasum

After leaving the omasum, the food moves into the last compartment of the ruminant's stomach, the abomasum. The abomasum is the only glandular (true stomach) stomach of the ruminant. The abomasum is located below the omasum and extends to the rear and to the right of the rumen. By the time food materials reach the abomasum, the fibers of the roughages have been broken down to the extent that they can be handled by the abomasum. The abomasum and the small and large intestines of the ruminant animal function much the same way as they do in monogastric animals.

THE RESPIRATORY SYSTEM

For bodily processes to take place and for the animal to live, oxygen must be taken into the systems for their use. The respiratory system takes oxygen from the air and places it into the bloodstream for distribution to the cells of the animal's body.

The respiratory system begins with the nostrils, which are openings on the face of the animal. Large amounts of air are brought in through these openings. One of the selection criteria in racehorses is that they have large, broad nostrils that can take in sufficient air when the animal is running (Figure 9–25). Air is also brought in through the mouth. The nostrils open into the nasal cavities inside the skull, and these cavities lead into the pharynx, which serves as a common passageway for the food, water, and air that the animal ingests. The opening is controlled by a valve-like structure called the epiglottis that closes when the animal swallows. This prevents water and food from entering the respiratory system. If these materials accidentally enter the airway, the animal coughs and expels it back into the pharynx. The pharynx also provides a junction for the esophagus and the larynx. The larynx, also called the voice box or Adam's apple, controls the voice of the animal and aids in preventing material other than air from entering the lungs. A large tube known as the trachea leads from the larynx to the chest cavity where it branches out into two tubes called the bronchi (Figure 9–26). Both the

Figure 9–25 *One selection criterion for racehorses is large nostrils that can take in sufficient amounts of air when the animal is running. Courtesy of James Strawser, The University of Georgia.*

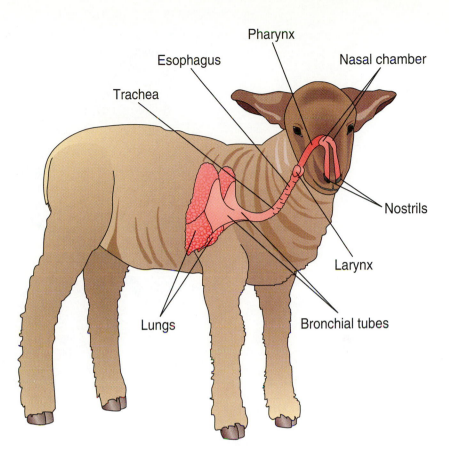

Pharynx

Esophagus

Nasal chamber

Trachea

Nostrils

Larynx

Lungs

Bronchial tubes

Figure 9–26 A large tube called the trachea leads from the larynx to the chest cavity where it branches to two tubes called the bronchi.

trachea and the bronchi are made up of ridged rings of cartilage that prevent the tubes from collapsing during breathing.

The bronchi branch out, and each branch divides further with each succeeding branch becoming smaller and smaller until they terminate in small sacs called alveoli. These tiny structures are so numerous that their combined surface area would be several times that of the total skin area of the animal.

The alveoli exchange gases with the bloodstream (Figure 9–27). The alveoli are surrounded by blood vessels that absorb oxygen from the ducts in the alveoli. Carbon dioxide is taken from the blood through the alveoli and expelled from the lungs. This process takes place through diffusion.

Breathing takes place by muscular contractions of the rib cage and the diaphragm. The diaphragm is a muscular structure that separates the chest cavity from the abdominal cavity. The inside of the chest cavity is a partial vacuum. When the muscles of the diaphragm contract and are drawn downward, the rib cage expands and air is drawn in. When the muscles of the diaphragm relax and return to their normal

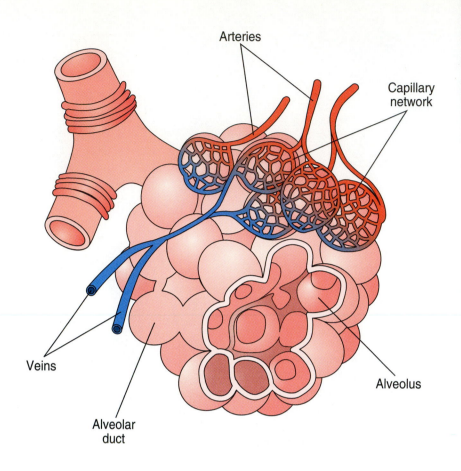

Figure 9–27 *Alveoli exchange gases with the bloodstream.*

upward arch, the ribs move inward, and the animal exhales. The process of exhaling expels carbon dioxide out of the lungs.

THE CIRCULATORY SYSTEM

The transportation of food nutrients, water, and oxygen is accomplished through the circulation of blood through the animal's body. In addition, the bloodstream cleanses the body by carrying toxic materials to the kidneys and sweat glands for excretion. Body temperature is regulated by the circulating blood, and disease agents are removed by cells in the blood.

The center of the circulatory system is the heart. This muscular organ operates continuously to pump blood throughout the animal's body. In fact, several thousands of gallons of blood are pumped each day through the body of a large animal. The heart is divided into four chambers (Figure 9–28). The heart is divided lengthwise by a thick wall

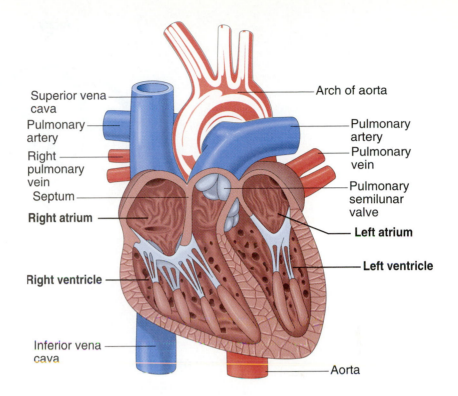

Superior vena cava

Pulmonary artery

Right pulmonary vein

Septum

Right atrium

Right ventricle

Inferior vena cava

Arch of aorta

Pulmonary artery

Pulmonary vein

Pulmonary semilunar valve

Left atrium

Left ventricle

Aorta

Figure 9–28 *The heart is divided into four chambers.*

of muscle called the septum, making right and left chambers. Each of these is divided into smaller chambers, the atrium and the ventricle. Blood returning from the body collects in the atriums and is pushed into the ventricles through valves that open and close. The sound of the heartbeat is the sound of these valves opening and closing. The walls of the ventricle are composed of thick, strong muscles that contract and force the blood through to the lungs and the other parts of the body. The right side of the heart pumps blood to the lungs through a large vessel called the pulmonary artery, and the left side pumps blood to the body through another large vessel, the aorta. These vessels branch out until they reach all areas of the body.

The blood vessels that take the blood from the heart are known as the arteries. Those that return blood to the heart are called veins. Veins and arteries are connected together by tiny, thin-walled vessels called capillaries that deliver the nutrients to the cells from the arteries and take away waste material through the veins (Figure 9–29). As the blood passes through the kidneys, the waste material is filtered out and is passed out in the urine. If all of the blood vessels of a large animal's body were laid out end to end they would reach around the world!

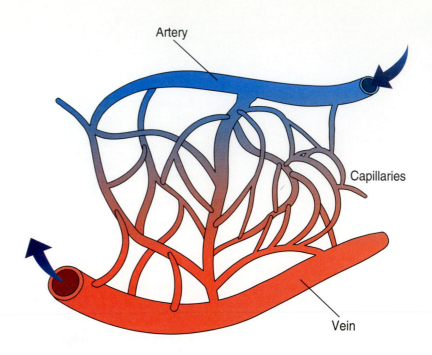

Artery

Capillaries

Vein

Figure 9–29 Blood is carried in veins, arteries, and capillaries.

Blood is an amazing fluid. More than half of its volume is composed of a tan-colored fluid called **plasma** that suspends several substances that help sustain life. The rest of the blood's volume is made up of three types of blood cells. Red blood cells give the blood its color and are shaped like donuts with holes that are not quite clear through (Figure 9–30). These cells are filled with a substance called hemoglobin that is composed of an iron-rich protein. The red blood cells bind with oxygen as they pass through the lungs. White blood cells are actually clear in color and destroy disease-causing agents (Figure 9–31). (More will be said about this process in Chapter 12.) The third type of cell is the platelets. These are actually only fragments of cells and contain bits of cytoplasm.

Red blood cell

Figure 9–30 Red blood cells are red and are shaped like donuts.

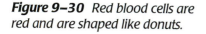

Viewed from above   Viewed from an angle   Cross section

Their function is to prevent blood loss when a vessel is injured. The platelets release chemicals called clotting factors that "patch" breaks in blood vessels.

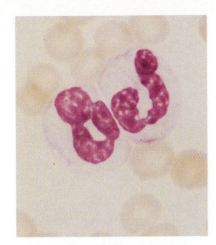

Figure 9–31 *White blood cells destroy disease-causing agents.* Courtesy of Rose Raskin, University of Florida.

THE NERVOUS SYSTEM

For all of the systems to function properly, the movements and processes have to be controlled by a central system. This is the function of the nervous system (Figure 9–32). The brain is the control center for the nervous system. It is an almost incomprehensibly complex organ. The most advanced computers in the world are simple compared to the brain of a higher-order animal. Messages are sent to all parts of the body from the brain through long fiberlike structures called nerves. The nerves that conduct impulses from the brain to other parts of the animal's body are called motor or efferent neurons (Figure 9–33). Nerves that send impulses from the body back to the brain are known as sensory or afferent neurons. All nerves are connected directly or indirectly to the spinal cord that runs through the center of the backbone. Different sections of the spinal cord receive and deliver messages to and from certain parts of the body.

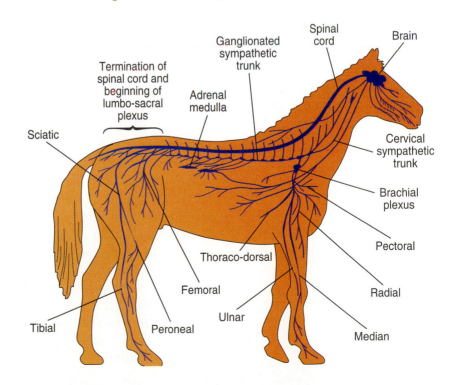

Figure 9–32 *The nervous system is the control center of the animal.*

Figure 9–33 *Neurons send messages to and from the brain.*

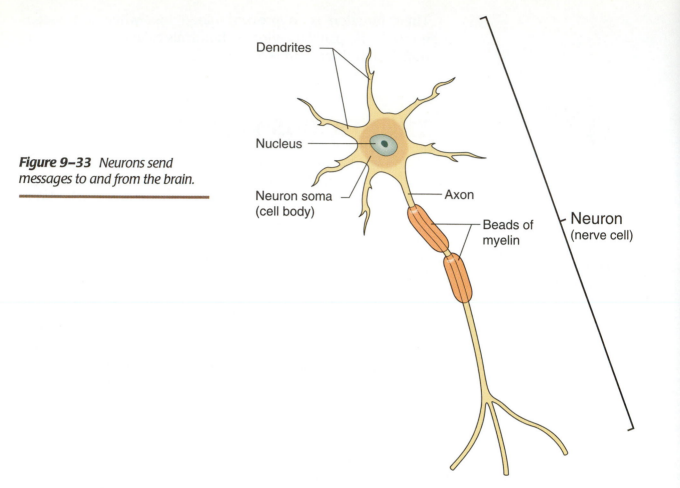

The upper portion of the spinal cord is attached to the brain. Impulses transmitted through the spinal cord are sent to or received from the brain and carried to or from the various parts of the body. The brain is divided into several sections, each of which controls certain body functions (Figure 9–34). The largest part of the brain is the cerebrum. This wrinkled, folded portion controls the thought processes of the animal. If an animal decides to move a foot, the impulses must come from the cerebrum. If the foot is placed on a sharp stone, the impulse is sent to the brain from the foot, and the brain sends a message back to the foot to lift off the stone.

The cerebellum acts as a coordination center for messages from the cerebrum. As an animal walks, many muscles are used to extend and contract the leg and back muscles. These muscles have to move in coordination with each other or the animal will not walk smoothly or may not walk at all. The cerebellum sifts through all of the messages to and from the

Anatomy of the Brain

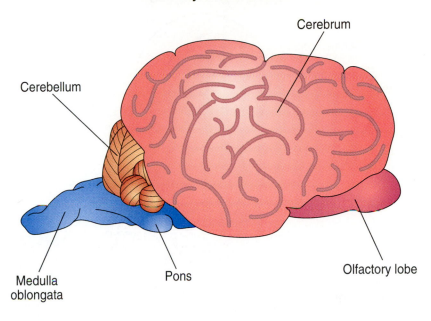

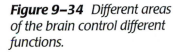

Figure 9–34 *Different areas of the brain control different functions.*

cerebrum and times and coordinates the movement of the muscles.

The involuntary activities of the body such as the beating of the heart or the breathing of the animal are controlled in the lower part of the brain called the medulla oblongata. This part of the brain responds to the need for changes in heart beat and blood pressure. If an animal begins to work harder (running, for instance), the breathing rate increases along with the heartbeat. This area of the brain also controls such bodily functions as the body temperature, the movement of food through the digestive system, and feelings such as fear or thirst.

THE ENDOCRINE SYSTEM

This system is composed of glands that secrete substances called **hormones** (Figure 9–35). Hormones are chemical agents that are sent to specific areas of the animal's body to stimulate or inhibit a response. Because hormones control such vital bodily functions as growth and reproduction, they have been the subject of a lot of scientific research. Producers of agricultural animals frequently use hormones or synthetic hormones to stimulate responses in animals. (Chapters 10 and 11 will discuss the uses producers make of hormones.)

Endocrine System of Domestic Animals

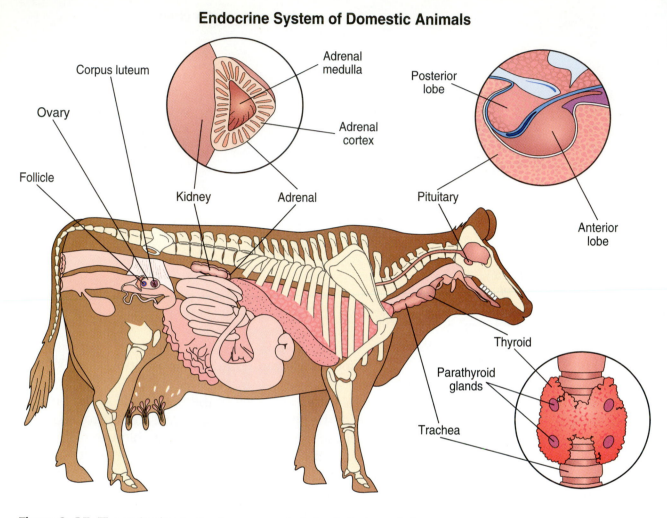

Figure 9–35 *The endocrine system is composed of glands that secrete hormones.*

The pituitary gland is located near the base of the brain. This gland is often referred to as the master gland because it controls all of the functions of the other endocrine glands through the hormones it secretes (Figure 9–36). It also manufactures and secretes hormones that control such processes as growth and reproduction.

The hypothalamus is located under the lower front part of the brain. It secretes hormones that control body temperature, hunger, sleep, and certain functions of the digestive process. Another major function is to manufacture hormones that stimulate the pituitary into producing hormones.

The adrenal glands are positioned on top of each kidney. They produce a hormone called adrenaline that stimulates

Functions of Endocrine Glands and Hormones

Endocrine Gland	Hormone Secreted	Primary Physiological Function
Hypothalamus	Gonadotropin-releasing hormone (GnRH)	Stimulates release of LH and FSH
	Corticotropin-releasing hormone (CRH)	Stimulates release of ACTH
	Thyrotropin-releasing hormone (TRH)	Stimulates release of TSH
	Growth-hormone-releasing hormone (GHRH)	Stimulates release of growth hormone
	Growth-hormone-inhibiting hormone (somatostatin)	Inhibits release of growth hormone
	Prolactin-releasing hormone (PRH)	Stimulates release of prolactin
	Prolactin-inhibiting hormone (PIH)	Inhibits release of prolactin
	Oxytocin	Causes ejection of milk, in mammals expulsion of eggs in hens, and uterine contractions
	Vasopressin (antidiuretic)	Causes constriction of the peripheral blood vessels and water reabsorption in the kidney tubules
Anterior pituitary	Growth hormone (GH or somatotropin)	Promotes growth of tissues and bone in the body
	Andrenocorticotropin (ACTH)	Stimulates secretion of steroids (especially glucocorticoids) from the adrenal cortex
	Thyrotropin or thyroid-stimulating hormone (TSH)	Stimulates thyroid gland to secrete thyroxine
	Prolactin (Prl)	Initiates lactation and promotes maternal behavior
	Gonadotropic hormones	
	Follicle-stimulating hormone (FSH)	Stimulates follicle development in the female and sperm production in the male
	Luteinizing hormone (LH)	Causes maturation of follicles, ovulation, and maintenance of the corpus luteum in the female. Causes testosterone production by the interstitial cells of the testes in the male
Thyroid	Thyroxine, triiodothyronine	Increase metabolic rate
	Calcitonin	Lowers the concentration of calcium in the blood and promotes incorporation of calcium into bone

Figure 9–36 Hormones of the endocrine system stimulate or inhibit responses in the animal. (continues)

Functions of Endocrine Glands and Hormones

Endocrine Gland	Hormone Secreted	Primary Physiological Function
Parathyroid	Parathyroid hormone	Maintains or increases the level of calcium and phosphorus in the blood
Andrenal glands Cortex (shell)	Glucocorticoids	Make energy available; increase blood glucose level, have an antistress action
	Mineralocorticoids	Maintain salt and water balance in the body
Medulla (core)	Epinephrine (adrenalin)	Stimulates the heart muscles and the rate and strength of their contraction
	Norepinephrine	Stimulates smooth muscles and glands and maintains blood pressure
Ovaries, Follicles	Estrogens	Cause growth of reproductive tract and mammary duct system
Corpus luteum	Progesterone	Prepares reproductive tract for pregnancy, maintains pregnancy, and causes development of mammary alveolar system
	Relaxin*	Causes relaxation of ligaments and cartilage in the pelvis, which assists in parturition or birth
Testes	Androgens (testosterone)	Cause maturation of sperm; promote development of male accessory sex glands and secondary sex characteristics
Pancreas (islets of Langerhans)	Insulin	Lowers blood glucose
	Glucagon	Raises blood glucose
Placenta (in some species)	Gonadotropins (pregnant mare serum gonadotropin, Human chorionic gonadotropin), estrogens, and progesterone	Promote the maintenance of pregnancy

*Relaxin is produced by the corpus luteum in the pig but is produced by the placenta in the mare.

Figure 9–36 *Hormones of the endocrine system stimulate or inhibit responses in the animal. (continued)*

physiological responses in the animal in times of stress or danger. The heart rate and breathing rate increase to give the animal added oxygen and energy to meet a crisis.

The thyroid gland is located on the front of the windpipe and produces a hormone called thyroxin that aids the body in the use of energy from digested food. This substance controls the rate at which the body cells break down carbohydrates and sugars. If the body needs more energy, the thyroid releases more thyroxin, and the rate of breakdown increases. When the need is lessened, the production of thyroxin is slowed. The thyroid also secretes calcitonin, which causes the storage of calcium in the bones. The parathyroids are situated inside the thyroid glands and regulate the amount of calcium and phosphorus in the blood. If the calcium level of the blood is too low, the parathyroid hormone causes some of the bone tissue to break down and restores the proper level of calcium in the blood.

The pancreas is a gland situated below the stomach. It secretes two hormones, insulin and glucagon, that regulate the amount of glucose in the blood. Humans sometimes suffer from a disorder called diabetes that results from a lack of the proper amount of insulin. The condition is corrected through the daily intake of insulin.

Other organs of the body, such as the reproductive organs, also produce hormones. (Chapter 11 will deal with the hormones produced by the reproductive system.) Scientists often discover hormones they did not previously know existed. Animals are complex organisms composed of interdependent systems that still baffle our most knowledgeable scientists.

SUMMARY

Higher-ordered animals are complicated organisms. Many systems have to function properly, and at the same time function in coordination with each other. By understanding and using the naturally occurring processes of animal systems, agriculture is better able to produce the meat and other animal products so vital to our survival.

CHAPTER 9 REVIEW

Student Learning Activities

1. Make a chart of the different animal systems and their uses. Be sure to include food, pharmaceuticals, and cosmetics.

2. Create a chart contrasting plant and animal systems. List all of the similarities and differences in the systems.

Define the Following Terms

1. cartilage
2. periosteum
3. vertebrae
4. synovial fluid
5. myofibrils
6. striate
7. tendons
8. myoglobin
9. monogastric
10. ruminant
11. plasma
12. hormones

True/False

1. All systems in an animal's body are independent and do not depend on other systems for support.

2. The skeletal systems of immature animals contain a larger proportion of cartilage than mature animals.

3. Red marrow is the substance responsible for the formation of blood cells.

4. The purpose of the ribs is to provide a protective cavity called the sternum.

5. The main purpose of flat bones is support and protection.

6. Skeletal muscle makes up the largest portion of total muscles in the body.

7. Mitochondria is the iron-rich compound that gives blood its red color.

8. Monogastric digestive systems are often referred to as simple stomach systems.

9. Ruminants have no upper front teeth but have a dental pad instead.

10. Nerves that send impulses from the body parts back to the brain are known as sensory neurons.

11. The endocrine system is composed of glands that secrete hormones.

Fill in the Blank

1. The _____ system of animals provides the frame and support for all of the other systems and organs.

2. Bones are held together by bands of tough tissues called _____ that bind the bones at the joints.

3. There are three types of muscles: _____, _____, and _____.

4. There are two types of digestive systems in agricultural animals: _____ and _____.

5. Even though it serves little purpose in most monogastric animals, the _____, which is located in the large intestine, helps animals such as horses digest hays and grasses.

6. In the ruminant digestive system, the _____ is considered to be the "true stomach."

7. Air can be brought into the respiratory system through the _____ or the _____.

8. In the pharynx there is a valve-like structure called the _____ that prevents food or water from entering the respiratory system.

9. _____ are tiny sacs in the lungs that exchange gases with the bloodstream.

10. Breathing takes place by muscular contractions of the _____ and the _____.

11. The heart is divided lengthwise by a thick wall of muscle called the _____.

12. Red blood cells are filled with an iron-rich substance called _____.

13. The adrenal glands produce a substance called _____ that helps the animal in times of stress or danger.

14. The pancreas secretes two hormones, _____ and _____. Diabetes results from a lack of _____.

Discussion

1. How might producers ensure that all the systems of their animals' bodies function properly?

2. Name the types of bones found in the skeletal system. Describe their functions and give an example of each.

3. Describe the differences between white muscle and red muscle.

4. Name each compartment of the ruminant stomach and briefly describe the function of each.

5. Describe the pathway that food travels in the monogastric digestive system.

6. Identify the three types of cells found in blood and the function of each.

7. Name the parts of the brain and briefly describe their functions.

8. Which system is responsible for the production of hormones? Why are hormones so important to bodily functions? Why might producers use hormones or synthetic hormones in the production of animals?

CHAPTER 10

ANIMAL REPRODUCTION

KEY TERMS

zygote
meiosis
spermatogenesis
chromatids
synapsis
oogenesis
testosterone
follicle
copulation
artificial insemination
cloning

STUDENT OBJECTIVES

After studying this chapter, you should be able to:

◆ List and explain the steps involved in meiosis.

◆ Discuss the parts and functions of the male and female reproductive systems.

◆ Describe the functions of the hormones that control reproduction.

◆ Describe the phases of the female reproductive cycle.

◆ Explain the process by which fertilization takes place.

◆ Explain the procedures used in artificial insemination.

◆ List and explain the steps used in embryo transfer.

◆ Describe the advantages of estrus synchronization.

◆ Explain the process of estrus synchronization.

◆ Define and explain the process of cloning.

Perhaps the most important part of the animal industry is animal reproduction. Through this phase, new animals are brought into the production cycle for use by humans (Figure 10–1). By carefully controlling the selection of breeding stock, improvements can be made in the livestock produced.

Much of what we know about reproduction in humans is a direct result of research conducted on agricultural animals. Each year millions of dollars are spent researching more thorough animal reproduction. Through a better understanding of the process, producers are better able to care for their animals and produce them more efficiently.

Figure 10–1 *Through the process of reproduction, new animals are brought into the world.* Courtesy of James Strawser, The University of Georgia.

Not only must the animals reproduce, but they must reproduce efficiently. This means that cows must give birth each year, and sows must have a large number of pigs at each farrowing. Reproduction is complicated. In Chapter 9, the different systems of an animal's body were discussed. All of these systems must function efficiently for an animal to reproduce regularly. Reproduction involves the act of mating, the chemical processes of hormones, cellular division, nutrition, and many other processes. These processes involve all of the systems of the animal's body, and they must work in harmony for reproduction to occur.

Agricultural animals reproduce sexually. This means that the offspring receives genetic material from two parents. Both the father (sire) and mother (dam) of the animal contribute half of its characteristics. The beginning of the young animal is the union of the gametes or sex cells of the parents. Just as in plants, the male gamete is known as a sperm cell, and the female gamete is known as an egg cell. The union of the two gametes results in a new cell that contains genetic material from both parents. The new cell, called a **zygote**, results in the beginning of a new animal similar to the parents. The zygote divides by the same process of mitosis that was explained in Chapter 8 on plant growth.

THE PRODUCTION OF GAMETES

The reproduction process begins with the formation of the gametes in the male and the female. The testicles of the male form the sperm cells, and the formation of the egg takes place in the ovaries of the female. Through a process known as **meiosis**, cells divide into cells that contain only one set of chromosomes for the formation of the young animal (Figure 10–2). When the

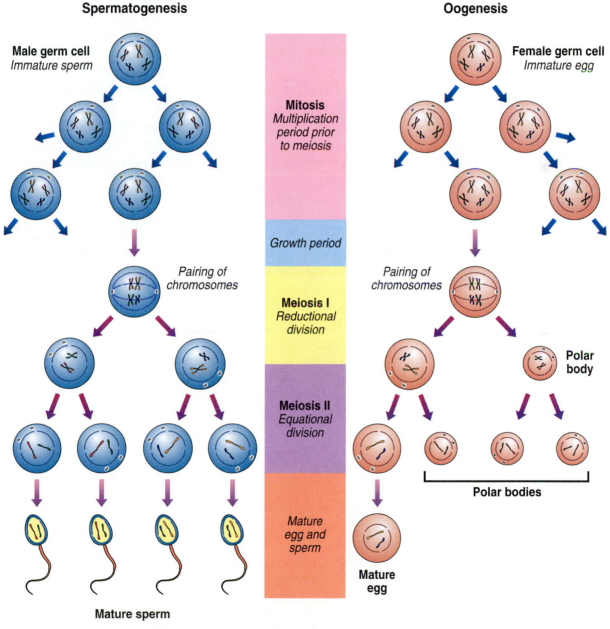

Meiosis
Maturation Process

Spermatogenesis

Male germ cell
Immature sperm

Pairing of chromosomes

Mature sperm

Mitosis
Multiplication period prior to meiosis

Growth period

Meiosis I
Reductional division

Meiosis II
Equational division

Mature egg and sperm

Oogenesis

Female germ cell
Immature egg

Pairing of chromosomes

Polar body

Polar bodies

Mature egg

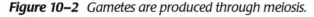

Figure 10–2 *Gametes are produced through meiosis.*

gametes unite, the full number of chromosomes is accomplished by each parent contributing half (see Chapter 4).

The production of the male gamete or sperm takes place through **spermatogenesis**, which occurs in the testes of the male. Cells develop into spermatozoa (sperm) through a four-step process. The first step involves a process called replication in which the chromosomes make an exact copy of themselves. The replicated chromosomes are called **chromatids**. To complete the second step, the chromatids come together and match up in pairs in a step called **synapsis**. In the third step, the cell divides, and the chromosomes separate with each cell receiving one of each chromosome from each pair. However, because each chromosome replicated itself and the chromatids are still attached together, in the fourth step both the cells and the chromatids separate, and the chromatids become chromosomes. These new sperm cells each contain only half of the chromosomes that the original cell contained. The result of this process is that four new sperm cells are produced from the original cell.

The female gametes (the eggs) develop in the ovaries in processes similar to the production of sperm. Gamete production in the female is known as **oogenesis**. There is, however, one important difference. In meiosis in egg production, only one egg cell is produced instead of four. In oogenesis, three of the newly divided cells become what is known as polar bodies, and only one cell becomes a viable egg. Because the egg is considerably larger than the sperm, it needs more nourishment. Most of the cytoplasm (cell material outside the nucleus) from the cell goes into the one cell that will become the egg. The function of the polar bodies is to provide sustenance for the egg until conception.

THE MALE REPRODUCTIVE SYSTEM

The male reproductive system consists of certain glands, the testicles, and the penis (Figure 10–3). The testicles produce the sperm through meiosis. They also produce hormones that aid in the reproductive process. The main hormone produced by the testicles is **testosterone**, an important hormone that controls the animal's sex drive or libido. It also plays an important role in the growth process and in the development of the sex characteristics of the male. Sex characteristics are those that make a male look like a male and a female look like

The Male Reproductive System

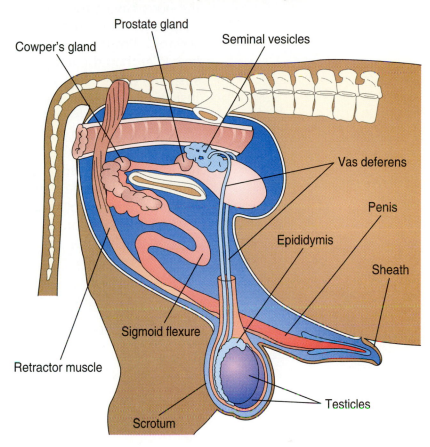

Figure 10–3 *The reproductive system of the male consists of the penis, testicles, and certain glands. This is a diagram of a bull's tract.*

a female. For example, bulls have broad, massive heads with a large portion of weight concentrated in the shoulder region (Figure 10–4). Females are more refined with smaller shoulders and a relatively larger pelvic area.

Figure 10–4 *Testosterone plays an important role in developing male sex characteristics such as a massive head and large shoulders. Females are more refined.* Courtesy of Lee Pritchard, Texas Limousin Association.

With the exception of poultry, which have testicles inside the body cavity, the testicles of most agricultural animals are suspended from their bodies in a saclike structure called the scrotum. Proper sperm production requires that the testicles be cooler than the temperature of the animal's body. By being suspended they can remain cooler in the summer months. Inside the scrotum are muscles that draw the testicles up against the animals body when the weather is cold to prevent them from freezing and to maintain a constant temperature. When the weather warms up, the muscles relax, and the testicles are suspended away from the body. In addition, the scrotum has little hair and no fat beneath the skin. This helps dissipate heat in hot weather.

Sperm are produced in the interior portion of the testicles (Figure 10–5). After the sperm are created through meiosis,

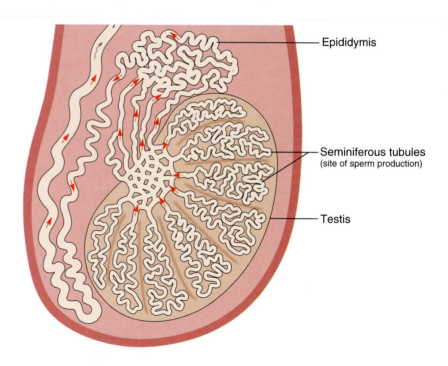

Epididymis

Seminiferous tubules
(site of sperm production)

Testis

Figure 10–5 Sperm are produced in the interior portion of the testicles.

they move to a region called the epididymis, which is composed of many tube-like structures that hold the sperm until they mature. The sperm are conducted through tubular structures called the vas deferens that lead to the seminal vessels on the urethra. The urethra are tubes that lead from the bladder to the penis and transport urine for expulsion from the body. The seminal vessels are a holding place for the sperm and secrete a fluid that is mixed with the sperm. This fluid protects the sperm and provides a medium in which the sperm can be transported. Two other glands also provide fluid for the mixture referred to as semen. The Cowper's gland secretes fluid that helps clean the urethra before the sperm is passed along the tube. Remember that the urethra also serves as a conduit for urine. The Cowper's gland secretion also acts to coagulate or thicken the semen. The prostate gland also secretes a fluid for the semen mixture that provides nutrients for the sperm. This muscular gland encircles the urethra and also serves to expel the semen during the mating process.

The penis of the male serves two purposes, expelling urine from the body and depositing sperm in the female tract. This organ is composed differently in different animals. The penis of the boar, bull, and ram is composed mostly of connective tissue. The upper end of the penis is S-shaped and flexes to extend the penis outward during mating. The penis of the stallion is made of vascular tissue that allows the organ to become engorged with blood. This causes the penis to become extended and is said to be erect. This allows penetration of the female. The external covering of the penis is called the sheath or prepuce. It protects the penis from injury and infection.

THE FEMALE REPRODUCTIVE SYSTEM

The female reproductive system is more complex than that of the male (Figure 10–6). The male system constantly produces sperm and provides millions of gametes at the time of mating. In contrast, the production of the fully matured egg of the female comes about only in carefully controlled cycles. The cycle produces the egg, places the egg in the proper place, causes the female to accept the male for mating (called estrus or heat), and ensures that the fertilized egg remains in place throughout the gestation period.

The female gametes are produced by the ovaries, which are located in the abdominal cavity and are supported by strong ligaments. The ovaries also produce the hormones estrogen and progesterone that play essential roles in the reproductive cycle.

The entire female reproductive cycle is controlled by hormones. The events in the cycle have to occur at the proper time, or fertilization (the joining of the egg and the sperm)

The Female Reproductive System

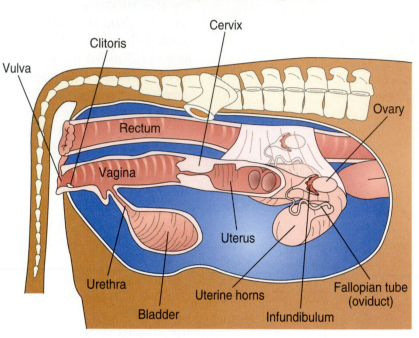

Figure 10–6 *The female reproductive system of a cow.*

will not take place. Remember from Chapter 9 that the pituitary gland controls the action of other glands. The ovaries are stimulated by a hormone called the follicle stimulating hormone (FSH) into producing a blisterlike projection called a **follicle**. Within the follicle, the egg or ovum grows and matures. At the same time, the follicle secretes a hormone called estrogen that stimulates the rest of the female reproductive system into preparation for the mature egg (ovum). When the egg is mature, the blisterlike projection ruptures, and the egg is expelled. This process is known as ovulation (Figure 10–7). Animals such as cattle and horses typically expel only one egg. Occasionally, two (and rarely three) eggs are expelled, and fraternal twins result. Pigs produce several eggs at one time and therefore have litters of young.

The newly matured egg leaves the follicle and enters a tube called the fallopian tube, which leads from the ovary to the uterus. It is usually in the fallopian tubes that fertilization occurs. The uterus (sometimes called the womb) is a muscular sac-like organ in which the fertilized egg (zygote) develops into an embryo and then into a fetus. The lower end of the uterus is sealed by a thick group of circular muscles called the cervix. It is through this opening that the sperm travel and the new animal emerges from the uterus. The cervix contains glands that secrete waxy material that seals the uterus.

Ovulation

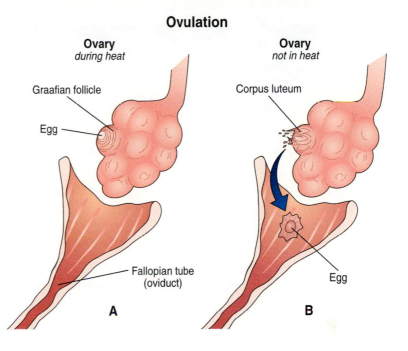

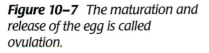

Figure 10–7 *The maturation and release of the egg is called ovulation.*

The tract that leads from the uterus and cervix is called the vagina. This sheath-like organ accepts the male animal's penis during mating. The new animal travels through this tract once it leaves the uterus. The outermost part of the female reproductive tract is called the vulva. This organ provides a closing to the vagina and also serves as the terminal end of the urinary tract. Within the vulva is a small sensitive organ called the clitoris that provides stimulation during the mating process.

THE MATING PROCESS

At about the time of ovulation, the estrogen produced by the follicle causes a condition known as estrus or heat in which the female will allow the male to mate with her. This condition may last from a few hours to two or more days depending on the species of animal. The female displays aggressive behavior such as grunting (pigs) or bellowing (cattle). They may also mount other animals and move around restlessly in search of a male (Figure 10–8). The males seek out females that are in estrus and may display belligerent behavior toward humans or other animals. During the act of mating, called **copulation**, the male's sperm is deposited into the female tract by a process called ejaculation.

B I O ✦ B R I E F

Vitrification Keeps Pig Embryos Viable

Consumers want high-quality food at lower prices, so producers are scurrying to meet price and quality demands of the marketplace. Interest in foods with enhanced qualities such as leaner, more tender meat is forcing livestock producers to look to science, particularly genetics, for answers.

Fortunately, Agricultural Research Service scientists at the Germplasm and Gamete Physiology Laboratory in Beltsville, Maryland, are helping to meet the challenge.

"In order to produce a better product for the consumer, we must develop strategies for maximizing the genetic potential of our domestic animals," says John R. Dobrinsky, an animal physiologist at the lab.

"Maintaining genetic diversity is the key to maintaining the most valuable genetic traits in animal populations that will provide the foundation for meeting the needs of future generations," he says.

Dobrinksy and his colleagues are working to develop technologies for preserving germplasm and embryos from today's genetically and economically superior animals. They are developing methods to preserve pig embryos indefinitely in liquid nitrogen at −320°F (−196°C). Preserved in this way, the embryos' biological activity all but ceases. After being removed from storage and transplanted into surrogate mothers, the embryos resume normal development.

Since the mid-1980s, the meat animal industry has been routinely cryopreserving embryos of several livestock species, especially cattle. However, the $11 billion-a-year swine industry has not had this technology available. Now that could change.

Scientists at the Beltsville lab are using a technique called *vitrification*. They cool a liquid medium so fast that ice crystals can't form and then store pig embryos in it. For a solution to vitrify, it must be instantaneously cooled in liquid nitrogen.

"Rapid cooling prevents ice crystals from forming in or around the embryos, and this is key to their safe storage and later development," Dobrinsky says. "Pig embryos are extremely sensitive to slow cooling below normal room temperature—about 59°F (15°C). This type of slow cooling is required during conventional embryo freezing methods, and this is why pig embryo survival after cooling or cryopreservation has been so poor," says Dobrinksky.

This problem, he explains, is that embryos suffer physiological and structural changes when going from normal body temperatures to cooler temperatures. "Hypothermic conditions can change normal cell function and skeletal structure, making the embryo incapable of normal development."

Rapid cooling during vitrification is through to outrace damaging effects evident during slow cooling. With vitrification, Dobrinksy and his colleagues achieved modest success rates—this is, about 40 percent of the

This sow's five pigs developed from cryopreserved and surgically transferred embryos. Courtesy USDA-ARS.

embryos survived. But, that was not good enough. Dobrinsky wanted to know exactly how the embryos' cell structures were being damaged during cryopreservation.

"We focused on the embryonic cytoskeleton," he says. The cytoskeleton is a network of microfilaments and microtubules that gives embryonic cells their shape and support—just as the human skeleton shapes and supports the body.

With a confocal microscope, which uses lasers, Dobrinsky took a closer look inside embryos during and after cryopreservation.

The microscope produces a digital reconstruction of all cells in the embryo.

"We could see the cell plasma membranes, and the microfilaments were being disrupted. We wanted to prevent these membrane disruptions," he says. "Our hypothesis was this: If we dismantled the microfilaments in an orderly way before cryopreservation, they might reform normally and support the plasma membrane after cryopreservation. To do this," says Dobrinsky, "we place from 10 to 20 embryos into small straws in a solution containing a compound called microfilament

continued

BIO BRIEF *(continued)*

inhibitor. We vitrified and stored those straws in a sealed canister filled with liquid nitrogen."

"We learned that they can be stored this way indefinitely," he says. "And, once they are warmed, the embryos can then be further cultured or transferred to surrogate mothers, where the microfilament network will reform and normal cell development will resume."

Dobrinsky has found his system increases the survival rate to more than 80 percent in the laboratory.

"From the laboratory to the barn, we warmed vitrified embryos and transferred them to surrogate mothers, producing the first live offspring from vitrified/warmed pig embryos," says Dobrinsky. "This is a first for maternal genetics. Until now, the swine industry could only preserve sperm from select males through semen cryopreservation—processes that were developed at Beltsville in the 1970s."

This high-tech cryopreservation approach is more than just a scientific advancement: Both swine producers and consumers will benefit. This process could bring a major increase in the efficiency of making pigs with important genetic traits available to breeders worldwide. Today, many pigs are transported by air freight from countries with breeder herds to those where new breeding herds are being established.

"The chance for global expansion of the swine industry is another potential advantage," says Dobrinsky. "Keeping separate breeding herds would be costly for producers. This technology will allow us to import and export valuable breeding stocks and unique germplasm—without the worry of shipping live animals. It has the potential for changing the way we produce future generations by minimizing risk of disease transmission or loss of valuable animals during transport."

Source: Agricultural Research Magazine.

FERTILIZATION

At the time of ejaculation, millions of sperm are deposited in the female vagina (Figure 10–9). The fluid in the semen provides the sperm with a medium in which to move and with nourishment for the trip. The sperm move through the fluid using their tails in a whiplike action. Sperm that move about freely are said to be motile. The sperm travel from the vagina

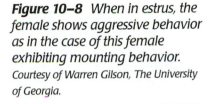

Figure 10–8 *When in estrus, the female shows aggressive behavior as in the case of this female exhibiting mounting behavior.* Courtesy of Warren Gilson, The University of Georgia.

Anatomy of Sperm

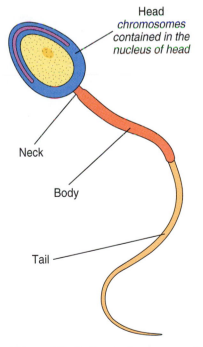

Head
chromosomes contained in the nucleus of head

Neck

Body

Tail

Figure 10–9 *Sperm is composed of tail, body, neck and head.*

through the cervix and uterus and into the fallopian tubes. Remember that in ovulation the egg was deposited from the follicle into the fallopian tube. Here, the sperm and the egg are united in the process of fertilization.

The egg secretes a chemical that attracts the sperm toward the egg. For many years, it was thought that only one sperm was needed for fertilization (Figure 10–10). However research has shown that many sperm are needed for the process. Although only one sperm fertilizes the egg, other sperm aid in the uniting. The egg is encompassed by a membrane that is surrounded by a jellylike substance that protects the egg. For the sperm to get inside the membrane, the protective layer must be dissolved. Each sperm releases a chemical that dissolves the membrane, and through the action of many sperm surrounding the egg, the layer is dissolved. When this happens, one sperm forms a tubelike connection with the membrane of the egg. The nuclear material of the sperm enters the egg, and the tail of the sperm is left outside the egg (Figure 10–11). When the nucleus of the sperm enters the egg, the egg releases proteins and carbohydrates that form a layer around the egg to prevent any more sperm from entering. This new layer is called the fertilization membrane. As soon as the nuclei of the sperm and egg combine, the correct number of chromosomes are established, and the fertilization process is complete.

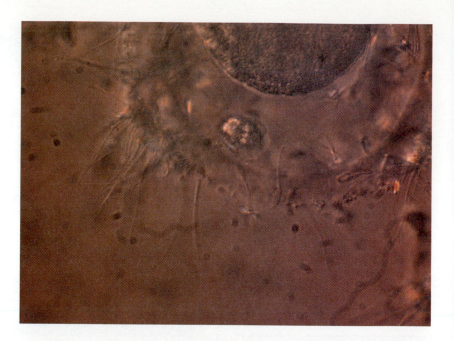

Figure 10–10 *Only one sperm fertilizes the egg, but many others aid in the process.* Courtesy of Richard Fayrer-Hosken, The University of Georgia.

Figure 10–11 *When the sperm enters the egg, the tail is left outside and a layer is formed around the egg to prevent the penetration of more sperm.* Courtesy of Richard Fayrer-Hosken, The University of Georgia.

If fertilization does not occur, the reproductive cycle of the female begins all over again. When fertilization does occur, certain processes take place to ensure that the female remains pregnant during gestation, which lasts from fertilization until the birth of the new animal. Gestation differs in length according to the species. To maintain gestation, the luteinizing hormone (LH) is produced by the pituitary gland. This causes a body called the corpus luteum to form on the follicle after the expulsion of the egg. The corpus luteum secretes a hormone called progesterone that causes the walls of the uterus to thicken in preparation for receiving the fertilized egg. After conception (the uniting of the sperm and the egg) occurs, the corpus luteum continues to produce progesterone, and the female remains pregnant. If conception does not occur, the corpus luteum recedes and the ovary returns to normal, and the cycle is begun again. In the embryo-development stage, the mass of cells begins to secrete a hormone that prevents the formation of a corpus luteum on the follicle of the ovary. As a result, the lining of the uterus remains intact, and the embryo can continue to develop.

ARTIFICIAL INSEMINATION

For many years livestock producers have used a technology known as **artificial insemination** (AI). In this process, sperm is collected from the male and deposited in the reproductive tract of the female by means other than natural mating. Artificial insemination has several advantages over natural mating. Producers can make use of superior males at a small fraction of the cost of owning and caring for the animals. They can select males with the specific characteristics they need for their females. This selection process is made much more reliable because of the number of offspring a male can have through AI. Data for the offspring are collected and give the producer information about the type of offspring the animal sires (Figure 10–12).

Artificial insemination also lessens the occurrence of diseases. Several livestock diseases are passed on through the mating process. Because the animals do not mate, the chances of passing diseases are kept at a minimum. Also, animals from all over the world can be used. Quarantine laws require that animals coming into this country be kept in isolation for a period of time to make sure that they are not diseased. Because this is an extremely expensive process, the use of animals from outside the country is not feasible except through AI.

Semen is collected from the male by several techniques. The most widely used method is the use of a dummy. The animals are trained to mount the dummy, and a technician guides the penis into an artificial vagina (Figure 10–13). Ejaculation occurs, and the semen is collected in an ampule at the end of the device. The semen is then examined for

Tri-State Breeders—1991 Beef Sire EPDs

Simmental		CE_H EPD	ACC	CE_C EPD	ACC	Birth Wt. EPD	ACC	Wean. Wt. EPD	ACC	Yearl. Wt. EPD	ACC	Mat. CE_H EPD	ACC	Mat. CE_C EPD	ACC	Mat. Wean. Wt. EPD	ACC	Mat. Milk EPD	ACC
14SM340	AF Redlands 35Y	-6	.07	-2	.07	+3	.16	+18	.15	+31	.14	+1	.06	+0	.06	+11	.08	+2	.08
14SM341	ASR Polled Pacesetter 413Z	-1.0	.16	+0	.16	-1	.21	+16	.20	+32	.18	+0	.15	+0	.15	+25	.17	+17	.17
21SM270	Bold Leader	-4.7	.68	-1.3	.68	+2.8	.92	+26.6*	.91	+47.4*	.88	-2.1	.68	-.5	.68	+24.5*	.78	+11.2*	.77
14SM334	F N Stamina	+4.5	.09	+1.2	.09	-.4	.18	+6.7	.17	+16.8	.17	+1.2	.09	+.4	.09	+1.0	.11	-2.3	.10
14SM338	Hancocks Pineview Regal	-7	.18	-2	.18	+2	.35	+13	.32	+28	.19	+0	.18	+0	.18	+12	.18	+6	.18
14SM326	HCC Prophet	+6.3	.28	+1.6	.28	-.5	.73	+.3	.65	+4.5	.58	+3.6	.19	+1.0	.19	+1.5	.24	+1.4	.23
36SM145	HF Phantom	-6.1	.37	-1.8	.37	+.8	.70	+18.6*	.66	+32.3*	.61	+5.0	.37	+1.3	.37	+4.7	.45	-4.6	.43
14SM327	HMF Gold Bar 304W	+3.4	.16	+1.0	.16	+.7	.53	+1.4	.46	+9.1	.40	+1.2	.14	+.4	.14	+8.5	.18	+7.8	.17
14SM330	HMF Polled Siegfried 230U	-3.2	.17	-.8	.17	+.9	.55	+9.0	.46	+25.0	.40	+2.3	.15	+.7	.15	+6.5	.20	+2.0	.19
14SM336	Keystone	-10.8	.53	-3.4	.53	+5.9	.66	+31.3*	.62	+50.5*	.61	+6.9	.53	+1.7	.53	+11.1	.55	-4.5	.55
14SM339	LCHM Black Baron 235X	+2.2	.17	+.7	.17	-.9	.45	+11.0	.40	+25.3	.35	+4.3	.17	+1.1	.17	+.8	.19	-4.7	.19
14SM337	Mr GF Train	-7	.17	-2	.17	+2	.20	+22	.20	+35	.18	-2	.17	-1	.17	+15	.18	+5	.17
14SM320	Paymaster	-1.5	.19	-3	.19	+.1	.75	+2.9	.68	+21.4	.66	-5.9	.21	-1.7	.21	+3.8	.41	+2.4	.39
14SM011	Pineview Apache	-7	.12	-2	.12	+1	.35	+32	.32	+57	.18	+1	.12	+0	.12	+23	.14	+7	.13
14SM142	Pineview Jazz	-5.5	.81	-1.5	.81	+.9	.93	+16.1*	.92	+38.8*	.91	-2.2	.80	-.6	.80	+12.7*	.86	+4.7	.85
36SM154	Pineview Presley	-4.8	.34	-1.3	.34	-.6	.77	-7.6	.72	-6.2	.65	+3.1	.23	+.8	.23	-.5	.32	+3.3	.30
14SM342	R&R Magician Z504	-2	.11	+0	.11	-2	.34	+14	.31	+31	.17	+0	.10	+0	.10	+14	.12	+7	.12
14SM332	R&R The Wizard 504X	+4.1	.14	+1.1	.14	-3.6	.42	+15.3	.34	+28.6	.30	-.9	.11	-.2	.11	+10.6	.14	+3.0	.14
14SM318	Royal Can Am	+.4	.16	+.2	.16	+2.2	.46	+9.1	.41	+10.9	.36	+2.6	.15	+.7	.15	+9.2	.19	+4.6	.18
36SM165	S&S Eclipse	-3.3	.27	-.9	.27	+.9	.67	+20.1	.59	+37.3	.57	-2.4	.20	-.6	.20	+9.8	.26	-.3	.24
14SM321	Sunny K Blackjack	+5.4	.18	+1.4	.18	-.4	.57	+2.6	.50	+9.4	.45	+.6	.15	+.2	.15	+1.1	.19	-.2	.18
36SM184	Super Light 19U	+.9	.18	+.3	.18	-.1	.37	+5.1	.32	+9.6	.30	+3.5	.17	+.9	.17	+5.0	.19	+2.5	.18
14SM324	The Greek	-8.7	.21	-2.7	.21	+6.0	.67	+36.7	.55	+58.9	.46	+7.7	.17	+1.9	.17	+3.0	.21	-15.4	.20
14SM357	TNT Mr T	-10.8	.60	-3.4	.60	+2.5	.90	+20.9*	.87	+41.6*	.85	-1.7	.55	-.4	.55	-7.9	.66	-18.3	.65
14SM325	Triumph	-13.8	.65	-4.6	.65	+5.6	.85	+19.9*	.81	+36.3*	.78	+3.7	.64	+1.0	.64	+.2	.68	-9.8	.67
14SM322	WRS Alien	-21.4	.15	-8.3	.15	+7.7	.70	+25.8*	.63	+51.9	.53	+6.9	.10	+1.7	.10	+10.5	.16	-2.4	.15
36SM156	WRS Enterprise	-2.2	.17	-.5	.17	-.2	.34	-.7	.31	+6.9	.29	-.3	.17	+.0	.17	+2.4	.20	+2.8	.19
14SM331	Y1 Yardleys P&B R248	+1.8	.25	+.6	.25	+1.9	.73	+7.1	.66	+24.9	.58	-1.4	.27	-.3	.27	-5.3	.46	-8.9	.45

*Trait Leader

Figure 10–12 Production data are used in selecting bulls for artificial insemination.

Figure 10–13 *One method of semen collection is using an artificial vagina.* Courtesy of James Strawser, The University of Georgia.

motility and morphology. Motility refers to the activity of the sperm, and morphology refers to the shape of the sperm.

Sperm must be active in order to reach the egg for fertilization. Sperm that are not shaped normally are usually defective and will not fertilize the female gamete (Figure 10–14). In addition to being normal and active, the sperm have to be in sufficient quantity in the semen. This is determined by an estimate of the number of active sperm in a milliliter of semen.

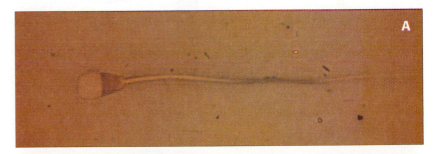

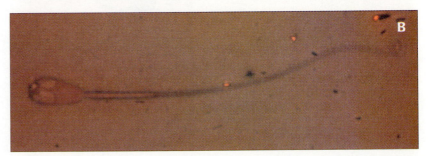

Figure 10–14 *Sperm have to be inspected for defects. (A) The sperm in this microscopic view is normal; (B) the sperm in this view is defective.* Courtesy of Ben Bracket, DVM, The University of Georgia.

After the semen has been inspected, it is processed using an extender. The extender may be such materials as milk, egg yolk, or glycerin. The extender dilutes the semen. One ejaculation may contain enough sperm to fertilize many eggs. The extender allows the semen from the ejaculation to be divided into units. The extender also protects the sperm when it is frozen and provides nourishment. At this time antibiotics may be added to the semen to help in disease prevention.

After the addition of the extenders, the semen is again examined for motility and is placed in straws. Each straw contains enough sperm for one mating. The semen is then frozen at a steady rate until it reaches a temperature of –320°F. The semen is stored and transported in liquid nitrogen to keep it at that temperature (Figure 10–15).

The procedure of AI begins with making sure the females are in estrus. Producers closely observe the animals for estrus. A technology that is widely used in the beef industry is estrus synchronization. In a large herd of cattle it is an advantage to the producer if all or groups of the cows come into estrus together. This allows an AI technician to work more efficiently than if the cows came into estrus naturally. Estrus synchronization is accomplished through the use of an artificial hormone that stimulates the ovary into follicle development.

Figure 10–15 *Semen is stored in liquid nitrogen. Courtesy of James Strawser, The University of Georgia.*

Artificial Insemination

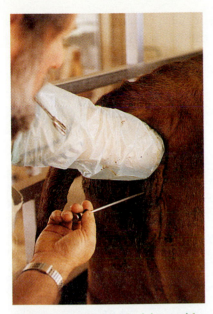

Figure 10–16 *A technician guides the straw of semen into the reproductive tract. Courtesy of James Strawser, The University of Georgia.*

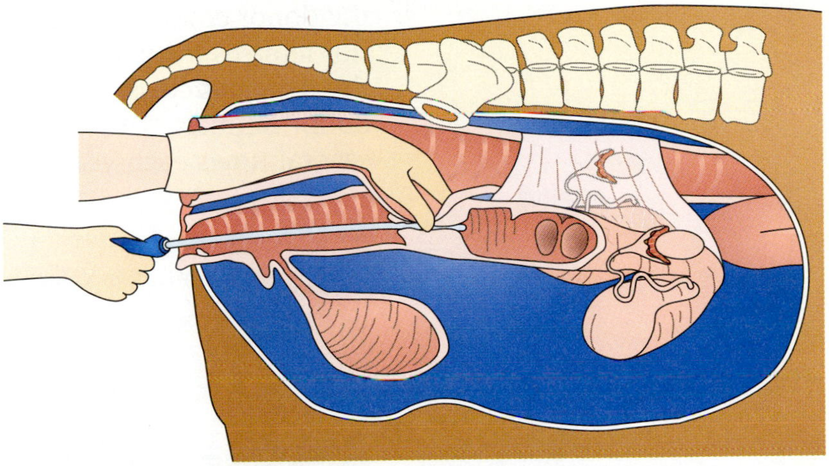

Figure 10–17 *The instrument must place the semen through the cervix and into the uterus in cows.*

When the follicle begins to develop, the reproductive cycle begins. If all of the females in the herd are injected with the hormone at the same time, they will come into estrus at about the same time. This aids in artificial insemination and causes the calves to be born at about the same time.

When the animals are in estrus, the technician begins the process of artificial insemination. The semen is thawed immediately prior to insemination. This must be done carefully in order to not damage the sperm. Too rapid or too slow thawing will harm the sperm. The semen is usually thawed by placing the straw in warm water for a few minutes. When the semen is properly thawed, the straw containing the semen is placed in a tubelike instrument. The technician carefully guides the instrument into the female reproductive tract (Figure 10–16). When the instrument passes through the cervix, the technician releases the semen into the uterus (Figure 10–17). From that point, conception takes place just as in natural mating.

EMBRYO TRANSPLANT

Another commonly used reproductive technology is that of transplanting embryos from one female to another. Instead of producing only one offspring a year, high-quality females (donors) can produce several offspring. Instead of allowing the donor cows to give birth, the embryos are carried by inferior females (recipients). Even though they are given birth by the recipient cows, genetically the young are offspring of the donor animals. This allows the superior cows to reproduce several times each year (Figure 10–18).

Figure 10–18 Through embryo transplant, high-quality females like this can produce several offspring a year. All of these calves are hers. Courtesy of James Strawser, The University of Georgia.

Embryo transplant is possible with most species of livestock but is most feasible with cattle. Embryos from superior dams may be purchased from companies that specialize in providing high-quality genetic material to producers. Just as artificial insemination has allowed producers to select high-quality sires, they may also order embryos from high-quality dams. The selection is based on production data on the animals ancestors and progeny. In this way producers can look for those traits that will be of most use in their herds. By combining a superior male with a superior female, the offspring should be of high quality. Recipient cows are usually selected based on their ability to have calves without difficulty and to provide milk.

A producer may wish to transfer embryos from the donor to the recipient during the same operation. Using estrus synchronization, the donor and recipient are at about the same phase of their estrus cycles. This allows for the proper transfer of the embryo from one reproductive system to another. The only difference is that the donor animals undergo a process known as superovulation. A follicle-stimulating hormone (FSH) causes the ovaries of the donor cow to release as many as 12 to 15 eggs at one time.

Two to three days after injection for superovulation, both the donor and the recipient are injected with prostaglandin to cause them to come into estrus. About 48 hours later, the females should both be in estrus or heat. When this occurs, the donor cow is artificially inseminated or is bred naturally.

After fertilization occurs, the fertilized eggs (embryos) are allowed to grow for about a week and then are collected. The embryos are removed in a process called flushing in which a long thin rubber tube called a catheter is passed through the cervix and into the uterine horn (Figure 10–19). The catheter has an inflatable bulb about two inches from the end that fills like a balloon and seals the entrance to the uterus. A solution is then injected through the catheter into the fallopian tubes. When the fallopian tubes and the uterus are filled with solution, the flow of the solution is stopped, and the solution is drained into a collection cylinder. The fertilized eggs (embryos) are carried out of the uterus with the solution.

After the embryos are collected, they are examined under a microscope to determine their quality (Figure 10–20). Only embryos that are in the proper stage of maturity and that appear normal and undamaged are used for transferring. The embryos are either transferred immediately or are frozen for storage and shipment.

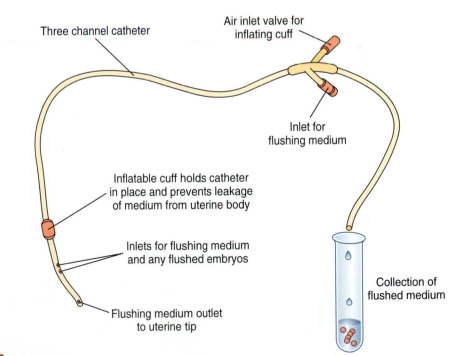

Three channel catheter

Air inlet valve for inflating cuff

Inlet for flushing medium

Inflatable cuff holds catheter in place and prevents leakage of medium from uterine body

Inlets for flushing medium and any flushed embryos

Collection of flushed medium

Flushing medium outlet to uterine tip

Figure 10–19 *A catheter like this is used to collect embryos.*

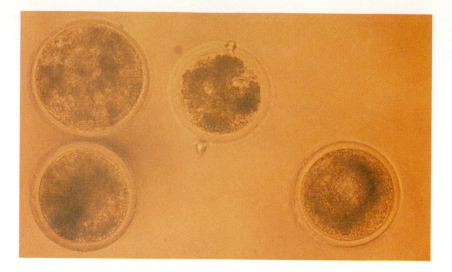

Figure 10–20 *Embryos are examined to determine defects. The embryo on the right is normal. The three on the left are defective.* Courtesy of Ben Bracket, DVM, The University of Georgia.

The recipient cows are brought into estrus in order to receive the embryos. This prepares the reproductive system of the cows to receive the embryos and to carry out a normal pregnancy. The embryo is placed in the uterus of the recipient, where it attaches to the uterus and begins to grow as in a natural conception (Figure 10–21).

Scientists have made it possible to determine the sex of the embryos prior to implantation. This allows the producer to choose the sex of the animal before the pregnancy begins,

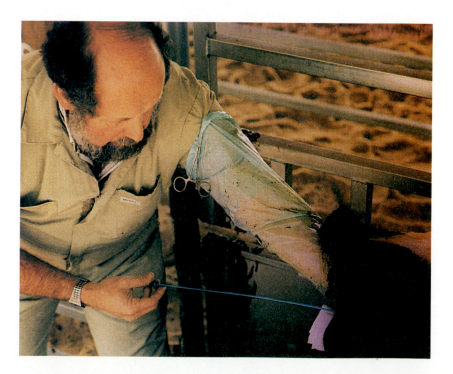

Figure 10–21 *Trained technicians implant embryos into the female.* Courtesy of James Strawser, The University of Georgia.

a tremendous benefit to livestock producers. For example, dairy producers would like for a majority of the calf crop to be female so they can be used as replacements in the herd. If the embryos are separated before implantation, only female embryos can be used. Breeders who need to produce mostly males to sell as herd sires also benefit.

The process involves the use of fluorescent dye that sticks to the DNA of the sperm cell. Remember from the chapter on genetics, sperm that produces a male has a Y-shaped chromosome, and a sperm that produces a female has an X-shaped chromosome. Female-producing chromosomes contain between 2.8 and 7.5 percent (depending on species) more DNA than male-producing chromosomes because of the larger X-shaped chromosome. When the sperm are dyed and subjected to a laser beam, the female-producing X chromosomes always glow brighter because of the greater amount of DNA. The sperm cells then are sorted using a cell sorter developed by medical researchers studying cancer cells.

The technology already exists to allow scientists to determine the sex of an embryo. The challenge is to develop a method that can be done at a low enough cost to be affordable to producers.

CLONING

One of the newer technologies is **cloning**. This process involves producing genetically identical individuals (Figure 10–22). A few years ago this technology was considered to be science fiction. Now several companies routinely perform the procedure and offer cloned embryos for sale.

When cloning first began, two methods were currently used to clone animals. The first method involved the use of an instrument called a micromanipulator that enabled the technician to hold and split the embryo into two halves. Each half then was implanted and developed into a normal newborn animal.

The other method used an unfertilized egg in which all of the genetic material was removed. One cell was removed from an embryo, and this cell was fused into the unfertilized egg. The genetic material from the embryo cell controls the growth and development of the egg into an embryo that was then transplanted. Using this method, several embryos could be developed from one embryo.

Nuclear Transfer Procedure

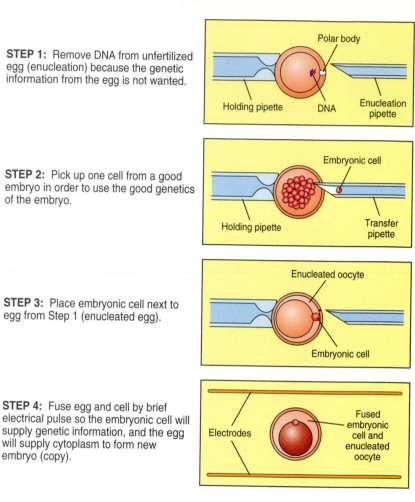

STEP 1: Remove DNA from unfertilized egg (enucleation) because the genetic information from the egg is not wanted.

STEP 2: Pick up one cell from a good embryo in order to use the good genetics of the embryo.

STEP 3: Place embryonic cell next to egg from Step 1 (enucleated egg).

STEP 4: Fuse egg and cell by brief electrical pulse so the embryonic cell will supply genetic information, and the egg will supply cytoplasm to form new embryo (copy).

STEP 5: Repeat Steps 1–4 until each cell of the embryo has been used. This is done because each cell has potential to form a copy of the original embryo. A 32-cell embryo could produce up to 32 copy embryos (clones) during one cycle of the cloning process.

STEP 6: Culture clones for 5 to 6 days until they reach the morula or blastocyst stages. Transfer or freeze clones. The best stages of embryo development for embryo transfer or freezing are that of morula and blastocyst.

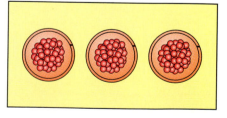

Figure 10–22 *Cloning is done using material from the nucleus of the egg.*

In 1996 a dramatic development in cloning was accomplished. A team of Scottish scientists successfully cloned a sheep using a cell from the udder of an adult sheep. The cloned sheep, Dolly, represented the first time an animal was cloned using cells other than sperm and egg.

The scientists first removed a cell from the udder of a ewe and made the cell multiply in a laboratory dish. Then they took an egg from another ewe and removed the nucleus and all the DNA. A cell from the culture was inserted into the egg with no DNA. A jolt of electricity joined the egg and the udder cell, and another electric impulse made the fused cell begin dividing. After the cell began dividing, it was transferred into yet another ewe. The embryo grew into a normal fetus and produced a healthy lamb. The procedure took a lot of patience and persistence on the part of the scientists. The successful experiment was the 277th time they had tried.

The cloning of an animal using technology other than splitting embryos has tremendous implications for the animal industry. If animals can be produced that are genetically alike, many production problems can be solved. For example, animals could be fed a ration that would be the most efficient for all in the group. Because the genetic makeup would be the same, all animals would mature at the same time and could be efficiently marketed at the same time. Ethical dilemmas, however, are involved with cloning of animals, and these issues must be dealt with before the public will be willing to accept the mass cloning of animals.

SUMMARY

Unless animals reproduce, their existence on earth will not last long. Scientists are now beginning to understand the tremendously complicated process of animal reproduction. By understanding this process, they can provide means to increase the efficiency of reproduction in agricultural animals, and this means a more efficient food system.

CHAPTER 10 REVIEW

Student Learning Activities

1. Research the origins of artificial insemination. Report to the class. Include in the report the milestone discoveries that made the process economically feasible.

2. Research and report to the class on the latest research on cloning, genetic engineering, and biotechnology associated with animal reproduction.

Define the Following Terms

1. zygote
2. meiosis
3. spermatogenesis
4. chromatids
5. synapsis
6. oogenesis
7. testosterone
8. follicle
9. copulation
10. artificial insemination
11. cloning

True/False

1. Agricultural animals may reproduce both sexually and asexually.
2. The union of two gametes results in a new cell called a zygote.
3. Through mitosis, cells divide into cells that contain only one set of chromosomes.
4. The penis of the male is responsible for sperm production.
5. The female reproductive system is more complex than the male reproductive system.
6. The egg and sperm are united in the process of fertilization in the fallopian tubes.
7. The hormone estrogen is mainly responsible for the female remaining pregnant after conception.
8. Sperm that are not shaped normally are usually defective and will not fertilize an egg.
9. Embryo transplant is most feasible with cattle.
10. The recipient cow of an embryo transplant must also be a superior dam to ensure the quality of the offspring.

Fill in the Blank

1. The production of sperm takes place through a process called _____, and the production of eggs takes place through a process called _____.
2. Replicated chromosomes are called _____.
3. The products of oogenesis are an egg and three _____, which provide sustenance for the egg.
4. The main hormone produced by the testicles is _____, which controls _____.
5. The _____ is the sac-like structure that suspends the testicles from the male animal's body.

6. The female gametes are produced by the _____.

7. The _____ is also called the womb and serves as a chamber for the development of the fetus.

8. After the egg is fertilized, it forms a layer to prevent any more sperm from entering. This layer is known as the _____.

9. _____ is the hormone that causes the corpus luteum to develop on the _____.

10. After semen is collected from the male for use in artificial insemination, it is checked for_____ (activity of the sperm) and _____ (shape of the sperm).

11. The process of _____ involves dividing an embryo to produce genetically identical individuals.

Discussion

1. Name some of the processes involved in the reproduction process.

2. How does oogenesis differ from spermatogenesis?

3. Name three parts of the male reproductive system and describe the function of each.

4. Briefly describe the ovulation process.

5. What occurs in the female after fertilization of the egg? What happens if the egg is not fertilized?

6. What are the advantages of artificial insemination?

7. Why would an extender be added to semen after collection for artificial insemination?

8. How is estrus synchronization accomplished? Why would a producer want to synchronize a herd?

9. Why might a producer choose to use embryo transplanting? What future technologies are being researched for use with embryo transfer?

ANIMAL GROWTH

STUDENT OBJECTIVES

After studying this chapter, you should be able to:

✦ Explain the importance of animal growth to agricultural producers.

✦ Analyze the factors that influence animal growth.

✦ Cite examples of the advances made as a result of research on growth.

✦ Distinguish between prenatal and postnatal growth.

✦ Explain the stages in prenatal growth.

✦ Explain the stages in postnatal growth.

✦ Describe the aging process in animals.

Producers of agricultural animals earn their living through the growth of animals. Whether the animals produce eggs, wool, meat, or perform work, growth of the animals is of primary concern to the producer. Through the natural process of growth, an animal increases in size and puts on weight. By efficiently managing the animal so that the maximum amount of growth occurs at the least cost, producers earn their living. Over the years, an enormous amount of research has gone into understanding the process of animal growth. Although this may sound like a relatively easy area to research, the process is actually quite complex because of the many factors that play a part in the ability of an animal to grow. **Heredity** plays an important role (Figure 11–1). How large an animal ultimately becomes and how long it takes to reach its mature size is determined to a large degree by the genetic code passed from its parents. If the genetic code dictates that the animal has the potential to reach a certain size,

Figure 11–1 *Heredity plays an important role in the growth rate of an animal.* Courtesy of Polled Hereford World.

not much can be done to cause that animal to get larger. Much of the advance in the growth efficiency of agricultural animals is due to the results of research on selective breeding.

Nutrition has an enormous effect on the growth rate of an animal. The nutrients taken in by an animal through its feed go into the growth process and into maintaining the animal's bodily processes. The proper nutrients in the proper amount and ratio allow the animal to grow to its full genetic potential. Without the proper balance of nutrients, the animal will not grow very well (Figure 11–2).

Environmental factors also affect the rate of growth for an animal. If the animal is kept in an environment where it is comfortable, it will grow better because any type of stress causes the growth process to slow. Producers work hard to ensure that animals are kept in an environment that allows for the maximum amount of growth (Figure 11–3). This means that the animal lives where the temperature is comfortable for that animal. Some animals may do well outside during cold weather. For example, cattle grow thick coats of hair to protect them from the elements. Producers make sure that the animals receive enough feed to keep fat on their bodies to act as insulation against the cold and to provide energy for metabolism. On the other hand, pigs do not tolerate the cold very well and are generally kept indoors where the temperature can be controlled.

Animals must also be kept healthy and free from disease. Energy that is used to fight sickness cannot be used for the growth process. Also, the animal should be kept free from parasites. Parasites rob animals of food nutrients and energy

Figure 11–2 *Without the proper amount of nutrients, an animal will not grow properly.* Courtesy of Ben Bracket, DVM, The University of Georgia.

Figure 11–3 *Producers keep animals in a controlled environment to obtain the maximum amount of growth.* Courtesy of James Strawser, The University of Georgia.

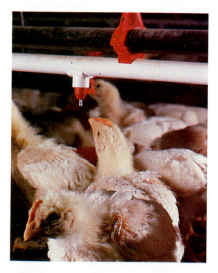

Figure 11–4 *Research has made a dramatic impact on the growth of poultry.* Courtesy of James Strawser, The University of Georgia.

that could otherwise be used to grow and put on weight. Diseases also slow the growth rate because sick animals do not eat well. Obviously, an animal that does not eat much will grow little.

As already pointed out, research has been conducted for many years on the growth of animals. As a result, a great amount of progress has been made in the rate at which agricultural animals grow. One of the most remarkable success stories of this research is the poultry industry (Figure 11–4). Over the past 70 years, the time for boilers to reach market weight was cut in half, from 15 weeks to 7.5 weeks. The weight at which the birds are marketed has increased from an average of 2.8 pounds to 3.38 pounds. In addition, the amount of feed required to put on a pound of live weight was cut from 4 pounds to 2 pounds. This means that we now can produce a heavier broiler in half the time on half the feed than we could in 1925!

Since 1950, the amount of feed required to produce a 200-pound hog has been reduced by about 50 pounds, and the time required to produce the hog has been reduced by almost a month. Since 1925, the live weight of calves marketed per breeding female has more than doubled. The tremendous increase in the efficiency of raising livestock has amounted to not only more profit for producers but less expensive food for consumers.

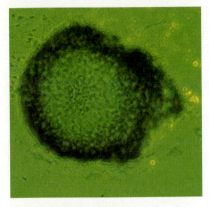

Figure 11–5 *All animals begin as a microscopic cell called a fertilized egg. Courtesy of USDA-ARS.*

Figure 11–6 *The first phase of animal growth is called the ovum stage. Courtesy of Richard Fayrer-Hosken, The University of Georgia.*

THE GROWTH PROCESS

Growth is the increase in the size of animals. This comes about as a result of either an increase in the size of cells (**hypertrophy**) or as a result in the increase in the number of cells (**hyperplasia**). The growth of young animals is mostly the result of the increase in the number of cells. As an animal matures, most of the increase in size and weight is a result of the increase in the size of the cells.

All animals begin as a single microscopic cell (Figure 11–5) known as a fertilized egg (zygote). From this microscopic beginning, an animal may increase in size to weigh more than a ton. The growth occurs in two major phases: **prenatal** and **postnatal**. Prenatal refers to the growth before the animal is born, and postnatal refers to the growth after the animal is born.

Prenatal Growth

Chapter 10 explains the process of how animals reproduce. The growth process begins with the fertilized egg. In Chapter 3 the process of cell division, called mitosis, is discussed. This same mechanism causes the growth due to the increase in the number of cells in the animal's body.

The first phase of an animal's prenatal growth is called the ovum stage (Figure 11–6). During this stage, the fertilized egg begins to divide and form a mass of cells that attach to the wall of the mother's uterus. As the cells divide, the genetic coding in the DNA is passed on to each succeeding cell through a replication of the DNA (see Chapter 4). As the cells divide and group together, they form into a ball-shaped mass called the morula (Figure 11–7). The process of the cells dividing and clumping into a mass is called cleavage. As the cells of the morula begin to increase, they form a spherical shape with an outer layer and a central core. This group of cells develops into a hollow cavity that is filled with fluid. This formation is called a blastula and is the final stage of the ovum phase of development (Figure 11–8). During this phase, which lasts about 10 days in sheep and 11 days in cattle, all of the cells are very much alike.

The Embryonic Stage

As the blastula grows and develops, the cells begin to change and take on different characteristics. They form different layers

Morula

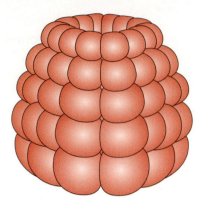

Figure 11–7 The morula is a ball-shaped mass of cells.

that develop into the organs of the body. The tissue that develops bone is different from the tissue that develops blood; the tissue that develops muscles has its own unique characteristics, and so on. This process is called cell **differentiation**. Scientists still do not fully understand what causes the cells and tissues to differentiate.

The blastula develops into three distinct layers. The outer layer is called the ectoderm, and from this layer such tissues as the skin, hair, and hooves are formed. Also, the central nervous

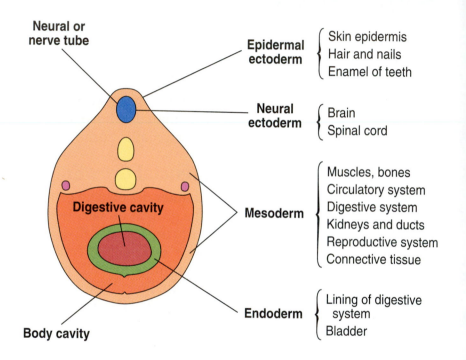

Figure 11–8 The blastula develops cells that differentiate. This structure develops three layers.

system, including the spinal cord and the brain, develop from this layer. This system begins with the formation of a hollow structure called the neural tube that allows a rod-like structure called the notochord to develop (Figure 11–9). This later develops into the spinal cord.

The middle layer is called the mesoderm. This layer develops into muscles, bones, the circulatory system, digestive system, reproductive system, renal system (the kidneys), and connective tissue. This layer develops the structural portion of the animal's body that allows the animal to stand and move about.

The innermost layer is called the endoderm and is responsible for the development of internal organs such as the bladder, the lungs, and the endocrine glands. The layering of the cells allows these organs to develop in a logical manner. The process of differentiated tissue developing into organs is

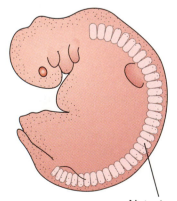

Notochord

Figure 11–9 The notochord develops into the vertabrae.

called **morphogenesis**. This process continues throughout the embryonic stage, which lasts from the 10th to the 44th day in sheep and from the 11th to the 45th day in cattle. At the end of the embryonic stage, the animal begins to take on a recognizable shape.

Also during the embryonic stage, the embryo becomes attached to the mother's uterus through a saclike pouch that forms around the embryo. This sac, called a placenta, protects the developing embryo by means of the amniotic fluid that fills the placenta (Figure 11–10). This fluid absorbs shock and keeps the embryo in a protective suspension. The umbilical cord passes through the placenta and attaches the embryo to the mother. Through this tube, the embryo receives nourishment. Waste material from the embryo is absorbed by the placenta and is passed out through the mother's excretory system.

Anatomy of the Fetal Horse

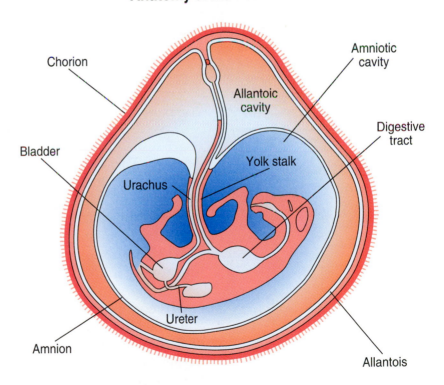

Chorion

Amniotic cavity

Allantoic cavity

Digestive tract

Bladder

Yolk stalk

Urachus

Amnion

Ureter

Allantois

Figure 11–10 A fluid-filled sac protects the fetus during development.

The Fetal Stage

The development of the organs begins in the embryonic stage with the differentiation and layering of the cells (Figure 11–11). During the next stage, the fetal stage, the organs fully

Prenatal Development from Conception to Birth

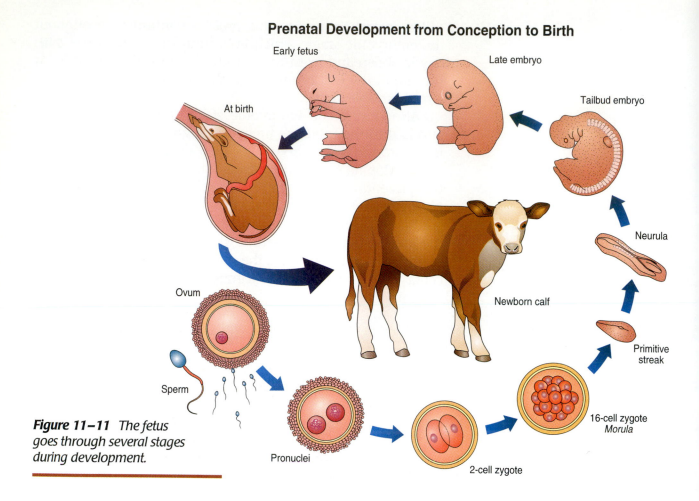

Figure 11–11 *The fetus goes through several stages during development.*

develop. The first are the liver, heart, and kidneys. These organs begin to function early in the development of the unborn animal. Although the fetus is attached to the mother by the umbilical cord, these organs aid the fetus in its maturation by circulating blood through the developing animal. Also, after the animal is born, these organs have to begin functioning immediately in order for the newborn to live.

Through the process of mitosis, the cells in the individual organs continue to divide and multiply. The process of differentiation allows the specialized cells to form organs that play vital roles in the animal's systems. Through the passage of oxygen and nutrients from the mother, the animal receives the essential elements needed to grow and develop. The organs develop at different rates, and the fetus becomes less and less dependent on the mother. At the point where the fetus is able to survive on its own outside the uterus, it is expelled in the birth process (Figure 11–12). A powerful hormone called oxytocin is produced by the mother's endocrine

Figure 11–12 *When the fetus is able to survive on its own, it is expelled from the uterus.* Courtesy of Warren Gilson, The University of Georgia.

system. Oxytocin causes the muscles of the uterus to contract and force the fetus out. The contractions also aid in forcing fluid from the fetus's lungs. The cervix and the birth canal are stimulated by the hormone and are caused to relax enough for the fetus to pass through. The newborn animal leaves the birth canal and falls to the ground. This abrupt contact with the ground stimulates the lungs to begin functioning. Although the heart of the fetus begins beating very early in the gestation period, the animal begins breathing when it leaves the uterus. The digestive system is stimulated into operation by the first milk it gets from its mother. This milk, called colostrum, is very rich with nutrients and contains substances that clear the digestive system and begin the processes of digestion (Figure 11–13). In addition, colostrum

Figure 11–13 *The first milk from the dam (mother) is rich in nutrients and imparts immunity to the young animal.* Courtesy of James Strawser, The University of Georgia.

BIO BRIEF

Cattle Gain on Pasture-Finishing

Ranchers who put their beef cattle out to pasture can produce animals that are ready for market, say Agricultural Research Service (ARS) scientists at the Grazinglands Research Laboratory at El Reno, Oklahoma. They have found evidence that adding a lot more grass to cattle's diet will still produce high-quality beef.

Animal nutritionists William A. Phillips and Samuel W. Coleman have been comparing performance of cattle from similar herds finished for market in two ways—either fed on grass with limited grain or fed a high-grain diet, a traditional feedlot practice.

"In the usual system, grain constitutes at least 95 percent of the diet," says Phillips. "In the system we developed, we use as much grass as we can and decrease the amount of grain." Findings from the three-year study show beef cattle can be finished as efficiently on grass pastures, with some grain, as they can with mostly grain.

A high-energy diet composed mostly of corn is provided in a covered feeder to give cattle additional energy for fattening. In the grain-on-grass system, cattle make their own dietary choices, deciding how much grain they need, depending on the grass supply. The ARS scientists finished cattle using wheat pasture and perennial grass pastures, such as Old World Bluestem, millions of acres of which grow in the Southern Great Plains region. They stocked the grass pastures with twice as many cattle as they would normally, to ensure that most of the grass would be consumed. "As the grass supply dwindled, the cattle ate more of the high-grain diet. Cattle fed grass plus grain needed less feed to reach market weight than herdmates fed in the feedlot," says Phillips. "Less feed means lower production costs. Under the grain-on-grass system, feed saving were around $25 per animal. With four animals per acre, the producer's grass pasture is worth $100 per acre for finishing cattle. That's a lot more dollars per acre than could be anticipated from other uses of the grass."

"And, carcass measurements have been similar between the two systems." says Coleman. "Cattle finished in the pasture

imparts immunity to the young animal while its own immune system develops.

The Postnatal Phase

Soon after the animal is born and its systems begin to operate independently of its mother, the postnatal growth phase

reach about the same end weight as those finished in feedlots, but they have about three percent less fat."

Research shows that cattle can be finished as efficiently on grass pastures with some grain, as they can with mostly grain. Courtesy of USDA-ARS.

"Finishing cattle under either system would bring the producer the same amount of money," Coleman says, "but production costs are lower under the grain-on-grass system."

Regulations require farmers to capture, store, and dispose of the animal waste they generate. Phillips says in their system the cattle distribute the manure over the pasture, where it can be incorporated into the soil and used to fertilize the grass for future growth.

"From an ecological standpoint," says Phillips, "the grain-on-grass system reduces the concentration of animal waste and allows some producers to finish their own cattle without incurring the added cost of waste disposal."

Phillips and Coleman say their system needs further refinement, but they see great opportunities for producers in the Southern Great Plains region to market their cattle more efficiently.

Source: Agricultural research Magazine.

begins. Initially the animal grows slowly as the organs fully develop. The organs develop in relationship to their importance to sustaining life. At birth the animal's head is proportionally larger than at any time thereafter. This is because the head encases the brain, which controls the functioning of all of the other organs and systems, including the processes that cause the animal to grow. For this reason the brain must be among the first of the organs to develop.

The legs are also a larger portion of the animal's body at this time. Strong legs are essential for the animal to move about almost as soon as it leaves the uterus. This ability was developed to be able to escape predators and to be able to reach the mother's udder (Figure 11–14).

Figure 11–14 *Strong legs are important to be able to escape predators.* Courtesy of Progressive Farmer.

When the organs develop, the young animal begins a phase of rapid growth. This growth continues until the animal reaches sexual maturity. This is the period during which the bones and muscles grow and develop. The animal grows more rapidly on proportionally less feed than at any other time. It is during this stage that producers achieve the necessary growth for the animals to be profitable. Animals that are raised for meat are marketed after this stage is completed.

Most of the weight of an animal is from muscle and bone. For animals being raised for food, an adequate amount of muscle is desirable because meat comes from muscle. This is the reason producers place so much emphasis on selecting breeding stock that have good muscling (Figure 11–15). The amount of muscle is a highly heritable trait.

Muscles that move the skeleton are composed of long thin cells that contain many nuclei. Other muscle cells, such as heart muscles, have only one nucleus per cell. All the cells of the muscular system are present at birth. Muscle growth is due to the increases in the size of these muscles.

Throughout the growth period until the animal matures, bone grows through cell division and multiplication. Most of the growth takes place at the end of the bone. At this site are tissues called cartilage that are softer and more pliable than bone. As the animal grows, the cartilage solidifies into bone tissue, and the bone grows longer. When the growth period ends and the animal is mature, all of the cartilage solidifies, and the

Figure 11–15 *Producers select animals for breeding that will produce a lot of muscles. This animal is well muscled.* Courtesy of Limousin World.

bone stops growing. This is called **ossification** of the bone. Mature bones are composed mostly of minerals and protein.

Fat tissue (adipose cells) helps the animals maintain its internal heat and stores energy. It also helps to cushion the internal organs and muscles. The animal first begins to deposit fat in the abdominal cavity. After the cavity is sufficiently layered with fat, the deposits are made between the muscles and under the skin. This gives the animals a sleek, smooth appearance. The last place an animal deposits fat is inside the muscles between the muscle cells. This is called marbling and is desirable in retail cuts of meat. It is used as a measure in quality grading beef.

The Growth Control System

As mentioned earlier, the brain controls all of the processes of the animal's body. The brain does this by stimulating the endocrine system to release hormones. Hormones are chemical substances that stimulate organs into operation or cause chemical reactions. How rapidly an animal grows depends to a large degree on the hormones released into the bloodstream by the glands of the endocrine system. The main glands are the pituitary, thyroid, testicles or ovaries, and the adrenal glands (see Chapter 10). An absence of the proper amount of these hormones will cause a condition known as dwarfism in which the animal remains small. This condition is genetic and may be passed on to the next generation. If too many hormones are released, the animal will be malformed or grow abnormally large.

One of the benefits of growth research has been the development of artificial growth hormones that cause the animal to grow more efficiently. During the past 20 years, these substances have been used extensively in the beef industry. These hormones or hormone-like substances are placed in the ear of the animal, just under the skin. They aid the animal in making more efficient use of nutrients. Concern has been expressed over the use of hormones in animals that are used for slaughter. Years of extensive research have gone into studying the effects of these materials in meat. The Food and Drug Administration (FDA) sets allowable amounts that may be present in meat. These allowances are so small that no harmful effects can come from them. Producers are required to withdraw the hormones a specified time before the animals are slaughtered.

Most male animals that are raised for slaughter are castrated. In this process both testicles are removed in order to affect the behavior of the animal, the growth pattern, and the quality of the meat. The major hormone produced by the testicles is testosterone. Although the major role of this hormone is to stimulate the development of male characteristics and the production of sperm, it also affects the lean-to-fat ratio. In other words, it helps determine the amount of fat cells compared to the amount of muscle when the animal is mature. In a castrated male, the fat in the body compared to the lean is relatively higher. This means that the castrated males will mature faster and have the desired amount of fat at an earlier age. Also, the meat from the animal will have a more desirable flavor.

Physiological Age

As animals grow older, they reach a point known as maturity when their bodies stop growing taller and larger. At this point they begin to use the nutrients taken in for depositing fat instead of growing. After the animals mature, their bodies begin certain processes associated with aging. One of the most noticeable changes is the solidification of bone. Remember that in the process of bone growth, bones turn to solid tissue as the bone tissue matures. New cartilage then turns to bone, and the animal grows taller or larger. When the animal stops growing the bone does not develop new cartilage, but the bone continues to solidify. The major method used to determine the age of an animal that has been slaughtered is by examining the bones of the rib cage and the vertebrae. The younger the animal, the more soft cartilage the animal will have in these areas; whereas older animals have more solid bone. USDA graders use these areas to determine the physiological age of an animal. The physiological age is the age an animal appears to be by the amount of bone solidification; chronological age is the actual age of an animal. The physiological age is used in setting the grade for a particular carcass.

As an animal ages, other physical characteristics change also. For example, the reproductive system secretes a smaller amount of hormones, and as a result breeding animals becomes less productive as they age. The solidification of the bones along with a lessening of muscular strength make it more difficult for an animal to move about. The aging process also brings about a gradual disintegration of the nervous system, and the animal has problems functioning. Because of all these changes, animals are more susceptible to disease and parasites as they become older. Most producers cull out aging animals as they slow down and their productiveness diminishes. If left to natural occurrences, all animals will eventually die of old age or disease brought about by advanced age. It is through this process of growth, reproduction, maturity, decline, and death that life is perpetuated on earth.

SUMMARY

Animal growth is controlled by many factors. Agriculturists try to understand and control these factors in an effort to produce animals efficiently. Animals that are healthier and grow faster on less feed are more economical. This translates into better products at a lower cost to the consumer.

CHAPTER 11 REVIEW

Student Learning Activities

1. Go to the library and research how artificial hormones are used to enhance animal growth. Find out what concerns consumers have about the use of the hormones. Decide whether or not you think the concerns are well-founded. Debate the issue with a student in your class who takes the opposite stance.

2. Visit an animal producer and compile a list of the management techniques he or she uses to ensure the efficient growth of the animals. Compare your list with the lists of others in the class.

Define the Following Terms

1. heredity
2. hypertrophy
3. hyperplasia
4. prenatal
5. postnatal
6. differentiation
7. morphogenesis
8. ossification

True/False

1. Producers earn their living by managing an animal so that the maximum amount of growth occurs at a moderate cost.

2. Proper nutrition can often overcome a genetic code for poor growth in animals.

3. Growth occurs as a result of an increase in the size or number of cells in the body.

4. The growth process occurs in two phases: embryonic and postnatal.

5. At the end of the embryonic stage, the animal begins to take on a recognizable shape.

6. The embryo receives nourishment through the umbilical cord.

7. The organs of an animal's body fully develop during the fetal stage.

8. Because bodily functions and organs must be regulated during the first few months of life, the head and chest regions are proportionately larger than at any time thereafter.

9. How rapidly an animal grows depends on the hormones that are released into the bloodstream.

10. Because physiological age and chronological age are so difficult to distinguish, they are often not important in the carcass grading system.

11. The reproductive abilities of an animal increase with its chronological age.

Fill in the Blank

1. Muscle growth is due to an increase in the size of the muscle cells. This type of growth is also known as _____.

2. The first phase of an animal's prenatal growth is the _____ phase. During this phase, the fertilized egg begins to divide and form a mass of cells that attach to the wall of the uterus.

3. The process of cells dividing and clumping into a mass is called _____.

4. The blastula develops into three distinct layers: the _____, _____, and _____.

5. The placenta is filled with _____ to protect the developing embryo.

6. The first milk produced by the mother, which aids in starting digestion and imparts immunity to the young animal, is called _____.

7. The _____ is the organ that controls all the processes that cause the animal to grow.

8. The solidification of cartilage into bone is known as _____.

9. _____ is the process that most male animals raised for meat undergo. In this process, the testicles are removed to alter behavior and growth as well as meat quality.

10. _____ age is determined by the amount of bone solidification and is used in grading carcasses.

Discussion

1. Several factors influence an animal's ability to grow. Name these and briefly discuss their effect on an animal's growth.

2. Briefly outline the phases of prenatal growth and the processes that occur with each phase.

3. What are the first organs that develop in the fetus? Why are these the first to develop?

4. Discuss the endocrine system, what organs compose the system, and the effects of hormone manipulation on animal growth.

5. Discuss the effects that aging has on animals, their growth, and their bodily processes.

PLANT AND ANIMAL DISEASES

STUDENT OBJECTIVES

After studying this chapter, you should be able to:

◆ Define disease.

◆ Explain the agricultural problems that are caused by plant and animal diseases.

◆ Distinguish between infectious and noninfectious disease.

◆ Describe and compare the different pathogens that cause animal diseases.

◆ Explain how pathogens cause disease.

◆ Discuss how antibiotics were developed.

◆ Explain how antibiotics work.

◆ Explain how an animal's immune system works.

◆ Discuss how vaccines work.

◆ Analyze the importance of plant diseases to producers.

◆ Tell how plants resist disease.

◆ Describe how disease is prevented

KEY TERMS

pathogens
bacteria
antibiotics
viruses
protozoa
antigens
antibodies
active immunity
passive immunity
vaccination

Perhaps the worst calamity that can happen to an agricultural producer is disease. Whether the producer raises animals or grows plants, disease can be devastating to the operation. Whole herds of animals and huge fields or forests can be wiped out if disease organisms are allowed to invade

Figure 12–1 *A disease destroyed vast numbers of American chestnut trees.* Courtesy of Ed Brown, The University of Georgia.

the animals or crops. For example, in Africa during the 1800s, a disease known as rinderpest wiped out huge herds of buffalo that covered the grassy plains. The disease spread to domesticated cattle and almost wiped out the industry. Similarly, a plant disease called blight attacked chestnut trees in the United States during the early part of the 20th century (Figure 12–1). Previously expansive forests of these huge trees blanketed much of the eastern part of the country. After the blight, the losses were so great that it is now very rare to find one of these native trees growing in this country.

Even though we have gained a large amount of knowledge about diseases through research and development, this problem still causes billions of dollars in losses to agricultural producers each year. Producers not only suffer losses because of the toll taken from diseases but also in the amount of money that must be spent in the prevention and cure of disease.

Agriculturists disagree on an exact definition of disease. Many factors can be considered when describing what a disease is and what it does. However, the most widely accepted definition is that a disease is any condition that causes the systems of a plant or animal to not function properly. The two broad categories of diseases are infectious and noninfectious. Infectious diseases are those caused by other living organisms, called pathogens. They are contagious and can be transmitted from one plant or animal to another. This is because the organisms that cause the disease reproduce and can go from one host to another.

The noninfectious diseases are caused by genetic defects, malnutrition, or other types of diseases that are not caused by other organisms that live within the plant or animal. Both plants and animals can suffer from noninfectious diseases. For example, lack of proper nutrition can cause the malfunctioning of the systems of either plants or animals and is therefore considered to be a disease. Because many of the causes of noninfectious diseases are covered elsewhere in the text, this chapter will focus on infectious diseases.

ANIMAL DISEASES

Infectious diseases of animals are by far the most costly in terms of animal losses and the most expensive in terms of prevention and cure. These diseases are caused by microorganisms that invade or infect the animal's body. Several types of microorganisms cause disease. These organisms are called **pathogens**, and their study is known as pathology.

Bacteria

The most common pathogens are **bacteria** (Figure 12–2). At any given moment, literally billions of bacteria are all around us. They live on our bodies, in our bodies, and on most objects in our environment. More than 2,000 species of bacteria exist. Some are helpful; some are neither helpful nor harmful; and some are harmful because of the diseases they cause.

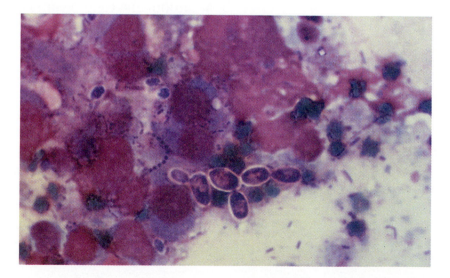

Figure 12–2 *The most common pathogen is bacteria. Courtesy of Dr. Ramos, University of Michigan.*

Some of the most devastating diseases to both humans and animals are caused by bacteria. During the Middle Ages, the population of Europe was almost wiped out by a disease called the bubonic plague, which was caused by bacteria spread by fleas that fed on infected rats. The bacteria were passed to humans by the fleas. No one knew the cause, and few effective prevention measures were taken. The terrible disease rinderpest, which was mentioned at the beginning of this chapter, is also caused by bacteria.

Disease is usually caused when the bacteria invade an animal's body and get inside the body cells. They may feed on the cells or may secrete a harmful substance called a toxin that acts as a poison to destroy the cells. A dreaded disease called tetanus (lockjaw) is caused by a toxin released by the bacteria *Clostridium tetani*. This toxin is 100 times as toxic to animals as the deadly poison strychnine. If a large enough number of bacteria are feeding on the cells of an animal's body or are secreting toxins, one or more of the animal's systems may not function well or may cease to function. When this happens, the animal becomes sick and may die.

Several types of bacteria are classified according to their shape (Figure 12–3). Cocci bacteria are shaped like a ball and may be bunched together like a bunch of grapes. These are called staphylococci and cause diseases such as mastitis in cattle. Other cocci bacteria may be strung together like a chain and are known as streptococci. These cause diseases

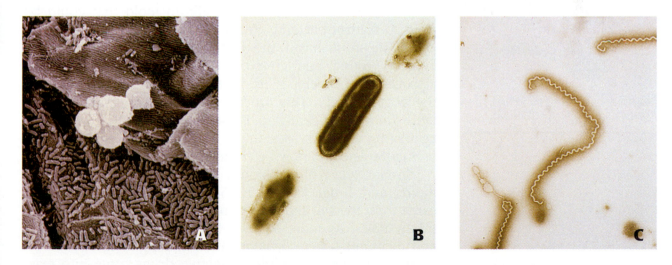

Figure 12–3 *Bacteria are classified by their shape. Photomicrographs of (A) cocci; (B) bacilli; (C) spiral bacteria.*
Courtesy of (A&B) Domingo, Barcelona, Spain and (C) Stephens, The University of Georgia.

such as distemper and meningitis. Rod-shaped bacteria are known as bacilli and move about by means of small whiplike projections called flagella. One of the most dreaded of all livestock diseases, anthrax, is caused by a species of these bacteria. Other bacteria are shaped like spirals or corkscrews. These cause many reproductive diseases such as leptospirosis, vibriosis, and others.

Scientists often take cultures to determine the type of bacteria infecting an animal. To do this, a sterile swab is wiped across the surface of the animal's skin, inside the mouth or other body areas. The swab is then rubbed onto a sterile growing medium that has been placed inside a sterile petrie dish. The growing medium (a common one is called agar) contains nutrients on which bacteria feed. The dishes are then placed in a chamber that is carefully controlled to allow the correct temperature for the growth of the bacteria. Within the next few days the bacteria transferred from the swab grow and multiply, and the scientist is able to obtain enough of the bacteria to examine under the microscope. He or she can then identify the bacteria.

Figure 12–4 *Penicillin comes from mold. Courtesy of USDA-ARS.*

Bacterial infections are usually treated with a class of drugs known as **antibiotics**. These drugs often originate from living sources. In the late 1800s, an Englishman named William Roberts noticed that bacteria often stopped growing or were killed by fungus. He observed a mold, *Penicillium glaucum*, that was especially effective in arresting the spread of bacteria (Figure 12–4). In 1928, Alexander Fleming developed a substance from the penicillium mold that could be used to control bacteria. He called the new substance penicillin. The problem with the new substance was that it was difficult to produce enough to be effective. During World War II, a process was developed using fermentation technology from the beer industry. This process allowed the manufacture of large amounts of penicillin. Penicillin revolutionized the fight against bacterial diseases. Cases that were once considered to be incurable could now be completely cured using penicillin.

Similar types of antibiotics have been developed using molds and fungi (Figure 12–5). One, called streptomycin, came from the soil and was tremendously effective in fighting tuberculosis. As more bacteria-fighting molds were isolated and techniques were developed for making synthetic antibiotics, many new and highly effective drugs came into existence. These have been successful in fighting a wide variety of bacterial diseases in both animals and humans.

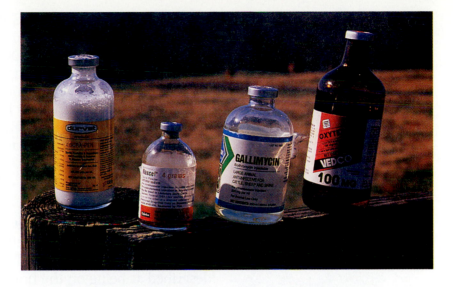

Figure 12–5 *An array of different antibiotics has been developed to fight bacteria.* Courtesy of James Strawser, The University of Georgia.

Antibiotics fight bacteria by several methods. Some prevent the cell walls of the bacteria from growing. Penicillin belongs to this group. Another group prevents the flow of nutrients through the cell walls of the bacteria. An example of this type of antibiotic is polymyxin. Antibiotics such as streptomycin and tetracycline prevent protein from being synthesized in the bacteria. Still other types disrupt the process of gene replication in the bacteria and thus prevent the bacteria from reproducing.

Livestock are given antibiotics through their feed to help them fight diseases associated with bacterial infections (Figure 12–6). One problem that has developed is that bacteria have

Figure 12–6 *Animals are sometimes given antibiotics in their feed.* Courtesy of James Strawser, The University of Georgia.

developed immunity to some antibiotics. This is a logical outcome because after an animal is treated, the bacteria that survive the antibiotics reproduce. This is a type of natural selection that develops strains of bacteria that are tolerant of antibiotics. This is why producers often use different types of antibiotics. A large industry has developed to produce and distribute these drugs. The federal government closely monitors the amount of drug residue in animal products intended for human consumption.

Viruses

Viruses are the most difficult disease-causing agents to treat. The antibiotics that are so effective against bacteria are not effective against viruses. These pathogens are sometimes described as being on the borderline between living and nonliving because they have characteristics of both living and nonliving things (Figure 12–7). For example, like all living things, viruses reproduce. However, they cannot reproduce outside of a living cell of another organism. Unlike other forms of life, virus cells are not complete because they have no nucleus and lack other cell parts characteristic of living things. Because they have no nucleus, they must rely on strands of nucleic acid to replicate in the reproduction process, and this can take place only within the cells of a living host.

Viruses cause disease by using up material in the cell that it needs to live and function (Figure 12–8). Cells may also be ruptured when the viruses reproduce inside them. Because

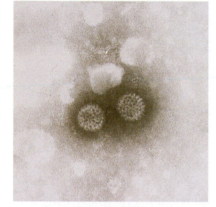

Figure 12–7 *Viruses are sometimes described as being on the borderline between living and nonliving. Courtesy of Eloise Styer, The University of Georgia.*

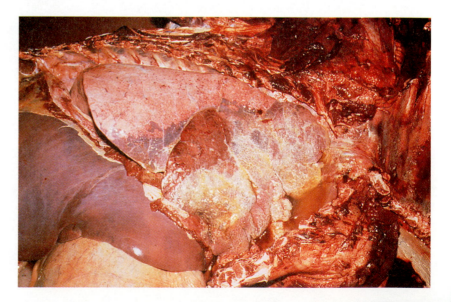

Figure 12–8 *The damage to this animal's lungs was caused by viruses. Courtesy of Dr. James Wright and Dr. Dawn Duncan, North Carolina State University.*

the viruses often become a part of the cell by interacting with DNA, they are very difficult to kill without destroying the host cell. Common diseases such as the cold in humans and animals are caused by viruses. Also some of the most dreaded diseases of animals and humans are caused by a virus. A good example in humans is AIDS. The virus that causes AIDS is extremely difficult to isolate and treat because of the interaction with the cells in the body. Animal diseases such as influenza, cholera, and Newcastle are all caused by viruses.

Often the virus itself may not cause the death of an animal but may weaken the animal. When this happens, the animal is much more susceptible to bacterial infection. This is known as a secondary infection, and it is actually the bacteria that causes death.

An exciting possibility for the treatment of viral diseases is a substance known as interferon. Scientists have known for years that cells infected with viruses manufacture and release a substance called interferon. This substance is absorbed by noninfected cells, where it inhibits the growth of viruses. Research scientists hope to develop ways in which the animal's own body can effectively fight viral diseases.

Protozoa

Protozoa are one-celled organisms that cause disease among animals (Figure 12–9). They invade the animal's body and feed on the cells. As with the other disease-causing agents, enough cells can be destroyed that a particular system of the animal's body can cease to function. One of the most troublesome diseases of the poultry industry, coccidiosis, is caused by protozoa. Usually these diseases are treatable with drugs.

Figure 12–9 Protozoa are one-celled animals that can cause disease. Courtesy of David Tyler, The University of Georgia.

ANIMAL IMMUNE SYSTEMS

As mentioned previously, disease agents are constantly present in the environment. Every day animals and humans come in contact with a wide variety of bacteria and viruses, some of which can cause disease. Animals are protected from these pathogens by the body's defensive or immune system. In order for the disease agent to cause trouble in the animal's system, it must first gain entry into the animal's body. This can come about through several means. Any of the body openings—the eyes, ears, nose, mouth, anus, or urinary/reproductive tract—can be a path of entry. Also, biting or blood-sucking external parasites can inject disease organisms directly into the animal's body when they pierce the skin (Figure 12–10). The parasites move from a sick animal to a healthy one and carry the disease agents to the healthy animal.

Figure 12–10 *Diseases are often transmitted through the bite of an insect.* Courtesy of Maxey Nolan, The University of Georgia.

The body openings have mechanisms to help prevent the entry of pathogens. Most of the openings are lined with membranes that secrete a viscous watery substance called mucous that traps and destroys bacteria and viruses. The digestive and the respiratory tracts take in large numbers of microorganisms each day as they eat and breathe. Most of the organs of these systems are lined with mucous membranes that trap and destroy pathogens. Foreign matter is kept out of the nostrils by hairs that line the opening (Figure 12–11). Foreign materials, such as dust that harbor pathogens, are caught in the hair and expelled when the animal sneezes. The largest organ of the body, the skin, also prevents the entrance of these agents. Cuts and other breaks in the skin allow the entrance of bacteria and viruses. Once inside the body, these pathogens enter the bloodstream and may be carried throughout the body.

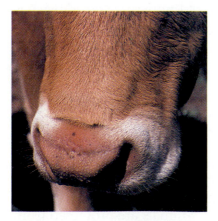

Figure 12–11 *Hair in an animal's nostrils can be a line of defense against disease.* Courtesy of James Strawser, The University of Georgia.

If a pathogen gets past the body's outer defense, the second line of defense comes into play. Chemicals and special cells in the bloodstream attack the disease agents. Remember from Chapter 11 that an animal's blood is made of both red and white blood cells. The function of white blood cells is to rid the body of wornout or dead cells (Figure 12–12). White

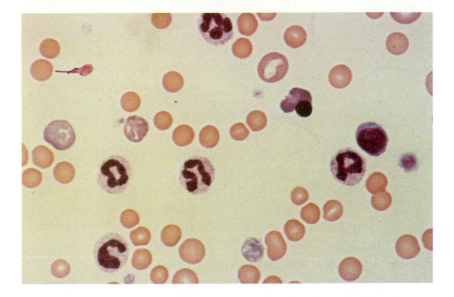

Figure 12–12 *White blood cells in the animal's circulatory system destroy pathogens.* Courtesy of Dr. French, Cornell University.

blood cells called phagocytes trap and destroy disease-causing agents in the blood. Foreign materials are called **antigens**, and they stimulate the white blood cells into releasing chemicals called **antibodies** that destroy the antigens. Also, the blood cells secrete a chemical that induces the production of more white blood cells to overcome an invasion of disease agents. In fact, one way a veterinarian can diagnose disease is by counting the number and type of white blood cells in an animal's blood (Figure 12–13). When the pathogens have been overcome, white blood cells called lymphocytes, which are produced by the lymph system, create "memory cells" that become activated when the same type of pathogen again enters the bloodstream. This is known as immunity.

Immunity

Animals possess a certain degree of immunity to all diseases. As disease organisms enter the body, the defensive system is activated. If the disease organisms are totally new to the animal's system, they may reproduce so rapidly that the defense system is not effective. If the disease agents are "familiar" to the defense system through a previous encounter, the immune system quickly activates and overwhelms the pathogens before they can build up.

Mammals get their first immunity from their mothers. The first milk they receive (called colostrum) is a nutrient-filled substance that is also rich in antibodies (Figure 12–14). This substance helps the newborn ward off disease until its own immune system can develop by being exposed to disease

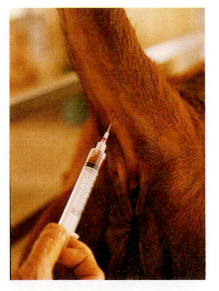

Figure 12–13 *Blood samples are taken, and white blood cells are analyzed to diagnose disease.*
Courtesy of James Strawser, The University of Georgia.

B I O   B R I E F

Making Coccidia Less Cocky

Each year, U.S. poultry producers raise about 7 billion broilers. Chickens are routinely given drugs to prevent infection by tiny, single-celled protozoa known as coccidia. These organisms invade cells in a chicken's gut, where they reproduce and make it harder for the bird to absorb feed and to gain weight quickly. Coccidiosis is among the top five chicken diseases that prevent weight gain, says Agricultural Research Service (ARS) microbiologist Harry D. Danforth. He is with ARS's Parasite Biology and Epidemiology Laboratory at Beltsville, Maryland. Each year, the protozoa cost producers worldwide an estimated $600 million in treatment and low carcass weights. And, it could become worse, because the protozoa are developing resistance to standard drugs. That possibility has Hyun S. Lillehoj and Mark C. Jenkins, who are based at the ARS Immunology and Disease Resistance Laboratory at Beltsville, working on ways to use the birds' own immunity against coccidia.

"We have short- and long-term goals," says Danforth, who is the agency's scientific liaison with the poultry industry. The short-term goal is a gamma-irradiated vaccine Danforth tested this year at Perdue Farms, Inc., in Salisbury, Maryland. First, however, Jenkins had to determine the radiation dose needed to weaken the live oocysts—the infectious stage of coccidia. This prevents them from developing or reproducing. Next, he determined the dose of weakened oocysts needed to produce immunity in the chicks. Then, Danforth put the oocysts in a gel delivery system he and researchers in the vaccine industry had developed earlier to get live vaccine into chicks at the hatchery. The gel was added to the feed of chicks destined to become Cornish hens.

Danforth says it takes about 1.6 pounds of feed for each pound of bird. With the gamma-irradiated oocysts, "feed conversion was 3 points better than with the anticoccidial drugs," he says. "That means the treatment reduced, by three hundredths of a pound, the feed needed to raise a two-pound Cornish hen."

The chicken's immune system is very complicated, and launching an immune response is walking a fine line. This type of response can protect a bird—or destroy it, if it goes too far. So researchers first have to unravel the complex inner workings in the bird's gut before they can get the optimum immune response without an overreaction. Jenkins approached this problem by identifying proteins in the oocysts that mark it as an intruder and elicit an immune response. He, Lillehoj, Danforth, and Michael D. Ruff have a patent on the recombinant DNA for two promising proteins from the oocysts' outside coat. Jenkins inserted the recombinant DNA for those two coat proteins and another promising protein separately into DNA loops, called plasmids, take from *E. coli* bacteria. In small-scale tests at the Beltsville lab, he shot the plasmids straight into chicken legs using a jet gun, like the ones dentists use to numb their patients' teeth and gums.

"We got the best protection with mixtures of al three plasmids," Jenkins says. Weight gain in the immunized chicks was significantly better than in the unimmunized birds. But, it was not as good as in the birds that never got close to a coccidia oocyst. "We have a little way to go because we want to have complete protection," he says. A more efficient system for delivering such an inoculum in an industry setting also is needed.

Scientists have developed a new vaccine that makes use of recombinant DNA. Courtesy of USDA-ARS.

Lillehoj's lab identified another promising protein, which enables the stage that emerges from the oocyst, called a sporozoite, to invade the bird's T cells. Lillehoj patented a monoclonal antibody to the protein that she says "consistently blocks this invasion in culture dishes." She is now collaborating with scientists in Japan and Korea to find the DNA that directs production of that protein.

Lillehoj's main focus is on the substances immune cells generate to communicate with one another. Animals produce these natural, hormone-like chemicals, called cytokines, during an infection. They are potent and function at low levels. Some cytokines enhance the immune response, but others can cause disease symptoms. "So, you have to know which ones are protective," says Lillehoj. "Understanding how this works may be a way to control infection without introducing anything unnatural."

She is looking for umbrella protection against the six or seven chicken coccidia species because, after cytokines are produced, they aren't picky about the species. They may also be given along with vaccines to increase their effectiveness. The challenge is that more than 20 different cytokines regulate immune response, Lillehoj says. "We're just beginning to understand how they work."

One all-purpose cytokine that has proved effective is interferon gamma. Interferon gamma activates macrophages—cells that behave like the Pac Men of the immune system, gobbling up invaders. Interferon gamma inhibits coccidia multiplication, so the birds lose less weight. Lillehoj says the birds' immunity level correlates with their interferon gamma level.

Source: Agricultural Research Magazine.

Figure 12–14 *The first immunity comes from the mother's colostrum. Courtesy of James Strawser, The University of Georgia.*

agents and building up antibodies. As an animal matures, immunity to specific antigens is developed.

Immunity can be either active or passive. **Active immunity** means that the animal is more or less permanently immune to the disease. **Passive immunity** means the animal is only temporarily immune. Naturally acquired active immunity is obtained by the animal having the disease and recovering. The memory phagocytes react quickly when the antigen that causes the disease enters the animal's body, and the antigens are overwhelmed.

Animals (and humans) are made to be artificially actively immune to certain diseases by a process called **vaccination**. In this process the animal is injected with antigens that cause the phagocytes to react without making the animal seriously ill (Figure 12–15). This in turn causes the buildup of memory phagocytes that attack the antigens whenever they enter the animal's system.

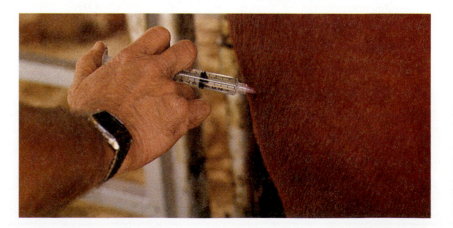

Figure 12–15 *In the vaccination process, antigens are injected into the animal's body. Courtesy of James Strawser, The University of Georgia.*

An Englishman named Edward Jenner developed this process in the late 1700s. One of the most devastating diseases of the time was smallpox. In some areas, this disease killed more than half of the people that contracted it. Those who survived were permanently immune to the disease. People also came down with a disease known as cowpox, which, as the name implies, was a disease of cattle. Jenner noticed that those who contracted cowpox never came down with smallpox and reasoned that a connection must exist between the two diseases. In humans, cowpox was usually mild and people recovered. Jenner collected material from sores that developed on people with cowpox and injected healthy people with it. The people injected became ill with a mild case of cowpox but were then immune to smallpox. The Latin word for cow is "vacca," and the word vaccination was coined because this technique was associated with cows.

Another scientist, Louis Pasteur, used this concept to develop several vaccines. All of the modern vaccines, whether given to humans or animals, work on basically the same principles. When he used materials from cowpox sores, Edward Jenner injected live viruses. This worked well in this particular case, but as more vaccines were developed, it was discovered that often the vaccination could cause the disease. Also, many of the viruses can live for long periods of time in the soil. If bottles of live vaccine were dropped and broken, the soil became contaminated. Research has proven that weakened or killed viruses can be effective vaccines against many diseases. These materials act as antigens in stimulating the production of antibodies in much the same way as live viruses, but without the dangers.

Producers regularly vaccinate animals against disease. Even though there may not be any known cases of disease anywhere in the area, the cost of the vaccine is small compared to the risk of disease.

PLANT DISEASES

Plant diseases can be just as destructive to agriculture as animal diseases. A good example is a disaster that happened in Ireland during the middle 1800s. A staple of the Irish diet was potatoes. The climate of the country was right for growing the crop, and many people depended on potatoes for their food.

Figure 12–16 *Potato blight caused widespread starvation when entire crops of potatoes were destroyed in the 1840s.* Courtesy of USDA-ARS.

Figure 12–17 *Plant diseases may appear as brown spots or wilted leaves.* Courtesy of Progressive Farmer.

In the 1840s, a disease called potato blight almost wiped out all of the country's crop of potatoes (Figure 12–16). As a result, thousands of people starved to death and more than a million other people left Ireland.

Plant diseases can have different symptoms depending on the type of disease. Plants may appear wilted or have brown spots on the leaves, fruit, or other parts (Figure 12–17). The disease may cause a powdery white substance to appear on the leaves or other plant surfaces. Bacterial or viral diseases may cause the plant to have a discolored or multicolored appearance, or the plant may be stunted. Regardless of the symptoms, plant diseases damage the plants and make them less valuable to both the producer and the consumer.

The same types of organisms that cause plant diseases can also cause animal diseases. Bacteria, viruses, and fungi all cause plant disease. However, with few exceptions, different species of the organisms cause disease in plants than those that cause disease in animals. Just as in animals, these disease-causing agents attack the cells of plants and interfere with their functioning. Cells may die, and as a result the plant may die or be seriously damaged.

Disease may attack different systems in the plant. The root system may be affected, and the plant may not be able to take in water and other nutrients. If the leaves are affected, photosynthesis may be impaired, and the plant may not be able to manufacture food. If the flowers are attacked, the plant may not be able to reproduce or produce fruit. Stems damaged by disease may not be able to translocate water and food nutrients to other parts of the plant.

In addition to the decrease in productivity, disease may cause another serious problem. The quality of the produce may be substantially lowered by diseases. Fruits may develop soft brown spots, or leafy vegetables may develop spots on the leaves. Root crops such as carrots or potatoes may have spots of disease growth in them. Not only do these problems cause the produce to be unsightly, but the taste may be significantly changed (Figure 12–18). In addition, removing fruit or vegetables that show signs of disease can be time-consuming and costly to food processors.

Some diseases may even render crops unfit for consumption by animals or humans. A good example is ergot, a disease of grain that is caused by a fungus. If animals or humans consume grain infected by this disease, they may go into convulsions, have hallucinations, and may even die. A powerful

Figure 12–18 *Disease causes plants to be unsightly. The spots on these tomatoes were caused by a disease called blossom end rot.*
Courtesy of James Strawser, The University of Georgia.

hallucinogenic drug called lysergic acid diethylamide (LSD) is derived from plants that are infected with ergot.

Diseases influence the industries of ornamental horticulture, landscaping, and turf management. All of these industries are based on producing attractive plants. If the plants have diseases, they will become unacceptable. For example, a golf course that has brown spots from a grass disease will not appear green and lush (Figure 12–19). Flowers that are diseased will not have vivid colors.

Pathogens enter plants through natural openings such as stomates, through the plant surfaces, or through injuries to the plant. Another important method of infection is through insects that feed on the plants. By going from plant to plant and feeding on each plant, disease organisms are transmitted. Fungal diseases may begin by spores landing on leaves or other surfaces of a plant (Figure 12–20). Spores are the one-celled or multicelled units that are like seeds in higher plants. The spores send tubes called hyphae down into the leaf to extract nutrients. As a result, a portion of or all of the leaf dies from lack of nutrients.

Plant Defenses Against Diseases

Like animals, plants also have defense systems that fight disease. The first line is the covering on the surface of the plant. The plant hairs or pubescence on the surface collect pathogens before they reach the surface of the plant (Figure 12–21). The plant may also have a thick waxy covering called the cuticle that prevents the penetration of the pathogens. Plant stomates close at night, and this action helps prevent

Figure 12–19 *The Turf Disease called "brown spot" on lawns or golf courses may appear as dead spots.*

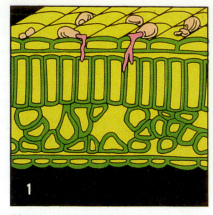

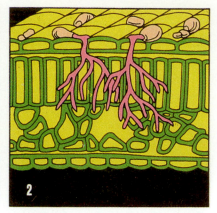

Disease spores, blown by the wind, land on leaf surfaces.

These spores send germ tubes through the leaf tissue.

Hyphae from the germ tube weave through the leaf tissue to take nourishment from the plant.

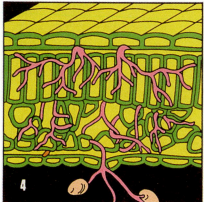

As the hyphae mass spreads, the leaf wilts, leaf cells die, and the food supply is cut off from the other parts of the leaf.

What you see from the outside is spore-bearing branches emerging from within the leaves.

And then—spores are carried by the wind from these infected plants to other plants, and the disease is on its way to spreading all over the field.

Figure 12–20 *Fungal spores with hyphae extending into a leaf.*
Courtesy of Ciba Agricultural Division.

the entrance of pathogens during the time when they are most active. If a plant is injured, a thick gummy substance may appear around the wound (Figure 12–22). This substance surrounds and destroys pathogens. Many fruit trees are particularly adept at gum formation around wounds.

After pathogens gain entry to a plant, biochemical processes combat the disease agents. Certain plants secrete chemicals that are toxic to pathogens. Concentrations of these chemicals on the leaves of the plant inhibit the growth

Figure 12–21 *Plant hairs help by collecting pathogens before they reach the plant surface.* Courtesy of Progressive Farmer.

Figure 12–22 *If a plant is injured, a gummy substance may appear around the wound. These wounds were caused by insects.* Courtesy of USDA-ARS.

Figure 12–23 *Keeping fields weed-free helps prevent disease outbreaks.* Courtesy of Case-International.

of such pathogens as fungi. Some plants exude the chemicals more than others and are therefore more resistant to disease. Also, certain chemicals are produced inside the plant as a result of pathogen invasion. These chemicals help to rid the plant of the infections. Scientists are conducting research to learn more about these chemical compounds.

Prevention of Plant Diseases

One of the most effective means of preventing disease in plants is through the use of disease-resistant varieties. All of the defense mechanisms of a plant are the result of genetic coding. Plant breeders concentrate on developing the characteristics of the plant that make them more resistant. For example, plants with more pubescence (plant hairs) or a thicker cuticle have an advantage in keeping out pathogens. Plants that secrete more pathogen-destroying chemicals will also have an advantage. The challenge is to breed plants that have a natural immunity and that yield high-quality produce. Scientists have come a long way in producing grains, fruits, and vegetables that are resistant to diseases.

Cultural practices can also help in preventing disease. A clean, weed-free field will eliminate plants that can harbor disease (Figure 12–23). Care in cultivation and other management practices can help prevent injuries that can let in pathogens. The removal of debris such as clippings from pruning or the stalks of harvested crops can lessen disease by eliminating places for disease to grow. Crops are usually

Figure 12–24 *Fungicides are applied to prevent and cure plant diseases.* Courtesy of James Strawser, The University of Georgia.

watered during the daylight to prevent the spread of pathogens that grow better at night. Also, producers do not disturb crops when they are wet. A person moving through wet plants can spread disease by carrying bacteria, viruses, or fungi from plant to plant through the water droplets picked up from the plants.

Producers also use chemicals that destroy disease organisms. The fruit industry regularly sprays crops with fungicides and bactericides (Figure 12–24). These chemicals prevent a buildup of pathogens, and the result is healthier, faster growing plants that yield better products.

SUMMARY

Diseases have been around almost as long as life on earth. They cause misery and destruction to animals and cause problems with the plants we raise. Many methods are used to control diseases. In the future, we will see some different and exciting new methods of controlling diseases.

CHAPTER 12 REVIEW

Student Learning Activities

1. Write a report on one plant disease and one animal disease. Include the plants or animals affected, the cause of the disease, how it is spread, the symptoms, the damage caused, the treatment, and prevention. Share the report with the class.

2. Interview a veterinarian about the most important animal diseases in your area. What methods are used to prevent and treat the diseases?

3. Collect five different plants that have been affected by a disease. Try to identify the disease, the cause, and the importance of the disease. Share your collection with the class.

Define the Following Terms

1. pathogens
2. bacteria
3. antibiotics
4. viruses
5. protozoa
6. antigens
7. antibodies
8. active immunity
9. passive immunity
10. vaccination

True/False

1. A disease is any condition that causes the systems of a plant or animal to not function properly.

2. Genetic defects and malnutrition are examples of infectious diseases.

3. Noninfectious diseases are by far the most costly because they are the hardest to prevent.

4. Bacteria are the most common pathogens of animals.

5. Antibiotics that are used to treat bacterial infections are also effective against viral infections.

6. Viruses can reproduce inside and outside of a living organism.

7. Viruses cause disease by invading and mutating living cells.

8. Mammals get their first immunity from their mothers by means of the mother's first milk.

9. Animals can be made artificially immune to certain diseases with vaccines.

10. Live viruses are more effective in vaccines than weakened or dead viruses.

11. The same categories of pathogens that cause animal diseases can also cause diseases in plants.

12. One of the most effective means of preventing disease in plants is by using preventative chemicals.

Fill in the Blank

1. Microorganisms that cause disease are called _____, and their study is known as _____.

2. Several types of bacteria can be classified according to shape. _____ bacteria are shaped like a ball; _____ are rod-shaped bacteria.

3. A substance released by cells that inhibits viral growth and is being researched as a treatment against viruses is _____.

4. _____ are one-celled organisms that cause disease among animals.

5. White blood cells called _____ trap and destroy disease-causing agents in the blood.

6. Immunity is either _____ or _____. _____ immunity means that the animal is more or less permanently immune to the disease. _____ immunity means the animal is only temporarily immune.

7. _____ is a plant disease that resulted in the starvation of thousands of people in Ireland.

8. Plant hairs, also called _____, and the waxy coating of a plant, called the _____, help protect the plant against disease.

Discussion

1. List the two categories of disease, discuss the differences between the two, and give examples of each.

2. Discuss the types of microorganisms that cause disease in animals and how they cause disease.

3. Discuss the methods by which antibiotics fight bacteria.

4. Some argue as to whether viruses are living or nonliving organisms. Give arguments for each point of view.

5. Describe the lines of defense the animal body has against disease pathogens.

6. Discuss the symptoms plants may exhibit when diseased.

7. Describe the various ways in which plant diseases may cost producers and how these diseases may be prevented.

WEED SCIENCE

STUDENT OBJECTIVES

After studying this chapter, you should be able to:

◆ Distinguish between weed and nonweed plants.

◆ Explain the different ways weeds can be pests.

◆ Explain how weeds are classified.

◆ Analyze the characteristics of weeds that make them difficult to control.

◆ Cite examples of imported weeds.

◆ Evaluate the various means of controlling weeds.

◆ Discuss how chemical herbicides work.

◆ Explain how herbicides are selective in the plants they kill.

◆ Discuss how biological agents can be used in weed control.

◆ Discuss the concept of integrated pest management.

Weeds have been a part of civilization for as far back as recorded history. Many ancient documents speak of humans battling weeds in the crops they grew. Throughout history, where there has been agriculture, weeds have also been there.

There are many definitions of a weed. The most commonly accepted definition is that a weed is a plant growing where it is not wanted (Figure 13–1). This unwanted plant usually causes problems for humans. Almost any plant can be a weed if it is not wanted. For example, millions of acres of oats are grown each year for human food and livestock feed. Oats are a very desirable plant, but if oats are growing in a wheat field, the oats are considered a weed. On the other hand, plants that

Figure 13–1 *Weeds are plants that are growing where they are not wanted. Courtesy of* Progressive Farmer.

are commonly described as weeds can be desirable if they are growing on a hillside and controlling erosion.

Weeds are undesirable and are considered to be pests just as insects and disease organisms are considered to be pests. They can be found everywhere that humans live or work. They infest our lawns, clog waterways, fill up ponds, invade flower beds, ruin golf courses, and grow in our forests (Figure 13–2). Each year millions of dollars are spent in the attempt to lessen the effect of weeds on agricultural production. Weeds damage agricultural plants and animals in several different ways.

Weeds compete with crops for materials essential to growth. If a field is infested with weeds, no matter how much fertilizer or water is applied to a crop, weeds get a share of the

Figure 13–2 *Weeds may clog waterways or fill up ponds. Courtesy of James Strawser, The University of Georgia.*

nutrients. Because these nutrients go to plants that are non-productive, these valuable inputs are wasted. Also, the weeds can compete with the crop for space and sunlight. Weeds that grow taller than the crop will shade it and reduce the amount of sunlight the crop plant can use for photosynthesis. This causes the crop to grow more slowly and to produce less.

Weeds may serve as a host for insects and disease organisms. Many weed species attract pests that are harmful to the crops. With weeds growing among the crop plants, control of insects and diseases can be more difficult because the producer must deal with more than one type of plant. One of the tools in the fight against insect pests is the interruption of the insects' life cycle. This is more difficult if the insects have a different plant they can live on or in. Likewise, disease organisms can be spread to crop plants through weeds that serve as hosts to disease organisms.

Weeds cause impurities in agricultural products. Cows that graze in pastures where wild onion or wild garlic grow will produce milk that has an offensive odor and taste. This lowers the quality of the milk and reduces its value as an agricultural product. Soybeans that contain the seed of the coffee weed plant are less valuable than pure soybeans (Figure 13–3). The weed seeds are not only poisonous to livestock, but are so close to the size and shape of soybeans that it is difficult to sort them out. Cotton that is picked among weeds will have trash in the lint. This reduces profit because the trash has to be removed from the lint. Similarly, wool that

Figure 13–3 *Coffee weed lowers the value of soybeans because the seeds are poisonous to livestock.*
Courtesy of Steve Reeve, Soybean Digest.

Figure 13–4 *Cockleburs can be toxic to livestock.* Courtesy of Progessive Farmer.

contains burs or other parts of weed plants is less valuable because of the added expense of removal.

Weeds can be poisonous to livestock. Some species of weeds can be deadly to livestock. A good example is the plant tansy ragwort that grows in the northwestern part of the United States. If a horse ingests the plant, death can occur in a short time. Cocklebur, nightshade, and bracken ferns are just a few examples of plants that can be poisonous to livestock (Figure 13–4). The removal of poisonous weeds from hay fields and pastures is both time-consuming and expensive.

Weeds cause problems to humans. Many people suffer from allergic reactions to the pollen produced by weeds (Figure 13–5). Ragweed and goldenrod are just two of the many plants that produce pollen that is the right size to cause allergic symptoms in humans during certain times of the year. In addition, weeds are unsightly and destroy the beauty of gardens, yards, landscaped areas, and many natural areas. Beautiful green lawns and golf courses are the result of many hours of labor and much expense to keep them weed-free.

Figure 13–5 *Pollen from weeds like these causes allergic reactions in some people.* Courtesy of William K. Vencill, The University of Georgia.

THE CLASSIFICATION OF WEEDS

Weeds are classified in several ways. One way is to group weeds as being common or **noxious weeds**. States determine which weeds are noxious—that is, the weeds that are the most difficult to control and those that create the biggest problem. Obviously, a weed may be declared a noxious weed in one state and not another. A federal law known as the Noxious Weed Act was passed in 1974 to help prevent the spread of noxious weeds. Among other provisions, the act provides for the inspection of crop seed for noxious weed seed. Any seed coming into this country or that crosses a state line must be inspected for noxious weed seed. If any are discovered, the label on the bag of seed must indicate the percentage of noxious weed seed (Figure 13–6). If certain types of noxious seed are in the crop seed, they may be condemned.

Weeds are also classified according to their life span. Annuals live only one year; biennials live two years; and perennials live more than two years. Control measures are different for the different life-span classifications.

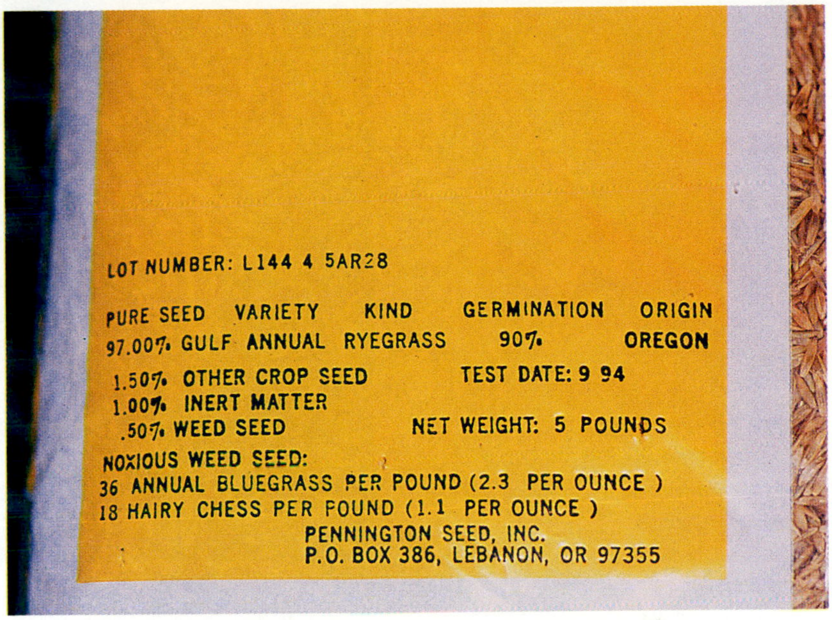

LOT NUMBER: L144 4 5AR28

PURE SEED	VARIETY	KIND	GERMINATION	ORIGIN
97.00%	GULF	ANNUAL RYEGRASS	90%	OREGON

1.50% OTHER CROP SEED   TEST DATE: 9 94
1.00% INERT MATTER
.50% WEED SEED   NET WEIGHT: 5 POUNDS

NOXIOUS WEED SEED:
36 ANNUAL BLUEGRASS PER POUND (2.3 PER OUNCE)
18 HAIRY CHESS PER POUND (1.1 PER OUNCE)
PENNINGTON SEED, INC.
P.O. BOX 386, LEBANON, OR 97355

Figure 13–6 *Seed tags indicate the percentage of noxious weed seed in the bag. Courtesy of James Strawser, The University of Georgia.*

CHARACTERISTICS OF WEEDS

Weeds generally have characteristics that make them problems in crop or animal production. The main reason particular plants become troublesome pests is that they grow and reproduce efficiently. If they did not, they would be easily controlled and not considered a pest.

Most weeds have at least some of the following characteristics. For producers to be effective in controlling weeds, they must understand the life cycle characteristics of the weeds. All of these traits have to be considered in a science-based program aimed at weed control.

Weeds produce a large number of seeds (Figure 13–7). Some weeds may have as many as a quarter of a million seeds from a single plant. Obviously, any plant that produces that many seeds will be difficult to control if all of the seeds mature and germinate. Fields that are completely covered in weeds generally have weeds that are prolific in seed production.

Weed seeds generally remain viable over long periods of time. This means that when weed seeds mature, they can remain living for a long time before they germinate. They go into a period called **dormancy** in which they remain inactive until the germination process is triggered. Dormancy may be

Figure 13–7 *Weeds often produce a tremendous number of seeds. Courtesy of James Strawser, The University of Georgia.*

controlled by conditions such as the absence of light, or a concentration of carbon dioxide may cause seeds to remain dormant. Every time a field is plowed, weed seeds are covered deeply in the soil. This can create conditions favorable for seeds being dormant. The same plowing can also bring dormant seeds to the surface, where they germinate. Some weed species have been known to lie dormant for 40 years and germinate at a rate of more than 80 percent! Eradication of weeds that produce seeds that remain dormant and viable for this long is extremely difficult.

Weeds grow rapidly. Some of the fastest growing plants in the world are weeds. The rapid growth of these plants causes them to be serious problems. As mentioned earlier, one problem with weeds is competition with the crop plant. If the weed grows more rapidly than the crop, the crop will be crowded out and receive less nutrients and sunlight (Figure 13–8). With rapidly growing weeds, it is essential that they not be allowed to get larger than the crop. Some weeds can grow so rapidly that they completely cover other plants and kill them.

Weed seeds are efficiently dispersed. If the seed of weeds could be contained in a relatively small area, they would be easier to control. However, many weed species have adapted efficient means of scattering seeds. Some seed are carried by the wind (Figure 13–9). Such weeds as milkweed, thistle, and dandelion have long featherlike structures that catch in the wind and propel the seeds over considerable distances. Other

Figure 13–8 *If weeds grow faster than the crop, the weeds can take over. Courtesy of William K. Vencill, The University of Georgia.*

Figure 13–9 *Some weed seeds are dispersed by the wind. Courtesy of James Strawser, The University of Georgia.*

seeds, such as cocklebur and beggar lice, have projections on their coats that catch in the hair of animals or the clothing of humans and are carried away. Seeds from the fruits of weeds may stick to the feet or bodies of birds and be carried to new locations. Also, seeds that are ingested by animals may be passed through their digestive tracts and remain viable. By the time the seeds make the journey through the animal's system, the seeds may be a long way away from where they grew. Weed seeds can be carried by farm machinery from one field to the next or may be transmitted on carts, wagons, or crates.

Weed seeds are also carried by water (Figure 13–10). Streams transport seeds that float on the water and are deposited downstream. One way weed seeds get dispersed is through the use of flood irrigation. Water is released onto fields from open canals. Water in these canals may have traveled for many miles and collected weed seed all along the way. When the water is released onto the fields, the seeds suspended in the water are scattered all across the field.

Weeds may reproduce by more than one method. Many weeds may reproduce both sexually and asexually. For example, a noxious weed, yellow nutsedge, propagates both by

Figure 13–10 *Flooding can disperse weed seeds through the water. Courtesy of University of California.*

seed and by rhizomes (Figure 13–11). Remember from Chapter 7 that rhizomes are underground stems that grow out from the plant and form new plants. Tubers are formed on the rhizomes that store food and nutrients for the plant. New shoots sprout from these tubers, and new plants are formed. If one form of reproduction fails, another may succeed. In developing a scheme to control weeds, a dual form of reproduction makes the task more difficult.

IMPORTED WEEDS

Many of the worst weeds we have in this country were imported into the United States from other parts of the world. In fact, of the 15 most serious weeds in this country, 13 of them were introduced from outside the United States. Many of these weed pests were introduced on purpose. At the time they were brought in it seemed a good idea to introduce the new plants because they were thought to serve a good purpose. Around 1840, Colonel William Johnson brought a new grass from Africa to Selma, Alabama (Figure 13–12). His idea was to introduce a high-quality forage grass. Indeed, the grass looked promising. It grew in conditions ranging from large amounts of rainfall to drought conditions. It could be easily established because it spread rapidly by either seed or through rhizomes. The new grass also grew extremely fast and came back rapidly when the upper part of the plant was

Figure 13–11 *Nutsedge is propagated by both seed and rhizomes. Courtesy of William K. Vencill, The University of Georgia.*

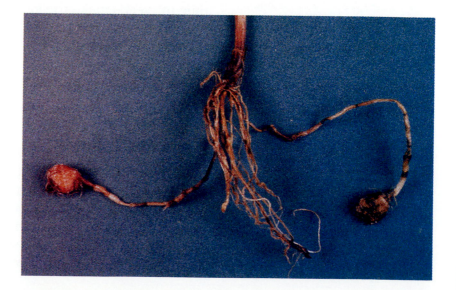

Figure 13–12 *Johnsongrass was brought in to provide grazing but developed into a serious weed pest.* Courtesy of Vann Cleveland, Progressive Farmer.

cut off from the roots. In fact, in one season's growing time, the grass could reach heights of greater than 10 feet. The problem began when the plant escaped and invaded other crops. Because it grew so rapidly and spread so efficiently, it soon became a serious weed pest of other crops. The grass was named Johnsongrass after Colonel Johnson. It spread rapidly to other areas of the country. Today, Johnsongrass has been found in most states and is considered one of the most difficult weed pests to control.

Another example is a vine called kudzu that grows throughout the southeastern United States (Figure 13–13). As explained

Figure 13–13 *Kudzu is a good example of an imported plant that is an extremely fast growing weed.* Courtesy of James Strawser, The University of Georgia.

in Chapter 5, this plant was introduced from Japan around the turn of the 20th century as a means of controlling erosion. In many parts of the South, large areas were so eroded that it was difficult to establish new plants on the thin soil. Kudzu grew extremely fast and established itself well on the eroded ground. Because the plant's dense foliage died back each year, valuable organic material was added to the soil. Also, the plant is a legume, so nitrogen was put into the soil, and the nutritive level of the soil was improved. In addition, the leaves of the vine are a highly palatable forage that is high in protein. This combination made the plant attractive to people who owned land that was susceptible to erosion and had a thin topsoil. The plant controlled erosion and provided high-quality forage, but as a weed pest kudzu has few equals. This vine grows so fast that in one year's time it can completely cover tall pine trees and kill them. At frost, the plant dies back to the ground, and the roots put out new runners the next year that can grow just as far as the vines of the previous year. Many trees across the South have been killed by this invading plant. Whole forest plots can be rapidly taken over by the vines, which can grow more than a foot each day. Its rapid growth and deep root system make this weed difficult to control.

Today, federal laws prohibit the indiscriminate introduction of plant materials into the United States. If plants are brought into this country, a long procedure of permits and qualifications has to be met before permission is granted. This helps to prevent the introduction of plants that will, like Johnsongrass and kudzu, become serious weed pests.

CONTROLLING WEEDS

The control of weeds parallels the development of civilization. Humans have been devising ways of controlling weeds for as long as they have been growing crops. This was a major obstacle to overcome before people could become efficient at producing their own food. As methods of controlling weed pests developed, growing crops became easier, more efficient, and more profitable. Billions of dollars have gone into researching new and better ways of controlling weeds. Today, a variety of methods can be used by producers in the battle against weeds.

Figure 13–14 *People learned to plant crops in rows that could be cultivated with animal power.* Courtesy of University of Illinois.

Figure 13–15 *Hoeing is still used as a weed control measure in many parts of the world.* Courtesy of Tom Dodge, Successful Farming Magazine.

Mechanical Weed Control

The first method was simply pulling the weeds from around the plants. This was time-consuming hard work. In addition, pulling the plants out by the roots often disrupted the roots of the crop plants, and damage was done to the plant. People learned to plant crops in long rows so the crop plants could be worked with less damage. This allowed the invention of plows that could be pulled by animals (Figure 13–14). Because the crop was in a row, the plow could be pulled down the middle between the rows, and the roots of the crop would not be harmed. The plow either cut off, up-rooted, or covered up the weed. Hoes were devised first of wood, then flint, and finally of metal and used to work weeds out from around the plants in the rows. This method was used in this country until the 1960s, when chemical herbicides began to be used. Hoeing is still widely used in many parts of the world (Figure 13–15).

Mechanical cultivation to rid crops of weeds is still used today (Figure 13–16). Plowshares have been scientifically designed to destroy the maximum amount of weeds while doing a minimal amount of damage to the crop. Several designs of cultivators are used with differing crops, weeds, soil, and climatic conditions. This method of weed control has four main disadvantages. First, the roots of the crop may suffer damage. Even with a careful operator and a well-designed implement, some root damage occurs. Any damage to the roots of the crop will result in damage to the plant. This damage will be repaired by the plant using cell growth that would otherwise increase the plant size. Second, mechanical control is expensive. The machinery is both expensive to buy and to maintain. Implements that plow through the ground wear out quickly and must be replaced at considerable cost. Also, the cost of operating the machinery is high. Diesel fuel and other oils and lubricants can be quite an added expense to the production of crops. Third, the stirring of the soil can lead to erosion problems. The loss of the topsoil is a serious concern, and any time the soil is loosened conditions exist for erosion. Wind and water can remove the soil. Programs are underway in several states to keep tillage of the soil at a minimum. A fourth disadvantage is that when the soil is stirred, moisture loss occurs. The soil has a higher moisture content beneath the surface because the sun cannot create as much heat to evaporate it. If the moist lower soil is brought to the surface,

Figure 13–16 *Mechanical cultivation of crops is still widely used. Courtesy of USDA-ARS.*

it is heated by the sun, and evaporation occurs. In areas where crops must be irrigated, cultivation can result in higher water costs.

Chemical Weed Control

One of the most efficient weapons in the war against weeds is chemical weed killers, known as **herbicides** (Figure 13–17). People have used chemicals for hundreds of years to kill

Figure 13–17 *Chemical herbicides are used to control weeds. Courtesy of Photo Disc, Skip Nail.*

unwanted plants. For example, salt has been used for centuries to kill plants. The obvious problem with using compounds such as salt is that it is residual in the soil and will not allow any plants to grow if the concentration is too high. Modern herbicides were first used around the time of World War II with the invention of a compound known as 2-4D. Progress on the development of herbicides moved rapidly through the 1950s, and they became widely used by the 1960s. Modern chemicals have to meet standards set by the Environmental Protection Agency to ensure that they do not harm the environment.

How Herbicides Work.   The purpose of a herbicide is to kill plants. How this takes place differs with the different chemicals that make up the active ingredient of the herbicide. Scientists still do not fully understand the exact mechanisms that cause certain herbicides to work, but different herbicides affect plants in several ways.

Interference with Photosynthesis.   Remember from previous chapters that photosynthesis is the process through which a plant uses sunlight to form energy for use in its metabolical processes. Certain types of herbicides inhibit the process of photosynthesis. It is thought that this causes the formation of a compound that destroys the cell membranes. When the cell membranes are destroyed, the contents of the cell leak out, and the cell dies. When a large number of plant cells die, the plant dies.

Inhibition of Amino Acids and Protein.   Amino acids are the building blocks for protein that are used by the plant to form structures and to grow. Also, chemical agents called enzymes, which cause all of the chemical reactions in a plant, are composed of proteins. The formation of the long chains of amino acids that make up protein is controlled by deoxyribose nucleic acid (DNA) and ribonucleic acid (RNA). Remember from Chapter 4 that RNA transports the genetic information from the DNA that determines how the amino acids are assembled. Certain herbicides interfere with the transportation of this information, and the ability to form protein is hampered. Because the formation of enzymes is decreased, the chemical reactions necessary for plant survival are decreased, and this results in the death of the plant.

Mitotic Poisons.   Remember from Chapter 8 that plants grow by a process called mitosis. This is the mechanism of cell

BIO BRIEF

Hard-To-Control Weeds Need a Mix of Measures

Farmers and land managers will have to become proficient at long-range planning for weed control—and also learn to use a mix of different techniques—to control weeds, according to Agricultural Research Service (ARS) scientists. Hemp dogbane, waterhemp, leafy spurge, cocklebur, and thistle are becoming more difficult to control with one-step approaches, such as using a single tillage method, annual crop rotations, or seasonal herbicide applications. The ARS scientists advocate the use of integrated weed management systems that increase cost efficiency for farmers, promote more ecologically sustainable production, and conserve soil and water resources.

Weeds and weed control cost U.S. farmers about $15 billion each year. Part of the problem is that the long-term use of specific herbicides has led to the development of weeds with herbicide resistance. Another problem is that shifts in tillage practices make it easier for different types of weeds to get established. Also, some cropping patterns have discouraged competition among weed species, causing certain ones to spread and crowd out crops.

One scientist tackling weed problems in new ways is ARS range scientist Robert A. Masters. He is in the ARS Wheat, Sorghum, and Forage Research Unit at Lincoln, Nebraska. Masters has combined herbicides, controlled burning, and replanting of native warm-season grasses, without tillage,

to supplant leafy spurge, a noxious weed that threatens Northern Plains grasslands.

"The productive capacity of Great Plains grasslands has been reduced greatly by invasive weeds like leafy spurge," he says. "These weeds have displayed desirable native forages as well."

Leafy spurge was introduced into the northern Great Plains from Eurasia in the late 1800s. It has no natural enemies in this country, and herbicides provide only short-term control. Unlike sheep and goats, cattle and horses won't graze land infested with leafy spurge, and they avoid eating forage grasses growing next to the weed. In field tests in cooperation with scientists at the University of Nebraska, Masters applied a three-pronged strategy to fight leafy spurge. The result was a 60-percent reduction in spurge populations and a surge in the production of warm-season grasses.

First, Masters applied a combination of the herbicides Arsenal and Oust to kill leafy spurge plants in the fall. He burned the dead plant residue the following spring. Then he planted native prairie species such as big bluestem, switchgrass, and indiangrass.

"Two years after planting," says Masters, "switchgrass and bluestem hay yields increased to over 4,000 pounds per acre, and indiangrass produced more than 3,000. Our goal is to develop economical, integrated weed-management strategies that enable land managers to convert marginal cropland and degraded grasslands to high-value grasslands that are resistant to noxious weed invasion," he says. "In our current research, we're refining this strategy by using a combination of the herbicides Plateau and Roundup, in place of Arsenal and Oust, to promote establishment of mixtures of native grasses and legumes."

Loyd Wax, an agronomist at the ARS Crop Protection Research Laboratory in Urbana, Illinois, is taking a similar approach to thwart waterhemp, a species similar to redroot and smart pigweed. Waterhemp biotypes have become increasingly resistant to

An integrated approach using a variety of methods was used to control weeds in the corn on the left.
Courtesy of USDA-ARS.

ALS-inhibiting herbicides that block the production of branched-chain amino acids, the building blocks of protein. Wax found super-resistant waterhemp biotypes that withstood up to 520 times the labeled rate of certain ALS-inhibiting herbicides.

To combat these ALS-resistant biotypes, Wax conducted studies showing that growers could markedly improve waterhemp control by combining non-ALS inhibiting herbicides with cultivation. This strategy gives the grower two very different methods to control ALS-resistant waterhemp populations.

"We controlled established waterhemp populations with non-ALS herbicides and then tilled the soil at a depth that was not favorable to waterhemp seed germination and emergence," says Wax. "We also found that we could combine sequential applications of non-ALS-inhibiting herbicides with the rapid canopy closure of close-drill soybeans for very good waterhemp control in no-tillage systems," he says.

"Farmers will greatly benefit from knowing which species of weeds are established in their fields and using this information when they plan crop rotations and farming operations. They should use a variety of farming techniques to control weeds. As farming systems change, weeds and weed populations change. So farmers can no longer expect one-shot solutions to weed problems."

Source: Agricultural Research Magazine.

division in which the cells multiply and the plant grows. Some herbicides prevent the formation of new cells by blocking cell division. If the cells cannot divide, growth cannot occur, and used up cells cannot be replaced. This in turn brings about the death of the plant. Because the herbicides prevent mitosis, they are referred to as mitotic poisons.

Blocking of Pigment Formation. Pigments are substances that impart color. The yellow colors in plants, known as **carotenoids** and the green colors, known as chlorophyll, are located in the chloroplasts of the cells (see Chapter 3). Both of these pigments absorb light during photosynthesis, and the carotenoids seem to function to protect the chlorophyll pigments by absorbing excess light that would be harmful. Some herbicides prevent the formation of carotenoids. Without the protection from excess light, the chlorophyll is broken down by the light, and the process of photosynthesis is stopped. Without the ability to produce energy, the plant soon dies.

Growth Regulation. Remember from Chapter 8 that plant hormones called auxins aid in the control of plant growth. These substances are produced naturally by the plant in very small quantities and are controlled by the plant. If large quantities of auxins are artificially applied to the plant, the actions of the auxins cannot be controlled by the plant. The result is uncontrolled enlargement and growth of plant cells, which blocks the tissues that transport food and water. Without this movement, the plant dies.

Herbicide Selectivity

Have you ever wondered how a chemical herbicide knows which plant to kill? If weeds are to be removed from a crop, the chemical must do little or no harm to the crop. It would do little good to kill the weeds if the crop plants were also killed!

Some herbicides are called nonselective or knockdown herbicides. These chemicals kill all plants they are applied to and are used where all vegetation is to be killed. Others are **selective herbicides** and kill only certain types of plants (Figure 13–18). It sometimes seems almost magical that a crop that is full of weeds can be sprayed, and only the weeds are killed. This selectivity is brought about by several methods.

For the herbicide to be effective, the chemical must penetrate into the plant. This can happen through contact with the leaves, stems, or through the roots. Contact herbicides are

Figure 13–18 *Selective herbicides kill only certain plants. Courtesy of William K. Vencill, The University of Georgia.*

sprayed over the top of the weeds after they have germinat-ed. This usually means that the crop plants are sprayed as well. The crop plant may have a thick cuticle or coating on the surface of the leaves or stems that prevents the absorp-tion of the herbicides into the plant, but the weeds may be easily penetrated by the chemical. Leaves may have a waxy material or be covered with plant hairs that may not allow the herbicide to wet the surface. Plants without these features allow the herbicide to contact and adhere to the leaf or stem surface, where the chemical is absorbed into the plant. Young weeds may be sprayed over the top by a directed spray that only contacts the woody stem of the crop plant (Figure 13–19). Physiological differences between younger and older

Figure 13–19 *A directed spray may keep the herbicide from the leaves of the crop plant. Courtesy of William K. Vencill, The University of Georgia.*

plants play an important part in the selective action of herbi-cides. Younger plants are more susceptible to herbicides than older plants because they are smaller and more tender. In addition, structural differences between monocots and dicots aid in selectivity. The two groups of plants have differences in the makeup of the stems, leaves, and vascular system (the system that transports food and water).

Chemicals are sometimes incorporated into the soil before the crop is planted (called a **pre-emergent**) or are sprayed over the soil surface after planting (called a post-emergent). These herbicides are taken into the system of the plants and are called systemic herbicides. Many weeds germinate in the top half inch of soil. Crops planted to a depth of more than an inch will have roots that grow well below the herbicide. If the prop-er amount of herbicide is applied to the top of the soil, the roots of the crop plant will be out of reach of the herbicide.

Different plants also have different chemical makeups and different chemical reactions. A chemical herbicide entering one plant may be altered by an enzyme within the plant that makes the chemical ineffective. A weed plant may not have this enzyme that neutralizes the chemical, and the weed will be killed. This phenomena allows for exciting potential in genetic engineering (see Chapter 5). If crop plants could be developed, which produce enzymes that neutralize herbicides, weeds could be more efficiently controlled. The key to development of such plants is the ability of scientists to splice in genes that carry the trait for production of the enzymes.

There are several disadvantages to using chemicals to control weeds. Because an enormous amount of time and money goes into the research and development of chemical herbicides, they are expensive. From the time a new herbicide is developed to the time it is released for use may be years. The chemical must undergo many tests and trials to ensure its safety. It must also have little impact on the environment (Figure 13–20). Chemical companies have to price new chemicals to offset these costs of development.

In addition, the cost of spray equipment and of operating and maintaining the equipment can be substantial. Herbicides can also harm the environment. Although modern chemical pesticides are safer to use than those of the past, some risk is incurred. If care is not taken during application, the wind may carry the herbicide to desirable plants that are susceptible to the chemical. It is possible that fields sprayed with a particular herbicide for a specific crop may not be used the following year with a different crop. These chemicals are said to be residual herbicides. Those with long-term residual effects have been banned by the federal government.

Figure 13–20 *Herbicides must not be harmful to wildlife or any part of the environment.* Courtesy of James Rathert, Missouri Conservation Department.

Biological Control of Weeds

The use of biological agents is another method that can be used in the efforts to control weeds. Biological agents include vertebrates, insects, and disease organisms. Remember that most of the serious weeds in this country were brought in from other countries. This means that the introduced weeds may not have natural enemies in this country. The location and introduction of organisms that feed on particular weeds can be an alternative means of controlling them.

Animals have been used to control weeds for many years (Figure 13–21). For example, goats may be kept in areas where sheep or cattle graze. The goats eat the coarser brushlike

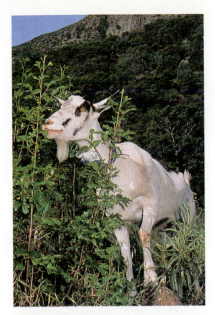

Figure 13–21 *Animals have been used to control unwanted plants.* Courtesy of University of California.

Figure 13–22 *Insects may be used to destroy weeds if they do not also feed on the crop.* Courtesy of USDA-ARS.

plants and make more room for the grasses eaten by other animals. Geese have been used in cotton, mint, and strawberry fields to help control grass. The geese like the tender shoots of grass but will not eat the young crop plants. As mentioned in Chapter 16, some species of fish are important factors in the control of aquatic weeds in fish ponds and lakes.

Insects are another means of destroying weeds (Figure 13–22). Certain species of insects will eat only certain plants. If the insects like a particular weed and not the crop plants, they can be used to help keep the weeds under control. For example, the spread of tansy ragwort has been slowed in California by the introduction of the cinnabar moth that eats the leaves of the plant. In Hawaii a woody weed know as lantana has been successfully controlled by the introduction of several species of insects that eat the plants. One species eats the flowers and seeds, another eats the leaves, while yet another girdles the trunk of the bush.

Disease organisms can also be used to control weeds. Plants get sick the same as animals do (see Chapter 12). These organisms include bacteria, viruses, and fungi. This area is perhaps the most promising method of biological control, provided disease agents can be isolated that affect only the weed species. In fact, several products known as mycoherbicides are currently on the market for commercial use. These compounds are composed of strains of fungi that attack specific weed pests.

In the near future, genetic engineering may provide us with pathogens that attack only certain species of weeds. This method would be efficient and sound environmentally.

The use of **biological control** has disadvantages. It is relatively slow. By the time insects have consumed a weed plant, the weed may have done considerable damage to the crop. The introduction of insects, animals, or diseases can be expensive. All these agents have to be bought and cared for until they are used on the weeds. This adds to the cost of production. Producers also have to be extremely careful that the agents that are introduced do not themselves become pests. If these agents attack crop plants, problems can occur.

Cultural Methods

Several management operations can reduce weed problems. Producers should be sure that they are not responsible for spreading weeds. Equipment, especially cultivators and plows, should be cleaned of all debris before moving to another field.

Figure 13–23 *Equipment can scatter weeds.* Courtesy of James Strawser, The University of Georgia.

Figure 13–24 *Plastic mulches can be used to control weeds.* Courtesy of Progressive Farmer.

One way weeds are spread is by rhizomes that cling to the bottom of implements and are moved from field to field (Figure 13–23).

The rotation of crops can also be a huge benefit. Some weeds are more difficult to control with certain crops and less difficult with others. For example, weeds that are difficult to control in corn may be relatively easy to control in soybeans. By alternatively planting corn and soybeans different years, the life cycle of weeds (as well as insect life cycles) can be disrupted.

A relatively new method is the use of black plastic mulch (Figure 13–24). Mulch has been used for many years to control weeds. Traditionally, mulch has been organic materials such as straw and sawdust that are put around crop plants to prevent the growth of weeds. It has the added benefit of preserving moisture. The problem with these mulches is that they are time-consuming to apply and leave a residue after the crop is harvested. Modern mulching involves the use of black sheet plastic that is rolled down the rows and secured on the edges by soil. Holes are made at the proper intervals and plants are set out or planted. Very few weeds can grow through the plastic, and almost total weed control is achieved.

The use of cultural techniques, mechanical methods, herbicides, and biological control in a coordinated effort is called integrated pest management (**IPM**). This method is probably the closest to ideal. By using the best of all these methods, the producer can control weeds more effectively and efficiently and at the same time protect the environment.

SUMMARY

For as long as humans have been cultivating crops they have battled weeds. Although modern science has given us efficient means of controlling weeds, the war against weeds will probably never end. In the future, scientists will find even better ways of controlling weeds that will be friendlier to the environment.

CHAPTER 13 REVIEW

Student Learning Activities

1. Identify and collect 10 different weeds from fields in your areas. Find out what measures are used to control the weed. Develop a plan to rid the field of the weed pests.

2. Identify and collect 10 different weed seeds. Analyze how the seeds are transported. List some ways to prevent their spread.

Define the Following Terms

1. weeds
2. noxious weeds
3. dormancy
4. herbicides
5. carotenoids
6. selective herbicides
7. pre-emergent
8. biological control
9. IPM

True/False

1. If a cotton plant is growing in the middle of a cornfield, it cannot be considered a weed because of its economic value.

2. Weeds can harm crops by serving as a home for diseases and insects.

3. Weeds can be classified as noxious or obnoxious.

4. A weed may be considered noxious in one state but not another.

5. Weeds can be classified according to their life span.

6. Although weeds reproduce very efficiently, their seeds do not remain viable for long periods of time.

7. Some weed seeds have special structures that allow them to be carried over great distances.

8. Most of the weeds that afflict crops in the United States are native plants.

9. Chemicals are one of the most efficient means of weed control.

10. Some herbicides prevent cell division in weed plants.

11. Although certain plant pigments are necessary for photosynthesis, the plant can survive due to secondary food storage within the cell structure.

12. For a herbicide to be effective, the chemical must penetrate into the plant.

13. Herbicides affect young and old plants in exactly the same way due to the genetic makeup of the plants.

14. Disease organisms are used as a means of biological control of weeds.

Fill in the Blank

1. Weeds can be classified as _____ or _____. _____ weeds are the most difficult to control and create the biggest problems.

2. Weeds may be classified according to life span. _____ live one year; _____ live two years; _____ live more than two years.

3. The period that a seed goes through until the germination process is triggered is known as _____.

4. _____ and _____ are examples of weeds that have been imported into the United States and have become very difficult to control.

5. The invention of the compound _____ was the beginning of the modern herbicide.

6. The yellow colors in plants are called _____; the green colors are called _____. They are both located in the _____ of the cells.

7. Some herbicides are called _____ or _____. These chemicals are used when all vegetation is to be killed.

8. Biological agents used to control weeds include _____, _____, and _____.

9. Compounds called _____, which are composed of particular strains of fungi, are used to attack specific weed pests.

Discussion

1. Discuss the ways in which weeds damage agricultural plants and animals.
2. List and explain the characteristics of weeds that make them so difficult to control.
3. List the alternatives available for the control of weeds.
4. List the different ways chemicals control weeds.
5. List the pros and cons of chemical control.
6. What is the difference between selective and nonselective herbicides?
7. List the ways weeds can be controlled biologically.
8. What are some cultural methods that can be used to control weeds?

AGRICULTURAL ENTOMOLOGY

KEY TERMS

entomology
siphoning insects
exoskeleton
metamorphosis
larval stage
pupal stage
phytophagous insects
parasitism
DDT
integrated pest
management
pheromones

STUDENT OBJECTIVES

After studying this chapter, you should be able to:

✦ Discuss the relative size of the world's insect population.

✦ Cite examples of the impact insects have had on the history of humans.

✦ Describe how insects are beneficial to agriculture.

✦ Analyze the scientific classification of insects.

✦ List the insect orders that are harmful to agriculture.

✦ Describe the distinguishing characteristics of insects.

✦ Compare the different types of insect mouth parts.

✦ Compare the different types of insect metamorphoses.

✦ Describe the damage to crops and animals caused by insects.

✦ Evaluate the different methods of insect control.

✦ Describe the benefits of chemical pesticides.

✦ Describe problems encountered with chemical pesticides.

✦ Analyze the concept of integrated pest management.

One of the most fascinating aspects of the science of agriculture is **entomology**, the study of insects. Insects are all around us and have always been a factor in the way humans live. Scientists known as entomologists spend their lives studying these organisms (Figure 14–1). It is hard to imagine any other group of animals that have received so

Figure 14–1 *Scientists who study insects are known as entomologists.* Courtesy of James Strawser, The University of Georgia.

Egyptian Scarab Beetle

Figure 14–2 *The ancient Egyptians considered the scarab beetle to be sacred.*

Figure 14–3 *The anopheles mosquito held up the digging of the Panama Canal. It transmitted diseases to humans.* Courtesy of Beverly Sparks, The University of Georgia.

much study! Given the vast number of these organisms, it is no wonder they are intensely studied. More than 750,000 species of insects have been identified, and scientists estimate that another 2,000,000 species have not been described and classified. Estimates are that insects are so numerous that their combined weight would be 14 times as great as the weight of all the humans in this country. Yet with all the insects in the world, only about 10,000 of the named species are harmful to humans, and only about 500 are considered to be major pests in this country. However, this relatively small number of species are a considerable force in agriculture.

As far back as recorded history goes, people have written about the impact of insects on their lives. Many paintings and writings of the ancient Egyptians featured the scarab beetle as a sacred creature (Figure 14–2). In other ancient writings, swarms of insects have been referred to as plagues that have devastated crops and caused famine.

Insects have had quite an impact on history through both their destructive and beneficial characteristics. Most of the deadly disease epidemics of the past have been caused by disease organisms carried by insects (Figure 14–3). For example, the bubonic plague epidemic that almost wiped out the population of Europe in the 14th century was carried by fleas that infested rodents. Huge projects like the digging of the Panama Canal have been hindered by insects. The canal was first begun by the French in the 1880s, but the project was abandoned because of thousands of worker deaths caused by malaria and yellow fever. It was not until around the turn of the 20th century that scientists discovered that these illnesses were carried by mosquitoes. If it were not for these insects, the canal would have been finished much earlier and at a

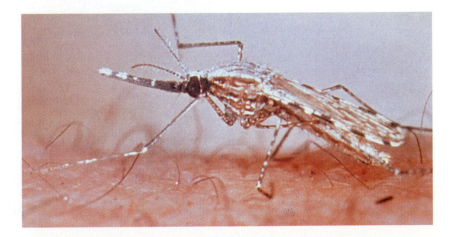

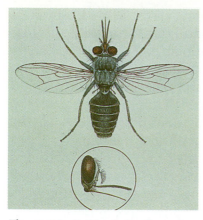

Figure 14–4 *The tsetse fly makes some parts of the world difficult to live in. Courtesy of Beverly Sparks, The University of Georgia.*

substantially lower cost in human life. For centuries, many places on Earth were uninhabitable because of insects, such as mosquitoes and tsetse flies that carry deadly diseases (Figure 14–4). The insects were so numerous that people living in the infected regions were almost certain to contract disease.

The discovery of America was indirectly tied to the larvae of a moth (*Bombyx mori*). The larvae of this moth feeds on the leaves of particular types of mulberry trees. They spin cocoons of long, smooth fibers created from body secretions. Humans use these fibers to create a cloth called silk. During the Middle Ages, this cloth was in such a demand that explorers searched for new and better trade routes to the eastern countries that made silk. Columbus was looking for a passage way to these countries when he ran into the New World.

In the town of Enterprise, Alabama, a statue to the honor of an insect sits in the center of the town (Figure 14–5). At the turn of the 20th century, the economy of the surrounding area depended on cotton production. The invasion of the boll weevil so devastated the crops that cotton was no longer profitable. A researcher named George Washington Carver developed

Figure 14–5 *In Enterprise, Alabama, a monument was erected to the boll weevil. Courtesy of Enterprise, Alabama, Department of Tourism.*

numerous uses for other crops that would grow in the area. One such crop was peanuts. Through the work of Carver (such as the development of peanut butter), peanuts became the main crop and turned out to be much more profitable than cotton. If the boll weevil had not intervened, the people around Enterprise might still be growing cotton instead of peanuts!

Many cultures in the world use insects for food. Termite larvae are considered to be delicious by some people who live in the tropics. Grub worms and other larvae have been eaten by humans for thousands of years. You may feel that eating insects is repulsive. However, remember that one of our common foods is a product of insects (Figure 14–6). That food is honey, which bees process from the nectar of flowers.

Figure 14–6 *The honey produced by bees has been a food for humans for thousands of years.*
Courtesy of James Strawser, The University of Georgia.

Insects are very much a part of our modern lives. They live in almost all of the land regions of the world, and they affect our lives every day whether we are aware of it or not. They invade our homes, spoil our food, destroy our belongings, and cause us discomfort. Yet they also pollinate the flowers of crops we grow for food and fiber (Figure 14–7). Without this function, our food supply would be far more limited and much more expensive.

Figure 14–7 *Insects pollinate crops as they gather nectar from flowers.* Courtesy of USDA-ARS, K-3652-12.

THE SCIENTIFIC CLASSIFICATION OF INSECTS

To identify and describe all of the millions of types of organisms in the world, scientists have devised a system of classification. All living organisms are organized into broad general groups having similar characteristics. These groups are then further divided into smaller groups with similar characteristics, and so on until very specific types are identified. The following is a list of the different levels of the classification system:

Kingdom
   Phyla
      Class
         Order
            Family
               Genus
                  Species

The highest level of classification is kingdom, and the most specific level is species. Organisms are usually identified by the genus and the species. This system is called a binomial nomenclature. For example, the honeybee is classified as *Apis mellifera*, with *Apis* being the genus and *mellifera* being the species.

Scientifically, insects are classified in the Kingdom Animallia, phylum Arthropoda, and class Insecta. The class is divided into more than 30 orders, but most harmful insects can be found in less than 10 orders. Six orders of insects damage plants (Figure 14–8). These are

1. Orthoptera, which includes the grasshoppers and locusts;

2. Hemiptera, which includes the true "bugs" such as leaf hoppers and plant bugs;

3. Lepidoptera, which includes the moths and butterflies;

4. Homoptera, which includes the aphids;

5. Thysanoptera, which includes thrips; and

6. Coleoptera, the largest of the insect orders, which includes the beetles.

Insects that feed on plants are grouped by how they feed. This classification is determined by how the mouth parts of the insects are structured and how food is taken into their mouths (Figure 14–9). Chewing insects tear off bits of plants

Figure 14–8 *There are six insect orders that damage plants. (A)* Orthoptera *(James Strawser, The University of Georgia.) (B)* Hemiptera *(James Strawser, The University of Georgia.) (C)* Lepidoptera *(Beverly Sparks, The University of Georgia.) (D)* Homoptera *(Beverly Sparks, The University of Georgia.) (E)* Thysanoptera *(Beverly Sparks, The University of Georgia.) (F)* Coleoptera *(James Strawser, The University of Georgia.)*

Egg Nymph Adult

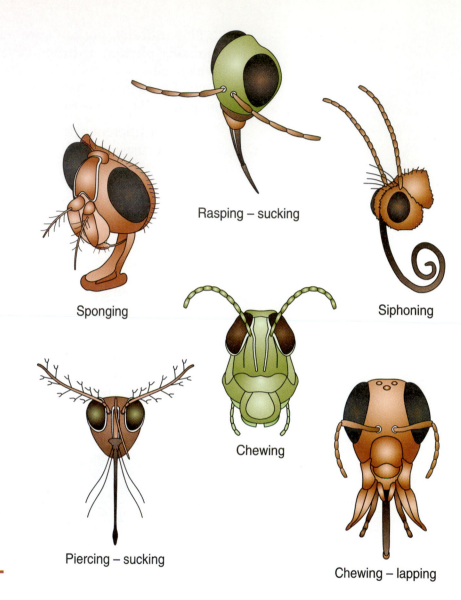

Rasping – sucking

Sponging

Siphoning

Chewing

Piercing – sucking

Chewing – lapping

Figure 14–9 *Insects can be grouped by the type of mouth parts.*

using mandibles (jaws) that are strong enough to bite off plant parts and grind the material up by chewing. These insects include most of the beetles and grasshoppers.

Another group of insects has structures called chewing and lapping mouths that are used to lap up liquid. Examples of this type of insect are bumblebees and honeybees. They have long hair-covered tonguelike projections that reach into blossoms and lap up nectar.

Rasping and sucking insects have mouth parts that scrape off cells and particles from a plant surface and suck the pieces into their mouths. Thrips feed on plants in this manner and belong to this group.

Some insects have mouth parts that are shaped like a drinking straw. This group, called **siphoning insects**, extends this mouth part, called a proboscis, to suck up nectar from plants. The moths and butterflies belong to this group.

Some of the most damaging insects use a proboscis that pierces a hole in the outer layer of a plant surface and sucks the sap from the plant. These mouth parts are called piercing and sucking. Scale insects and aphids belong to this group.

Flies use a mouth structure that serves as a sponge to absorb liquid nourishment. Nonbiting flies belong to this group, the sponging insects.

CHARACTERISTICS OF INSECTS

Insects are unique in many ways. Unlike many other animals, insects have no internal skeleton and rely on a hard outer coating called an **exoskeleton** that protects their inner organs and gives their bodies support. The exoskeleton is a noncellular layer composed of a material called chitin, which is secreted by an inner layer of cells. The chitin is coated with a waxy layer that helps the insect retain the fluid content of the body. The exoskeleton is tubular in shape and is very strong for its size.

All adult insects have three body parts: the head, the thorax, and the abdomen (Figure 14–10). The head has a pair of compound eyes and two sensory appendages called antennae. The thorax is divided into three segments from which are attached three pairs of legs. Some species of insects have wings, and these are attached to the last two segments of the thorax. The abdomen is attached to the thorax and contains several segments.

Insects are very diverse in size. They range in size from almost microscopic aphids and wasps to huge tropical beetles that grow to several inches in length. They come in almost every imaginable color from solid black to white to vivid red or blue. Some butterflies are among the most beautiful living things in the world. Other insects like the harlequin bug and the lady beetle have very attractive colors and patterns (Figure 14–11).

Life Cycles

Insects are different from other animals in that their physical form changes at different stages of their lives (Figure 14–12).

Parts of an Adult Insect

Figure 14–10 *All adult insects have three distinct body parts.*

Figure 14–11 Insects can have very colorful patterns. *Courtesy of USDA-ARS, K-4179-19.*

This process is called **metamorphosis.** Not all insects go through a metamorphosis; some, like the silverfish, merely go through a process of molting in which the insect outgrows the outer covering, sheds it, and grows a new one.

However, most insects go through some form of metamorphosis. In a complete metamorphosis, insects go through

Metamorphosis

Incomplete Metamorphosis   **Complete Metamorphosis**

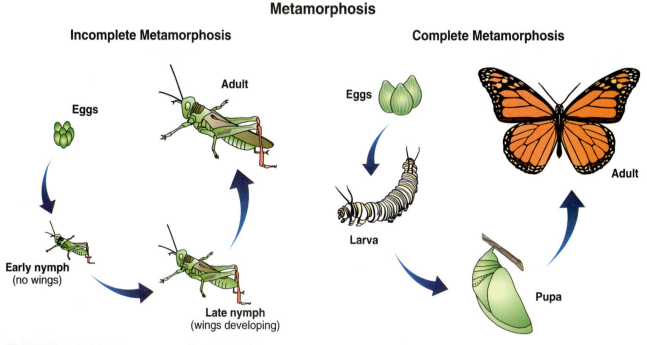

Figure 14–12 The life of an insect is divided into different stages.

Figure 14–13 *Larva are voracious eaters that can do a lot of damage to plants.* Courtesy of USDA Office of Information.

Figure 14–14 *A pupa is the intermediate stage between the larva and the adult.* Courtesy of USDA.

Figure 14–15 *Insects that undergo a gradual metamorphosis hatch from an egg into a nymph.* Courtesy of Beverly Sparks, The University of Georgia.

three stages from the time they hatch until they are mature adults capable of reproducing. At each one of these stages the insect looks completely different than the other three stages. When the young insect hatches, it is usually in the **larval stage** (Figure 14–13). This means that the young insect has a soft tubular body and looks very much like a worm. Larva usually are voracious eaters that can do a lot of damage to plants or to a host animal. A good example is the cotton boll worm that feeds on the bolls of cotton. The worm is actually the larva of a moth.

When the larva matures, it passes into the **pupal stage**, which is usually a relatively dormant stage (Figure 14–14). A pupa is the intermediate stage between the larva and the adult. During this stage the body tissues of the young insect convert from a larva to the adult. The last stage is that of adult. In this stage the insect lays eggs, and the cycle begins over again.

Some insects that spend at least part of their lives in water undergo an incomplete metamorphosis. In this process the insects go from an egg to a form called a naiad that looks and functions differently from the adult. When the growth process of the naiad is complete, it changes to a winged adult. Examples of this type of insect are damsel flies and dragonflies. Insects that undergo a gradual metamorphosis hatch from the egg into a nymph that looks very much like the adult (Figure 14–15). The nymph has different feeding habits and may live in a different place than the adult. From the nymph stage, the insect goes to the adult stage. Grasshoppers and cockroaches are examples of insects that undergo gradual metamorphosis.

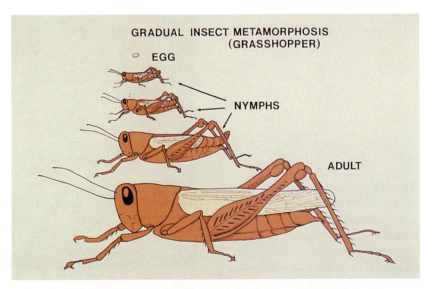

INSECT PESTS

The production of food and fiber for the use of people is a multiphase process. From planting crops to cultivation, harvesting, storing, and processing, insects play an important part. Even the raising of all types of animals involves combating insect pests.

When humans first began to produce their own food, the battle against insects took on new meaning. In the earliest era of agriculture, people accepted the fact that insects would devour a certain percentage of the crops and parasitic insects would damage their livestock. To compensate, larger and larger crops were planted to produce enough food above the amount destroyed by insects. As agricultural methods improved, ways were explored to prevent insects from damaging the crops or animals being grown. Only relatively recently have people been able to make progress in the battle against insects. However, millions of dollars are spent each year on control measures and insect research.

Insect Pests of Crops

Insect pests have to be dealt with from the time crops are planted to the time the food is consumed. As late as the 1950s, estimates were that insects destroyed around one-fourth of all agricultural produce in this country (Figure 14–16). Today, that figure is less than 10 percent due to

Figure 14–16 *As late as the 1950s, insects destroyed around one-fourth of all agricultural produce. Courtesy of James Strawser, The University of Georgia.*

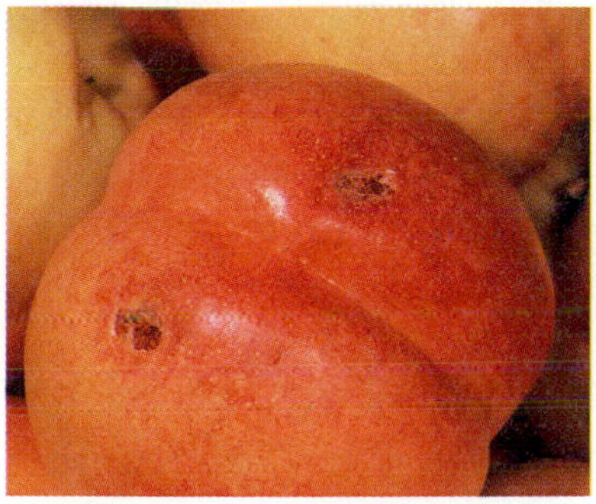

research that has helped us understand insects and how to control them. Even at this relatively small percent, the destruction caused by insects is significant. The monetary loss insects cause to crops and stored produce is estimated to be as high as $5 billion annually in the United States alone.

Insects damage crops in several ways. First, they damage the plant by feeding on it. Insects that feed on plants are called **phytophagous insects**. "Phyto" means plant, and "phagous" means eating. Some insects feed on the roots of plants; others may feed on the stems; while others eat the leaves. Wherever the plant is eaten, the plant is damaged in terms of growth and productivity.

Insects also damage the fruit of plants. Whether the crop is corn, soybeans, or peaches, the fruit is most often the reason plants are grown. Fruit that is eaten by insects is damaged even though only a small portion is consumed. Damaged fruit is made less saleable and less desirable for human consumption (Figure 14–17). Peaches, apples, pears, and grapes damaged by insects are almost impossible to market for human consumption. No one likes to eat an apple that contains or has been damaged by insect larvae.

Insects also attack produce such as grain that has been harvested and stored. Unless control measures are taken, insects can destroy a large portion of harvested grain and other produce.

Besides all the damage caused by the feeding habits of insects, another serious problem is caused by these organisms. As the insects go from plant to plant, they take with them disease organisms that may seriously damage or destroy plants. The disease problem in plants is covered in Chapter 12, but keep in mind that a major factor in the spread of disease is insects.

Figure 14–17 Fruit damaged by insects is less saleable and less desirable for human consumption. Courtesy of University of California.

Insect Pests of Animals

Some insects live off other animals in a relationship called **parasitism**. Parasitism is a relationship that is beneficial to one animal and harmful to the other. Almost all animals, including humans, are susceptible to parasites. The animal that lives off the other animal is called a parasite, and the animal that the parasite lives on or in is called the host. Parasites may live outside the host's body (external) or live inside the body (internal). Parasites cause almost a billion dollars worth of damage to agricultural animals each year (Figure 14–18).

BIO BRIEF

Trap Crops Prove Irresistible to Diamondbacks

Cabbage, broccoli, collards, kale, and other cole crops are an all-you-can-eat salad bar for diamondback moths, a pest names for the diamond-shaped markings embellishing its wings. Moth larvae, which chew on plant leaves, take a big bite out of cabbage and other crops worldwide, costing billions of dollars in control costs and losses.

Pesticide spraying can be costly, ranging from about $10 to $21 an acre for each application—depending on which pesticides are used—and typically costing growers $80 to $168 per acre, or more, each season to produce a crop. To the farmer's dismay, diamondback moths are becoming resistant to almost everything, including *Bacillus thuringiensis* (Bt)-based insecticides that are widely used to kill certain pests while preserving beneficial insects.

Now entomologist Everett R. Mitchell is taking another approach to spoiling the moth's meal. He says giving the pest a heaping serving of another vegetable—collard greens—spoils its appetite for cabbage. The moths can't resist the collards when planted completely around the edge of cabbage fields, a strategy called *trap cropping*.

"Invading diamondback moths stop and deposit their eggs on the collards, rather than on adjacent cabbage plants," says Mitchell. "Diamondback populations continue to recycle in collards as long as plants remain green and continue to grow." Mitchell heads the Insect Behavior and Biocontrol Research Unit, which is part of ARS's Center for Medical, Agricultural, and Veterinary Entomology in Gainsville, Florida.

Mitchell recently conducted experiments on nearby farms in northeast Florida. These experiment showed that the moths prefer to feed on highly fertilized collard plants. He tested this approach for more than two years. In all cases, he says, minimal cabbage damage was sustained from diamondback moth larvae. The quantity and quality of cabbage produced equaled that from conventionally sprayed fields. This simple, low-tech, cost-effective pest control method also

Parasites generally live out one or more phases of their life cycle at the expense of the agricultural animal. Because most parasites live off the blood of the host animal, a continual loss of blood occurs. If enough parasites are living off the blood of an animal, the blood supply may be greatly diminished. A condition known as anemia develops because not enough blood is left in the animal's body to provide body cells with oxygen and food nutrients. The animals are sluggish, feel poorly, and do not grow or perform as they should.

reduced pesticide use. "Cabbage fields surrounded by collards required 75 to 100 percent fewer sprays to control diamondback moths than fields treated conventionally with pesticides. That's a huge savings for farmers," notes Mitchell.

He also says *Diadegma insulare*, a naturally occurring parasitoid that attacks diamondbacks, builds in numbers in the collards and helps keep diamondback populations in check. The tiny *Diadegma insulare* wasp stings the larvae, preventing them from developing into adults and laying more eggs. When stung, a larva becomes sluggish and stops feeding within a few hours. The wasp doesn't attack other insects or humans.

"We established that there needs to be a threshold of 0.3 moth larva per plant before a farmer has to apply pesticides," Mitchell says. "We found that even though moth larval populations built up in collard planted around field margins, populations in cabbage generally remained well below the threshold."

—By Tara Weaver,
Agricultural Research Service Information Staff.

Source: Agricultural Research Magazine.

Trap crops can be effective in controlling insects such as the diamondback moth. Courtesy of USDA-ARS.

As a result of hosting parasites, animals that are in a weakened condition are more susceptible to disease. External parasites can also carry disease. Blood-sucking insects such as mosquitoes and fleas move from one animal to another. If blood from an infected animal is removed, disease organisms in the blood are also removed. When the insect carrying the infected blood moves to a healthy animal and pierces the skin, the disease organism is deposited in the healthy animal. As mentioned at the beginning of the chapter, disease spread

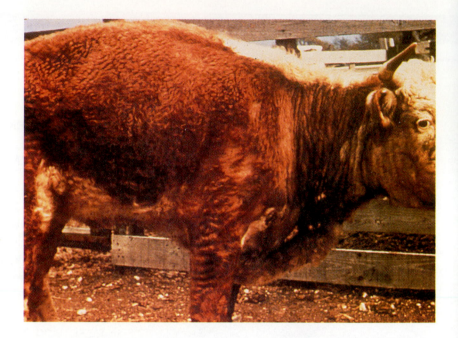

Figure 14–18 *Parasites cause almost a billion dollars worth of damage to agricultural animals each year.* Courtesy of Maxey Nolan, The University of Georgia.

by insects such as mosquitoes and fleas have caused an untold amount of death and suffering among humans and other animals.

CONTROL OF INSECT PESTS

Insects are hard to control for several reasons. One is that insects are so prolific. This means that they reproduce often and produce a large number of young. One insect female may lay millions of eggs during her brief life (Figure 14–19). Certain insects may have as many as 30 generations in a single year, with females laying thousands of eggs in each generation. Without some sort of control, the earth would soon be covered in insects.

Natural Control

To a certain degree, populations of insects are kept in check by nature. When left undisturbed, nature has always kept insects from becoming too dominant. One of the most effective ways of doing this is through predators that feed on insects. Many animals sustain their lives by eating large quantities of insects. Mammals such as anteaters and armadillos eat insects as the mainstay of their diets. Bats are particularly

Figure 14–19 *One female insect can lay millions of eggs during her life. Courtesy of James Strawser, The University of Georgia.*

adept at destroying insects. It is not unusual for certain types of bats to eat their weight in insects each day. Many birds are just as efficient as bats in consuming insects (Figure 14–20). Swifts, swallows, and flycatchers consume vast amounts of insects. Amphibians such as toads and frogs also eat a huge number of insects every day during the warm months. Look around a lamppost at night in the summer and you will generally find toads eating the insects that are attracted to the light. If it were not for insect-eating birds, amphibians, and mammals, agriculturists would have a much more difficult job in controlling insects.

Predatory insects destroy an unimaginable number of harmful insects. Without these predators, our lives would be much less comfortable. Some predatory insects eat the adults or larvae of other insects, and others eat the eggs. As a result, only a small percentage of the eggs laid by insects grow to be adults.

One of the world's most efficient hunters is the praying mantis (Figure 14–21). These insects are shaped like the branches of a plant and are green in color. This makes them blend in with vegetation and hide from their prey. Their large head is unique in that it can turn around and see insects behind them. This allows them to remain in one place and find insects to feed upon.

Lady beetles (sometimes called ladybugs) eat a large number of aphids and other pests that attack plants. Lacewing

Figure 14–20 *Many birds eat insects and are effective in controlling insects. Courtesy of Thomas Rosburg, Iowa State University.*

Figure 14–21 One of the world's most efficient hunters is the praying mantis. Here a praying mantis is feeding on a sulfur butterfly. *Courtesy of J.L. Castner, Florida.*

insects and damsel flies are also very efficient hunters that greatly aid in keeping the insect population under control.

Predatory insects are so good at destroying insect pests that people grow them for sale to gardeners to release on their crops (Figure 14–22). These beneficial insects are sent through the mail in special packages designed to keep them alive during shipment. Although these predators are useful in insect control, they are not entirely effective. To sustain a large population of predatory insects, the insects they feed on must be large in number. By the time this large population of harmful insects is built up, damage has already been done to the crops. In addition, while many of the insects will be eaten, enough will survive to damage the crop. It only takes one surviving insect to damage an apple enough to render it unacceptable to most consumers.

Figure 14–22 Lady beetles are raised commercially because of the large number of insects they eat. *Courtesy of James Strawser, The University of Georgia.*

Nature also helps keep insects under control by the weather. An unusually harsh winter can significantly lower the insect population of an area. Insects are generally dormant over the winter. Some may overwinter as eggs, some as pupa, and others may be adults. If the weather gets far below the normal temperature, freezing of the insects may occur. If a year is drier or wetter than normal, the life cycle of the insect may be sufficiently interrupted, and death may occur.

Disease is another way nature has of keeping insects in balance. They are susceptible to diseases like other animals, and large numbers of them may die during an outbreak of a disease. Overcrowding of the insects can contribute to disease outbreaks. The introduction of insect disease into a population of insects is a promising means of controlling them.

Chemical Control

People have used chemicals to control insect pests since ancient times. As far back 200 B.C., the Romans used an asphalt-like mixture called bitumen to control insects in vineyards. They also used extracts of a poisonous plant called hellebore to rid themselves of body lice. By the Middle Ages, the Chinese were using arsenic to control pests on crops.

During the 1800s, such pesticides as nicotine, copper sulfate, arsenic, and paris green (copper acetoarsenite) were widely used to control pests. These early pesticides offered some amount of control but were far from being completely effective.

From 1930 to 1950, modern pesticides were developed that offered the first truly effective weapons in the fight against insects. Perhaps the most important development was the substance dichlorodiphenyltrichloroethane. Popularly known as **DDT**, this chemical was discovered by a German scientist named Othmar Zeidler in 1874. In the 1930s, a Swiss scientist, Paul Muller, "rediscovered" the chemical and developed it as a powerful agent in controlling insects. He was awarded the 1939 Nobel prize in medicine for developing its effectiveness in controlling insects that carry some of the world's most dreaded diseases.

Some scientists say that DDT has saved more lives and alleviated more human suffering than any substance that has been invented or discovered. In the years since 1939, DDT has been used to spray areas in the tropics where mosquitoes carry such diseases as malaria, yellow fever, and typhus. As a result, malaria has been eradicated from more than 20 countries. The average life span of people in India rose by 15 years in a decade due mostly to the reduction in malaria-caused deaths. In our own country, cases of malaria numbered about 250,000 annually in the 1930s. That number has now been cut to around 10 cases per year.

As a pesticide used in crop protection, DDT probably has had no equal. The cotton industry of the South prospered as a result of the effectiveness of the chemical in controlling boll weevils and boll worms. In fact, almost all crops grown at one time used DDT as part of the effort to control insect pests. At

Figure 14–23 *At one time DDT was used widely as an insecticide. More than 100,000 tons were once used annually.* Courtesy of Progressive Farmer.

the peak of its production, more than 100,000 tons of DDT were manufactured and used annually (Figure 14–23).

Problems with Chemical Pesticides

In 1962, Rachel Carson published a book entitled *Silent Spring* in which she called attention to environmental problems caused by the use of chemicals and, in particular, DDT. In the book she visualized a time when songbirds were extinct due to the indiscriminate use of pesticides and other chemicals. The book was widely read and brought about widespread concern over the use of chemicals.

Although the book was controversial and criticized by some scientists, public attention was focused on the perceived problem. Studies indicated that the reproductive process of many bird species was interrupted due to deformed, thin-shelled, or infertile eggs. DDT was named as the main culprit. Also, a phenomena known as bioconcentration was pointed out. The gills of fish filter through a tremendous amount of water each day to remove oxygen. If the water contains a minute amount of pesticide such as DDT, the gills may also retain the chemical. Even though the water may contain only a very few parts per million (PPM), the continual filtering of the material by the fishes' gills causes a substantial buildup in the fish. Water birds and other predators (including humans) take in the chemical through the fish they eat (Figure 14–24). It should be noted

Figure 14–24 *Water birds take in chemical pesticide residues from the fish they eat. Courtesy of James Strawser, The University of Georgia.*

that after extensive research efforts, no evidence has yet been discovered that proves that humans have suffered ill effects from DDT residues. However, evidence was strong that the chemical was having an adverse effect on wildlife populations.

DDT belongs to a group of chemical pesticides called chlorinated hydrocarbons. Other pesticides in this chemical family, such as chlordane, heptachlor, and toxaphene, were developed to control insect pests in a wide variety of applications. Perhaps the worst characteristic of this chemical family was that they lasted so long in the environment. The chemicals can remain in the soil for many years, and after several years of application to crops, the residue of DDT and other chlorinated hydrocarbons in the soil collected and concentrated. With a concentration of the chemicals in the environment, they are passed up the food chain by animals that eat plants and in turn are eaten by other animals. These and other concerns brought about the ban of DDT in this country. However, chemical companies continue to manufacture DDT and other chlorinated hydrocarbons for use in other countries. Today, chlorinated hydrocarbons are rarely used in this country, and most have been banned.

Since the banning of DDT and similar pesticides, the federal government has passed laws and developed stringent regulations governing the release and use of new pesticides. A new pesticide takes about 10 years to gain approval (Figure 14–25). Many compounds can be used to kill insects, but they

Figure 14–25 *The development and testing of a new pesticide takes about 10 years. Courtesy of USDA-ARS, K-5562-04.*

have to meet certain requirements. An important consideration is that they must be safe to manufacture. Workers in the chemical plants must not suffer ill effects when pesticides are produced. The workers come in contact with the chemical and undergo safety measures to lessen exposure. Regulations require that the chemical must be safe to manufacture and that safety measures are used.

The chemical must not be harmful to the environment when used according to the manufacturer's directions. The Environmental Protection Agency (EPA) sets standards for the use of pesticides (Figure 14–26). As a rule, the pesticide is not

Figure 14–26 *The EPA monitors the manufacture and use of pesticides to ensure us that the environment is protected. Courtesy of James Rathert, Missouri Conservation Department.*

Figure 14–27 *All pesticides have to be labeled with a variety of information about usage. This pesticide can be used by the general public.* Courtesy of James Strawser, The University of Georgia.

allowed to persist in the environment. This means that it must break down rapidly and remain just long enough to control the pest.

Residues left on crops must not be harmful to humans or other animals that eat food from the crop. The government sets the maximum allowable level of residue for a food product. This level must prove to be of no harm to humans. More will be said on consumer concerns in Chapter 18.

All pesticides have to be registered with the EPA and must be labeled with a variety of information, including the use classification, which specifies who can apply the pesticide (Figure 14–27). Certain pesticides can be applied only by licensed applicators. This lessens the risk of indiscriminate use of the pesticide by people who do not understand the hazards of the chemical.

To replace the effective chemicals that have been banned, other pesticides have been developed that break down rapidly and pose less of a threat to the environment. Organo-phosphates, such as diazinon, parathion, and malathion, are quite effective in insect control and only last a short time in the soil. The carbamates, such as carbaryl (sold under the name, Sevin®), are used extensively and are not as harmful to the environment. In addition, the newer pesticides are less toxic to humans and other warm-blooded animals. Newer and safer chemicals are constantly being developed.

Integrated Pest Management

The most modern method of pest control is an integrated or multifaceted approach. Chemicals alone can be effective, but the potential damage to the environment cannot be overlooked. Combining chemical applications with nonchemical means of control lessens the need for high levels of chemical use.

The use of a variety of approaches in coordination with each other can be effective in pest control and in protecting the environment. Listed below are some of the tools used in integrated pest management.

Cultural methods. Insects attack a plant during certain phases of the plant's growth. Control measures can be used during the most vulnerable phase of the plant. Instead of spraying at a set interval, pesticides can be used only when needed. This requires close monitoring of the plants for insect buildup (Figure 14–28). This can be done by insect scouts who periodically check fields for insect damage. Also, plowing

Figure 14–28 *Close monitoring can determine when pesticides need to be applied. Courtesy of University of Illinois.*

under plant residues after crops are harvested can deny insects a convenient place to overwinter. The control of weeds that serve as hosts to insect pests can also deny the insects' food and an overwintering place.

Insect diseases and predators. Many insects and other animals feed on insects. Encouragement of the presence of birds and predator insects can lessen the population of insects. Also, as mentioned earlier, nature helps keep insects in control through diseases. The intentional introduction of disease agents into an insect population can be a helpful tool in lowering their numbers.

Pheromones. Insects communicate to a large degree through chemicals that are discharged from their bodies. These substances are known as **pheromones**. They attract insects of the opposite gender for mating. The male locates a female by detecting the chemical pheromone given off by her body. If a field is sprayed with this compound, the males have difficulty locating females, and the mating process is interrupted. Also, the pheromone can be used as an attractant to trap the males (Figure 14–29). An absence of males means that the reproductive process has been interrupted.

Figure 14–29 *Insects are trapped using pheromones. Courtesy of University of California.*

Release of sterile males. Another way to interrupt the reproductive process is the release of thousands of sterile male insects over the infected area. In a laboratory, male insects are treated with radiation or other treatments to render them infertile, but they are still able to mate. The wild females mate with the sterile males, and no offspring are produced. Over several years of saturating an area with sterile males, an insect pest can be eradicated. Several success stories have been accomplished using this method.

Insect-resistant plant varieties. A promising new technology is producing plants that resist insect damage. Insects feed on a plant because they like it. If plants could be produced that insects do not like and yet retain the qualities we grow them for, the insect control problem could be lessened. Through genetic engineering, plant species are being developed that resist insects (Figure 14–30). In the future this may be a key method to control insect pest damage.

Figure 14–30 *Insect resistance can be put into plants through genetic engineering. The plant on the right has been genetically altered and is more insect resistant.*
Courtesy of USDA-ARS, K-5258-1.

SUMMARY

Although insect pests represent one of the most challenging problems in agriculture, modern research has given us tools and techniques to help solve the problem and at the same time maintain a healthy environment. Also, we are assured of a nutritious, wholesome supply of food that is the envy of the world. As new problems arise, research will find new solutions to those problems.

CHAPTER 14 REVIEW

Student Learning Activities

1. Identify and collect 20 different insects that are economically important. This importance may be the result of damage done or insects eaten. Explain the relative importance of each and how the insect can be controlled or used.

2. Identify three harmful insects in your collection and diagram their life cycles, including feeding habits. Devise a way to interrupt the life cycle without using chemicals. Explain how the method could be made feasible for a producer to use. Share your ideas with others in the class.

Define the Following Terms

1. entomology
2. siphoning insects
3. exoskeleton
4. metamorphosis
5. larval stage
6. pupal stage
7. phytophagous insects
8. parasitism
9. DDT
10. integrated pest management
11. pheromones

True/False

1. Few of the deadly disease epidemics of the past have been caused by insect-borne organisms.

2. The Panama Canal project was delayed because of workers' deaths caused by malaria and yellow fever.

3. Insects that feed on plants are grouped by how they feed.

4. All insects go through metamorphosis.

5. More than 750,000 different species of insects have been identified.

6. The intentional introduction of disease agents into an insect population will only increase their numbers.

7. One way of interrupting the reproductive process is the release of sterile female insects over an infected area.

8. Less than 10 percent of all agriculture the U.S. produces is destroyed by insects.

9. Predatory insects are so good at destroying insect pests that they are raised and sold to gardeners to be released on their crops.

10. In Enterprise, Alabama, there is a statue honoring an insect.

Fill in the Blank

1. An organism that lives off another organism is called a _____.

2. _____ are particularly adept at destroying insects.

3. One of the world's most efficient hunters is the _____.

4. The most modern method of pest control is _____.

5. Scientifically, insects are divided into the Kingdom _____; phylum _____; and class _____.

6. Only about _____ species are considered to be major pests in this country.

7. The boll weevil devastated the _____ crop.

8. The _____ and _____ are used in many cultures for food.

9. One popular product that comes from insects is _____.

10. The _____ sets the standards for the use of pesticides.

Multiple Choice

1. The bubonic plague is carried by:
   A. bees.
   B. fleas.
   C. roaches.

2. An example of an insect with a lapping mouth is a
   A. bee.
   B. ant.
   C. beetle.

3. All adult insects have these three body parts:
   A. head, legs, and tail.
   B. legs, tail, and brain.
   C. head, thorax, and abdomen.

4. The worst characteristic of the chlorinated hydrocarbons is that they last so long in the
   A. plant.
   B. environment.
   C. insect.

5. George Washington Carver developed
   A. the cotton gin.
   B. peanut butter.
   C. the peanut crop.

6. The approval process for a pesticide takes about
   A. 10 years.
   B. 5 years.
   C. 2 years.

7. An insect that undergoes incomplete metamorphosis is a/an
   A. ant.
   B. bee.
   C. dragonfly.

8. Mosquitoes can carry illnesses such as
   A. malaria and yellow fever.
   B. cancer and malaria.
   C. hypertension and scarlet fever.

9. Siphoning insects have mouth parts that are shaped like a
   A. bottle.
   B. straw.
   C. spoon.

10. The thorax is divided into three segments from which are attached
    A. fingers.
    B. ears.
    C. legs.

Discussion

1. List several ways that insects are helpful.
2. Explain the classification system.
3. What are the six orders of insects that damage plants?
4. In what ways do insects do damage?

5. Why are insects hard to control?
6. How does nature keep insects under control?
7. Explain how genetic engineering could benefit plants.
8. What are the distinguishing characteristics of insects?
9. What were the problems related to DDT?
10. Explain current laws concerning pesticides.

THE SCIENCE OF FORESTRY

STUDENT OBJECTIVES

After studying this chapter, you should be able to:

✦ Discuss the extent of the U.S. forestry industry.

✦ Summarize the evolution of the nation's forestry industry.

✦ Debate the issue of harvesting old growth timber.

✦ Explain the concept of natural succession.

✦ Discuss the forest as an ecosystem.

✦ Discuss the importance of the forests to humans.

✦ Explain modern tree harvesting techniques.

✦ Discuss processing methods for different wood fiber products.

One of our nation's largest industries is forestry. This gigantic industry stretches from coast to coast and encompasses almost 500 million acres of trees (Figure 15–1). From these trees come thousands of different types of products that are manufactured using wood fiber. More than 1.5 million people are employed in the management, harvesting, processing, and marketing of these products.

Timber harvesting is often thought of as denuding our forests until there will soon be little left. This is erroneous. Because of the conservation efforts of the forestry industry and the U.S. Forest Service, today there are more trees in this country than there were 100 years ago. According to the American Forest and Paper Association, each year, close to 2 billion trees are planted in the United States (Figure 15–2). This represents more than six new trees a year for every American. Through the replanting of the forests, new plant

Figure 15–1 *The forestry industry encompasses almost 500 million acres of trees. Courtesy of Georgia Pacific.*

Figure 15–2 *Each year close to two billion trees are planted in the United States. Courtesy of Georgia Pacific.*

growth flourishes. Each year more than 27 percent more trees are planted than are harvested. Areas that were once used for row cropping are now used for growing trees. Hilly land not suited for farming with modern machinery is now planted in pine trees and is once again productive (Figure 15–3).

An added benefit to the planting and cultivation of trees is the reintroduction of wildlife. For example, the white-tailed deer has made a tremendous comeback in recent years (Figure 15–4). Some sources say that more deer are in this country now than before the first European settlers arrived. A few decades ago, the wild turkey was on the verge of extinction due to the loss of its habitat and overhunting. Today more than 4 million wild turkeys are living and thriving in our forests.

Figure 15–3 *Hilly land not suitable for farming can be productive by growing trees. Courtesy of Georgia Pacific.*

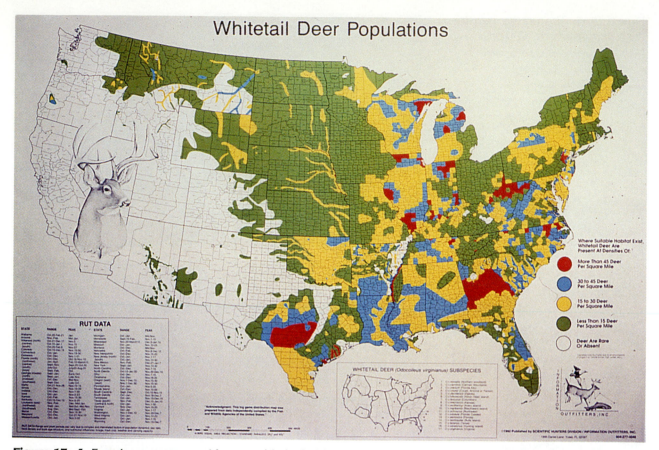

Figure 15–4 *Forestry management has provided a habitat that has dramatically increased white-tailed deer populations. Courtesy of Information Outfitters, Inc.*

One of the greatest natural resources this country has enjoyed has been its forests. When Europeans first arrived in America, a large part of the continent was covered with forests. Although these trees offered building material and an environment for game, the trees were often viewed as pests that stood in the way of cultivating crops. Often they were killed and burned to make room for crops (Figure 15–5). The forest seemed inexhaustible because mile after mile of the wilderness was covered with trees. The trees were cut and sawed into lumber to build homes, shops, and factories as the nation grew and moved westward. Even in the early days of our history, American timber was used to build ships that sailed all around the world. Lumber was cheap because of the large supply.

After the turn of the 20th century, the large trees of the virgin forest began to run out in areas where they were easily

Figure 15–5 *Pioneers cleared the forest to make way for the growing of crops.*

harvested. The mountainous areas and the swamps were then harvested. Even today, timber companies are cutting old growth timber in the Pacific Northwest (Figure 15–6). No finer construction timber exists than that taken from this area. Homes all over the world are made from lumber cut from trees in the northwestern United States. Many people object to cutting the huge old trees (Figure 15–7). A constant struggle seems to go on between preservationists and the logging industry. The conservation activists insist that as much of the old growth timber as possible must be preserved. The logging industry points out that more than four million acres of old growth timber are preserved in this country that, by law, can never be cut. They contend that the country's housing industry needs the material from the large trees in areas

Figure 15–6 *Homes all over the world are constructed of lumber from the Pacific Northwest. Courtesy of Georgia Pacific.*

where harvesting is permitted. They further argue that if logging is stopped, many jobs will be lost.

Another point of controversy is over wildlife **habitat**. A good example is the spotted owl that lives in the Pacific Northwest. Some biologists contended that the owl needed large tracts of old growth timber in which to breed and raise young. Logging operations were halted, and many people lost their jobs in the effort to preserve the owl. The issue was debated over whether or not preserving the owl was more important than providing jobs. It should be noted that later evidence showed that the owls could reproduce in second growth forests.

Figure 15–7 *Many people object to the harvesting of huge old growth timber. Courtesy of Jim Peterson, Evergreen Magazine.*

THE NATURAL FOREST

At one time, the entire wood industry was based on the natural growth of trees. Different types of forest grew in different regions of the country (Figure 15–8). Basically there are two classifications of timber, **hardwood** and **softwood**. A hardwood is a broadleaf **deciduous** (sheds its leaves in the winter) tree. A softwood is a cone-bearing tree or conifer that is evergreen. Actually, the classification has little to do with the hardness of the wood. Some hardwoods, like yellow poplar, are softer than softwoods like southern pine.

The upper regions of the Northeast consisted of vast hardwood forests. Throughout the Midwest and the upper regions of the South, hardwood forests also dominated. Along the coastal plains of the South, the predominant trees were pines. The Rocky Mountain region was covered with conifers such as lodge pole and ponderosa pine. On the West Coast, gigantic softwoods such as Douglas fir, spruce, and redwood covered the slopes.

Forest Succession

The reason certain types or species of trees dominate a region is because of a natural process called **succession**. When a forest first begins from bare ground as a result of

Major Forest Types of the United States

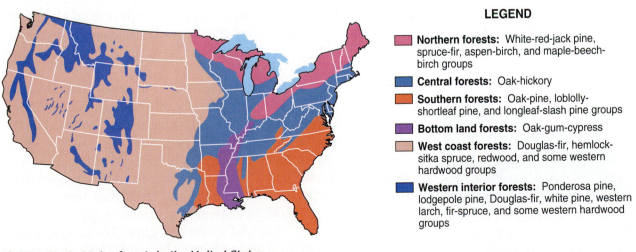

LEGEND

Northern forests: White-red-jack pine, spruce-fir, aspen-birch, and maple-beech-birch groups

Central forests: Oak-hickory

Southern forests: Oak-pine, loblolly-shortleaf pine, and longleaf-slash pine groups

Bottom land forests: Oak-gum-cypress

West coast forests: Douglas-fir, hemlock-sitka spruce, redwood, and some western hardwood groups

Western interior forests: Ponderosa pine, lodgepole pine, Douglas-fir, white pine, western larch, fir-spruce, and some western hardwood groups

Figure 15–8 *Major forests in the United States.*

fire, clear-cutting, or other causes, many species of plants emerge from the ground. Like all organisms, these plants compete for water, nutrients, and sunlight. The tree species grow taller than other plants. As the trees become taller than the assorted other plants, the trees cause a shade that covers the other plants growing beneath them (Figure 15–9). Plants

Figure 15–9 *Trees grow taller than the underbrush, causing a shade on the undergrowth.*

that do not receive enough sunlight generally stop growing and die. Trees that grow well in a particular climate are usually the ones that grow faster than the other species of trees. The limbs of the faster growing trees eventually unite to create a canopy or **overstory** above the other trees (Figure 15–10). This prevents trees of the same species from growing in competition. Also, trees that are not tolerant of shade will not grow under the taller trees that have created a canopy. However, trees that tolerate shade continue to grow at a slow, steady pace.

Eventually the trees that are shade tolerant may overtake and suppress the trees that first created the overstory. This is particularly true if the first trees mature and die. A good example of this process is the forests that covered the upper South. The first trees to dominate were the fast-growing pines such as the loblolly, slash, and shortleaf. These species created overstories that suppressed the species of trees that were not tolerant of shade. However, beneath this canopy, hardwood species such as red oak and white oak grew. These species tolerate shade fairly well and will grow beneath the pine canopy. The southern pine matures and dies at an earlier age than the

Figure 15–10 *Limbs of trees unite to form an overstory or canopy.*
Courtesy of Photo Disc, Russell Illig.

oaks. When a pine dies, the space in the canopy may be filled with a hardwood such as oak. When the hardwoods create a canopy, little sunlight will penetrate, and pines are intolerant of shade. The hardwoods eventually become the predominant species. This is called the climax vegetation because it is unlikely that other species will overtake the hardwoods (Figure 15–11).

Succession is a very slow process that may take 200 years to reach the climax vegetation stage. When the Europeans explored this region of the continent, they found a mixture of pine and hardwoods. The farther south they went, the more pine they found because the milder climate favored the pines. Also, the sandy soil of the coastal plains greatly favored pines over hardwoods.

The Forest Ecosystem

Forests play an important part in our lives because they are a part of our **ecosystem**. An ecosystem is any interdependent community of plants, animals, and their physical environment. The world as a whole is an ecosystem, and forests are instrumental in the support of life on the planet. Remember from Chapter 6 that plants use carbon dioxide and give off oxygen. Because animals use oxygen and give off carbon dioxide, plants support all animal life, including humans. Vast forests of

Figure 15–11 *A hardwood forest represents a climax vegetation stage.*
Courtesy of James Strawser, The University of Georgia.

Figure 15–12 *Young vigorously growing trees release relatively more oxygen than older mature trees.* Courtesy of Georgia Pacific.

trees provide essential filtering of the atmosphere, and young, rapidly growing trees produce more oxygen than older trees (Figure 15–12). As a tree produces new wood, carbon dioxide is removed from the air and oxygen is added. According to The American Forest and Paper Association, each year an acre of healthy trees produces 4,000 pounds of wood, uses 5,889 pounds of carbon dioxide, and gives off 4,280 pounds of oxygen. Because we have almost 500 million acres of trees, this represents a lot of oxygen that is put in the atmosphere.

Forests provide us with clean water. Roots of trees prevent water from running rapidly down a slope and instead cause the water to soak into the ground where it accumulates. This recharges streams, lakes, and wells without adding the soil from runoff (Figure 15–13). This process also helps keep the soil in place on the slopes. Were it not for trees that cover hills, lakes and streams would be muddy, and the slopes would eventually lose all of the soil. This would not only damage aquatic life in streams and lakes but would also keep plants from growing for lack of soil. In addition, our supply of clean drinking water would be lost.

Forests provide habitat for animals and other plants. Many species of plant life live beneath the canopy of trees where they are sheltered from direct sunlight. These plants would not thrive in the open sun. Many species of wildlife inhabit the forests and depend on the trees for their existence. Different types of forests support different types of wildlife. Animals and birds prefer particular types of foliage for food or to nest in. Without this habitat, certain species would become

Figure 15–13 *Trees help keep streams clean and free flowing.* Courtesy of Georgia Pacific.

Figure 15–14 *Some wildlife must have certain types of forest habitat in which to live.* Courtesy of Georgia Pacific.

Figure 15–15 *Forests provide areas for recreation.* Courtesy of James Rathert, Missouri Conservation Department.

extinct (Figure 15–14). The federal government has established laws and regulations regarding the protection of animal and plant species in the forests. Loggers and timber companies have to abide by the regulations in the production and harvesting of timber.

The United States is one of the most beautiful countries in the world. In large part this is due to our forests. Millions of acres of land are set aside by the state and the federal government to protect the natural beauty of our forests. These national and state forests preserve a portion of the virgin forests that were here when the country was founded. Today these areas provide a valuable resource for recreation (Figure 15–15). Hiking, fishing, hunting, and camping are favorite pastimes of many Americans, and our forests make these hobbies possible.

One of the greatest contributions our forests make is the large number of jobs associated with the forestry industry. Billions of board feet of lumber are grown, harvested, and processed each year. This requires a gigantic workforce to care for the trees, remove them from the forests, and convert the logs into finished products. This not only helps keep the nation's economy moving, but also supplies us with essential products that are used in almost all aspects of our lives.

THE PRODUCTION OF WOOD FIBER

Many people fail to consider the forest industry as part of agriculture. However, one of the largest segments of agriculture is the planting, management, and harvesting of trees. As mentioned previously, Americans at one time thought that the supply of timber was inexhaustible. Then around the turn of the 20th century, people began to realize that they were going to run out of forests to harvest if measures were not taken. Many saw that the methods used to harvest timber were wasteful and led to the depletion of the soil. As a result, in 1891 Congress passed the Federal Forest Reserve Act. This legislation gave the President the power to set aside timberlands to be managed so that people in the future would have forests to use and enjoy. Later large tracts of land were designated as federal timberlands, and the National Forest Service was established to manage these lands (Figure 15–16). Subsequent legislation during the last 100 hundred years has developed

Figure 15–16 *Large tracts of land are designated as National Forests.* Courtesy of Georgia Pacific.

Figure 15–17 *Throughout the Northwest, forests have been regenerated by planting seedlings.* Courtesy of Georgia Pacific.

the concept of the publicly owned land. Some land is designated for national parks and used only for recreational purposes. Other land was set aside as wilderness areas where the environment is carefully preserved to remain as close to its original natural state as possible. Still other public timberland is harvested by companies who buy the rights to cut the timber.

Forests were once looked on as being a natural resource that could be utilized. However, in recent years the timber industry has moved toward being an agricultural enterprise. Trees are planted, cultivated, and harvested in much the same manner as other agricultural crops. This means that instead of being a resource that will one day be exhausted, the timber industry can be sustained year after year through the careful planting and cultivation of trees.

For many years, areas **clear-cut** of all or most of the trees have been seeded or set out with tree seedlings (Figure 15–17). Throughout the Pacific Northwest, forests have long been regenerated by these methods. This allows forests to grow back in much the same way they originally did. In this area, for a forest to grow from seedlings to harvestable timber may take from 70 to 100 years. For this reason much of the timber industry has moved to the Southeast. In fact, the leading producer of wood is not Washington or Oregon but Georgia! In this region, the hot, humid weather allows trees to grow in much less time. A carefully managed stand of timber can go from seedlings to harvestable trees in as little as 15 years. It must be remembered, however, that the timber from

Figure 15–18 *Superior seedlings are produced in the nursery.*
Courtesy of Georgia Pacific.

the Pacific Northwest, while slow growing, provides a higher quality lumber. A large portion of the trees harvested in the South go for making paper or for plywood core material.

Tree Farms

On tree farms, trees are planted in rows and are given care throughout their growth period. At the proper time they are all harvested, and new trees are planted. Although there are tree farms in almost all states, most are located in the South.

The process of farming trees begins in the nursery, where superior seedlings are produced (Figure 15–18). Research goes into the selection and production of seeds from high-quality trees that are resistant to diseases and grow rapidly. Many of the seedlings are crosses between two species or strains of trees. For example, a longleaf pine may be crossed with a loblolly pine to produce a tree that is superior to either of the parent trees. This cross, known as a hybrid, benefits from a phenomenon known as heterosis or hybrid vigor (see Chapters 4 and 5). The hybrid grows more rapidly and is generally healthier than the purebred strains of trees (Figure 15–19).

After the trees are set out in large areas called plantations, they are managed using techniques that interrupt the process of succession. The underbrush may be cleared by mechanical means, chemicals, or by prescribed burning. The use of fire in prescribed burning destroys vegetation that competes with the trees for water and nutrients. The fire is carefully controlled to prevent damage to the trees and to prevent the fire from escaping.

Succession is also controlled by thinning the trees. As the trees grow and begin to compete with each other for sunlight and nutrients, some are harvested. Usually the trees cut are those that are slowest growing or poorest in quality. The best trees are kept. The trees harvested in the thinning process are used for fenceposts, poles, or paper.

Harvesting Trees

The techniques used in tree harvesting have changed over the years. The first trees were cut using a hand-operated saw or an ax. The logs were then "snaked" or dragged out of the woods using horses, mules, or oxen. Little thought was given to protecting the environment. In fact, until recent years, loggers were often destructive to the environment by building roads and cutting trees without regard to the effect on soil erosion. Trees were cut all the way to the edges of streams,

Figure 15–19 *These two seedlings are the same age. The one on the left is a superior hybrid.*
Courtesy of Georgia Pacific.

Figure 15–20 *This area has been clear-cut and is now growing back into forest.* Courtesy of James Rathert, Missouri Conservation Department.

causing the banks of the stream to erode and lowering the quality of the water. Today, timber harvesters follow guidelines that protect the environment as much as possible and still allow the removal of timber. Roads are kept to a minimum, and trees are left along streambanks.

Most modern operations use a method called clear-cutting, in which all of the timber is removed regardless of size (Figure 15–20). This management technique is used because the entire area can be reforested more efficiently than it could if some of the trees were left. After all of the old timber is removed, new seedlings can be set out that are genetically superior to those harvested. Leaving the small trees and removing the large trees is not considered to be the best practice because the smaller trees will most likely be slow growing. This is because they are either stunted or are genetically inferior.

Advances have been made in the way timber is harvested. Many of the large trees are still cut using chain saws, just as they have been for many years. Smaller softwood trees (less than 2 feet in diameter) can be removed using power machinery if the land is level enough. One type uses large hydraulic-powered cutting shears that cut trees off at the stump. Another type, called a felling head, grasps trees and holds it while a circular saw cuts the tree off close to the ground (Figure 15–21). The trees then may be limbed by a large machine called a tree processor that runs the entire tree through the machine and removes the limbs (Figure 15–22).

Figure 15–21 *A machine called a felling head is used to harvest softwood trees.* Courtesy of James Strawser, The University of Georgia.

Very large trees are delimbed with a chain saw. In the past, trees were cut into logs of a certain diameter and length. A lot of wood was left in the top of the tree. Now the entire length of the tree is hauled to the mill. The portion of the tree that is not large enough in diameter to saw into timber is chipped up for use in making paper or particle board.

Figure 15–22 *Trees can be limbed using a tree processor.* Courtesy of Dale Green, The University of Georgia.

BIO BRIEF

Putting Forests on Farms and Farms in Forests

Eastern Europe may have something to teach American about growing trees—includling American trees—on small farms. Soil scientist Charles M. Feldhake and horticulturist Carol M. Schumann at the Agricultural Research Service Appalachian Soil and Water Conservation Research Laboratory in Beaver, West Virginia, look enviously at the many varieties of American trees, like black locust, that Europeans have selected for livestock browsing, timber, and various other purposes.

When East Europeans think of farms, they see a picturesque scene with cows, goats, and sheep mixed in with forests, orchards, and beehives. Feldhake and Schumann think that's a much more suitable model for the Appalachian hill-lands than the large corn-soybean farms of the Midwest. The steep hills with shallow soils can't take any tillage, so perennials like trees and shrubs are a natural for the area.

Five years ago, Feldhake began growing 1,200 black locust trees on a steep, hilly pasture nestled near the mixed hardwood forest surrounding the lab. The trees are in rows about 30 feet apart in a 5-acre watershed. Currently, 2 sheep graze there.

Another 25 graze in an adjacent 5-acre watershed without trees.

"We want to find out whether the trees can reduce nitrogen losses by catching excess nitrogen from livestock urine and manure," Feldhake says. "To check that, we're measuring how much nitrogen leaves the two watersheds in subsoil water and surface runoff."

Feldhake says trees add diversity to a farmer's income. Black locust can be cut and sold for firewood or for pest-and rot-resistant fenceposts. The flowers are good nectar sources for honey bees. And, during a heat wave, sheep or other livestock can be rotated to the shady, forested pastures. With black locust, seedling quality has been a problem. Selecting for top-quality black locusts is a low priority with breeders in the United States.

"There isn't the commitment to this tree as a crop that there is in a place like Hungary," Feldhake says ruefully. But, the scientists are testing trees and shrubs with better breeding, such as black walnut, honey locust, and sea buckthorn on pastures in West Virginia and at Virginia Polytechnic Institute in Blacksburg, Virginia.

Lumber

Logs that are of sufficient quality for sawing into lumber are processed at a sawmill. The first step is to remove the bark from the logs with a debarker that rolls the logs around and grinds the bark off (Figure 15–23). The bark from softwood

Feldhake says more selections have been made for honey locust trees because of interest in their pods as a good source of sugar and protein for grazing livestock.

European farmers grow the sea buckthorn shrub for its berries, which have a high vitamin content and reputed medicinal value. They make jellies and juices from the berries for home use and roadside sales. How about growing berries and other fruits and nuts in forests? This past winter, Schumann and Feldhake literally cleared the way for that unique vision, opening up a 1.2-acre strip in the forest. In the spring, they planted red oak tree seedlings. They also planted faster growing trees and shrubs for the short term: white pine, Chinese chestnut, pawpaw, hazelnut, blueberries, and rasberries.

The red oak will be the climax tree in the long term, when it could be selectively cut for high-value veneer. The white pine will be ready for sale as holiday trees beginning December, 2006. The pawpaws and chestnuts will bear fruit and nuts beginning as early as 2008, Schumann says. "We need to find out if we can successfully grow marketable products from the shorter term plantings without negative effects on the red oaks," she says.

"In addition to meeting research objectives, this site is also designed to demonstrate to local farmers that they, too, can farm with trees and shrubs," Schumann says.

Trees such as the Chinese chestnut are evaluated for use in small-scale woodland agroforestry plantings. Courtesy of USDA-ARS.

Source: Agricultural Research Magazine.

trees is processed into small chunks and used for mulch in the landscape industry.

The logs are sawed into boards by a series of band saws that first cut the round logs into square or rectangular timbers. The outside layers that are cut off are called slabs. These

Figure 15–23 *At the sawmill, the bark is removed from the tree. Courtesy of Georgia Pacific.*

were once burned as waste but now are chipped into small pieces to be made into paper. In modern sawmills the arrangement of saws are set up to get the most efficient use out of the log. This process is controlled by a laser beam that "reads" the log and sends information to a computer that sets the saws at the proper spacing (Figure 15–24). This spacing allows the log to be cut into lumber of differing dimensions in order to eliminate as much waste as possible.

Figure 15–24 *Modern sawmills use laser beams to set up the sawing process. Courtesy of Georgia Pacific.*

The lumber that comes from the mill is graded and labeled for a particular use. The best quality lumber is used in the furniture industry or for making molding. Other lumber is graded for use as framing for buildings or for general use.

Remember from Chapter 6 that wood is composed of cells that form the xylem of the tree. The xylem is the tissue that transports raw nutrients throughout the tree. The light wood along the exterior portion of the tree trunk, called the sap wood, serves this purpose. As layers are added, the inner layers of xylem eventually cease to function and form the dark inner layers called **heartwood**. The heartwood gives support to the trees. The heartwood of species, such as cherry and walnut, provide beautifully colored wood that is used to make furniture.

Each year the tree adds a new layer or ring of sap wood (Figure 15–25). Under good growing conditions, the ring may be relatively thick. If growing conditions are less than ideal, the layer will be smaller, and the tree grows more slowly. A series of slow growing years results in closely spaced rings. Likewise, a number of good years results in widely spaced rings. Generally speaking, lumber from slow growing trees will be higher quality because it will have less tendency to warp or twist after it is sawed.

After sawing, the lumber is placed in large ovens called kilns to cure and dry the wood. The reason for this step is that water is bound up in the cells of the xylem. As the wood dries out, the cells (and the entire board) shrink. If the boards are used before they are properly dried, they will shrink and distort the building or other object that has been constructed. The natural removal of water from the xylem cells is a slow process that can take more than a year. By placing the lumber in a kiln, the proper humidity and temperature can be controlled in order to cure the lumber rapidly. The proper conditions in the kiln have to be maintained to ensure that the lumber dries evenly. Unevenly dried lumber splits and warps.

Veneer

At one time in this country, high-quality lumber was plentiful and cheap. However, as the supply of old growth timber declined, the higher grades of lumber became more scarce. This means that measures have to be taken to make the best use of the high-quality wood. One such measure is the use of veneers (Figure 15–26). A veneer is a thin sheet of wood that is less than .25 inch thick. To obtain veneer, a high-quality log

Figure 15–25 *Growth rings of a tree indicate how fast the tree grew. This tree grew rapidly.*
Courtesy of Georgia Pacific.

Figure 15–26 *Logs are cut into bolts for processing into veneer. Courtesy of Georgia Pacific.*

is cut to the proper length, debarked, and heated in hot water or steam vats to soften the wood. The log, called a bolt, is then placed in a lath where it is turned rapidly while sharp knives cut the wood into a long, thin continuous sheet (Figure 15–27). This sheet is then trimmed to size. A similar process is used to cut sheets from logs of lesser quality. The high-quality sheets are then glued to the lower quality sheets with high-pressure machines that compress the layers together. Obviously, the

Figure 15–27 *Using a large lathe, thin sheets of veneer are cut from the bolt. Courtesy of Georgia Pacific.*

Figure 15–28 *Sheets of veneer are laid at right angles to provide added strength.* Courtesy of American Plywood Association.

high-quality sheets are used on the outside. This process can make the quality wood go a lot further.

This product, called plywood, has other advantages. The sheets are placed together so that the wood fibers run at right angles to each other (Figure 15–28). This makes the panels stronger and also helps prevent warping. Plywood is used extensively in the construction industry as flooring, decking, and siding. High-quality veneer is used in cabinets and other furniture.

A variety of plywood is beginning to be used as framing timber. Instead of being glued into panels, the plies are glued and pressed into long beams that will carry a load (Figure 15–29). These beams are superior because of their strength and resistance to warping.

Figure 15–29 *The plies are glued and pressed together.* Courtesy of American Plywood Association.

Figure 15–30 *A variety of building materials are made from wood chips. Courtesy of Georgia Pacific.*

Another method of making building material is the use of wood chips (Figure 15–30). Low-quality wood is chipped into flat pieces. These chips are mixed with strong glue and compressed under extremely high pressure into sheets. To make stronger panels, the wood fibers are oriented or turned in the same direction. After being pressed together, the panels are glued and pressed with other panels that have the strands oriented at a right angle to the other panels in much the same way as plywood is made.

New products are constantly being developed that use wood fiber. A good example is the use of wood fibers with plastics (Figure 15–31). A product can be made that carves, saws, and nails much like wood but is stronger, will not warp, and resists decay.

Paper Making

Have you ever considered all the uses we make of paper? We print books, magazines, newspapers, and many other types of reading material. We use it for packaging everything from refrigerators to apples. It is used for filters in automobiles, water lines, and coffee pots. The construction industry uses paper from everything for insulation to roofing. There are literally thousands of different uses for paper products, and new uses are constantly being developed (Figure 15–32).

Figure 15–31 *Wood fibers can be combined with plastic to form new products. Courtesy of Mobil Chemical Co., Arthur Beck.*

Figure 15–32 *Paper is used to make thousands of products.*
Courtesy of Georgia Pacific.

Americans use a lot of paper. According to the American Forest and Paper Association, paper is used to print about 24 billion copies of newspapers and more than 2 billion books each year. We also produce and use more than 336 billion square feet of corrugated shipping materials annually. These are just a couple of examples of the tremendous amount of paper we need each year. This requires that a lot of wood fiber be grown, harvested, and processed.

Paper has been used by humans since ancient times. The Egyptians made a form of paper called papyrus by weaving reeds and pounding them together into thin sheets. Many ancient manuscripts were written on this material and rolled into scrolls. The word paper originated from this material.

The form of paper that is used today is generally thought to have originated in China around 105 A.D. The Chinese made paper using a combination of rags, tree bark, and hemp. To separate the fibers, this material was smashed into a pulpy mixture, pressed together to remove the water, and hung to dry in the sun.

Modern paper is made from wood fibers. The long fibers that form the xylem of trees are composed of plant cells. The cell walls are composed of a substance known as **cellulose** that remains after the plant cells die. It is this material that is used to make paper.

The first step in making paper is to convert the logs into a form called pulp in which the fibers of cellulose have been separated (Figure 15–33). This is done either by mechanically grinding the wood into fibers or by treating chips in a chemical process. The pulp process for some forms of paper may combine mechanical grinding and chemicals. In the chemical

Figure 15–33 *Wood is chipped into small pieces to begin the paper-making process. Courtesy of Champion Paper Co.*

process, the wood chips go into a tank called a digester that subjects the chips to high temperatures and pressures (Figure 15–34). Chemicals are added to the solution that dissolve the **lignin** in the wood fibers. Lignin is the substance that holds the wood cells together, and its dispersal results in the separation of the cellulose fibers.

After the wood has been reduced to a pulp slurry, the pulp is washed to remove the chemicals and screened to remove undissolved debris. It is then treated in a process called beating, in which the fibers are pounded to further separate them and to give them added flexibility. During this process several materials are added. Dyes may be added to color the paper,

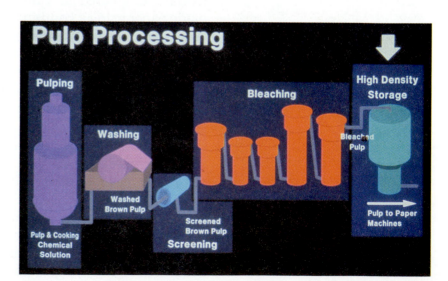

Figure 15–34 *Chemicals are added to wood chips in huge digesters that break the fibers apart. Courtesy of Champion Paper Co.*

or the pulp may be bleached to make it whiter. Fillers such as chalks or clays are added to fill in the pores left by the fibers. Rosins, gums, or other materials are added to make the paper water resistant to withstand ink or other moisture during printing or processing.

After coming from the beater, the pulp is about 99 percent water and only about 1 percent fiber. This material is then blown against a rotating screen cylinder (called a fourdrinier) that removes a lot of the water. The paper then goes onto a conveyor that travels at high speed and subjects the mixture to a suction that takes off additional moisture. At this point the mixture is in one continuous sheet. Next, the paper winds over a series of heated drums, and additional moisture is removed. This is called the drying process (Figure 15–35).

Figure 15–35 *The paper is dried in large sheets. Courtesy of Champion Paper Co.*

After leaving the dryer, the paper goes to finishing. The finishing process differs with the type of paper that is being produced (Figure 15–36). Some goes to a machine called a calendar where the paper is ironed between heavy rollers. This produces a smooth finish on the paper. Some papers may be coated with a type of clay to finish it. Other operations may finish the paper into a specialized paper product having particular properties for a special need. There are many finishes and product types, ranging from heavy cardboard to delicate writing paper.

Figure 15–36 *Paper is finished according to the purpose for which it is to be used.* Courtesy of Georgia Pacific.

Recycled Paper

For the past several years, people in this country have been recycling materials in an effort to help preserve our resources (Figure 15–37). Such items as aluminum, glass, and plastic are being recycled into new products. The material that is recycled more than any other is paper. Each year millions of tons of paper are recycled into new products. In fact, about one-third of all the raw material used for the production of paper comes from reclaimed paper. If wood such as tree tops, limbs, and sawmill slabs is added in, almost 60 percent of the paper produced comes from material that was once considered waste.

Figure 15–37 *Paper is recycled in an effort to save our resources.* Courtesy of University of Minnesota Agricultural Experiment Station.

Figure 15–38 *The amount of paper recycled continues to grow each year.* Courtesy of USDA-ARS, K-5259-17.

Recycling paper not only aids in the conservation of trees but also helps alleviate problems with waste disposal. We are quite literally running out of space to put garbage. Many large landfills are nearing capacity, and each truckload of garbage adds to the problem. Paper that is not recycled usually ends up in the landfill. Each year Americans are recycling more paper products (Figure 15–38). Since 1988, there has been a 28 percent increase in the amount of paper that is recycled, and the amount continues to grow.

SUMMARY

Our forests supply us with much of what we enjoy in life. Through scientific research and the development of new and more efficient management practices, we will always have trees to beautify our land and clean our air. We will also continue to have the necessary products that come from wood. The forest can truly be a renewable resource.

CHAPTER 15 REVIEW

Student Learning Activities

1. Collect branches from five trees that are of economic importance in your area. Classify them scientifically and determine whether they are softwood or hardwood. List all of the uses for the trees.

2. Look around your home and make a list of all the products that are made from paper. If possible identify whether or not they were made from recycled paper.

3. Visit a building supply store and identify wood products that are made from restructured wood. List the advantages of restructured wood over solid wood. Share your list with the class.

Define the Following Terms

1. habitat
2. hardwood
3. softwood
4. deciduous
5. succession
6. overstory

7. ecosystem
8. clear-cut
9. heartwood
10. cellulose
11. lignin

True/False

1. Timber harvesting is very destructive and is a major factor in the depletion of our forests.
2. Because of improved forests and increased acreage of forests, some species of wildlife on the verge of extinction have made a comeback.
3. There are less trees in the United States now than 100 years ago.
4. Georgia is the number one wood producing state in the United States.
5. Wood from the Pacific Northwest is the highest quality construction wood harvested in this country.
6. Trees are classified as hardwoods or softwoods based on the hardness of their wood.
7. After a canopy is established over the forest floor, succession is complete.
8. Because of its planting, management, and harvesting techniques, the forest industry is now viewed as an agricultural enterprise.
9. Trees can be planted only as seedlings on tree farms because hybrids do not grow well when planted directly from seed.
10. Hybrids are usually superior to parent trees.
11. Lasers and computers are being used in the wood processing industry.
12. High-quality wood is needed in the production of wood veneer.
13. The ancient Mayans of South America are credited with the invention of paper.

Fill in the Blank

1. A _____ is a broadleaf deciduous tree; a _____ is an evergreen tree.
2. A(n) _____ is an interdependent community of plants and animals and their physical environment.
3. _____ are operations where trees are planted in rows and given care throughout their growth period.
4. The superiority of a hybrid over its parent plants is due to a phenomenon known as _____.
5. Most modern tree harvesting operations use a method called _____ where all of the timber is removed regardless of size.

6. Logs that are of sufficient quality for sawing into lumber are processed at a
   _____.

7. The _____ of a tree is composed of nonfunctioning inner layers of xylem
   that give support to the tree.

8. After sawing the lumber, it is placed in ovens called _____ to cure and dry the
   wood.

9. _____ is thin layers of wood that can be glued and pressed together.

10. Today paper is made from wood _____.

11. The first step in making paper is to convert logs into _____.

12. _____ is the substance that holds wood cells together.

13. The product that is recycled more than any other is _____.

Discussion

1. Explain forest succession and why it dictates the dominant species in a region.

2. Discuss the contribution that forests play in the world ecosystem.

3. Discuss the evolution of the forestry industry from natural resource to agricultural
   enterprise.

4. Contrast the differences in harvesting trees over the years.

5. List and describe four products made from trees.

6. Briefly outline the paper-making process.

7. Discuss the importance of recycling paper products.

THE SCIENCE OF AQUACULTURE

KEY TERMS

carp
catfish
seine
raceways
milking
gills
thermoclines
oxygen meter
aerators
runoff

STUDENT OBJECTIVES

After studying this chapter, you should be able to:

◆ Analyze why the aquaculture industry is growing.

◆ Discuss the historical beginnings of the industry.

◆ Explain the biological aspects of fish that give them advantages over other agricultural animals.

◆ List the types of aquacultural enterprises in the United States.

◆ Distinguish between the ideal conditions for growing warm- and cool-water fish.

◆ Discuss the methods used in breeding and raising fish.

◆ Explain how the gills of a fish take in oxygen.

◆ Summarize the ways oxygen is dissolved in the water.

◆ Explain the concept of thermoclines.

◆ Summarize the ways in which oxygen depletion can occur in a pond.

Aquaculture is one of the fastest growing industries in agriculture. This form of production deals with growing animals that spend most or all of their lives in water. Aquaculture includes the growing of shrimp, oysters, crawfish, and prawns, but the most important part of the aqua-culture industry is the production of fish. Another segment of aqua-culture includes growing aquatic animals that are used as pets or ornamentals.

Fish have always been an important part of the diet of humans. Because water is essential to all animal life, humans are no exception. From the dawn of civilization, people have lived near sources of water. Not only did streams provide the people with water to drink and use, they also provided fish and other animals that could be used for food. Fishing became a science and an art as people devised ways of harvesting fish from the wild. However, fish were sometimes difficult to catch and at certain times were not available.

The growing of fish dates back to ancient times. Records indicate that fish were grown in captivity in ancient China, Egypt, and Rome. This allowed people to have a readily available supply of fish even when conditions were not favorable for catching wild fish. The culture of fish may have begun when people caught more fish than they could eat and needed a way to store them. One obvious way would be to keep them in captivity until they were needed and eat them fresh. When they saw that the fish could survive in captivity in a holding pond or in a cage, it probably occurred to them that they could raise their own fish as a ready source of food.

Although the culture of fish can be traced back thousands of years, until relatively recently, fish caught from the wild accounted for almost all of the fish consumed in the world. The annual amount of fish caught per person increased every year until around 1970, when it began to level off. Until then the oceans of the world seemed to have an unlimited reserve of fish available for harvesting (Figure 16–1). The demand for fish has become so high and the techniques of catching them

Figure 16–1 *The oceans have been a bountiful source of fish.* Courtesy of James Strawser, The University of Georgia.

Figure 16–2 *The oceans can no longer keep up with the demands for fish and other seafood.* Courtesy of James Strawser, The University of Georgia.

have become so sophisticated that the limit in the number of fish that can be harvested from the seas has been reached. Even with the Earth's vast expanses of ocean, there is a limit to the amount of fish that can grow (Figure 16–2). As the world population grows and the demand for fish increases, the resources of the world's oceans will be even more strained. Countries of the European Community have decided to reduce the number of fishing vessels in their fleet. Other countries have closed their waters to foreign fishermen in an attempt to limit the number of fish that are being caught out of the ocean.

In this country until the 1960s, the aquaculture industry consisted primarily of producing fish to be released into the wild for recreational fishing (Figure 16–3). As far back as the early 1800s, the raising of fish has been studied. Groups such as the American Fish Culture Association (later called the American Fisheries Society) promoted the rearing of fish to replenish the fish in waterways in the wild. Around 1960, producers began to raise fish for sale as food rather than for release. This came about as commercial fishermen began to reach the limits of the potential catch in the ocean. When the supply of a product falls below the demand, the price of the product usually goes up. Because the demand for fish could not be met from fishing the oceans, producers reasoned that a profit could be made from the commercial growing of fish. Since that time, the aquaculture industry has been continually growing. Not only is the price for farm-raised fish competitive, but the supply is constant, and we do not have to rely

Figure 16–3 *Until the 1960s, aquaculture was primarily for rearing fish to be released for sports fishing.* Courtesy of Georgia Pacific.

on having favorable conditions for harvesting in the wild. Currently, around 10 to 11 percent of the world consumption of fish comes from farm production.

FISH AS AGRICULTURE

Fish have several advantages over other agricultural animals. One of the major advantages is that fish are ectothermic animals. This means that they depend on the environment to provide the body temperature needed for life. Because of this, the animals do not have to use energy to produce body heat that must be kept to a constant temperature. Most agricultural animals must use a large portion of their feed intake to maintain body temperature. This is particularly true in very cold weather.

Another advantage is that fish move about and are suspended in water (Figure 16–4). The natural buoyancy of the water allows them to move about with less energy than animals that live on land. Because of these characteristics, fish

Figure 16–4 *Fish are suspended in and move about through the water. This uses less energy than an animal moving on land.* Courtesy of James Strawser, The University of Georgia.

are able to put on more body weight with less feed than most other farm animals. For example, a steer must take in around 8 to 9 pounds of feed for every pound gained, while a fish can gain a pound of body weight on less than 2 pounds of feed. However, this is not as great an advantage as it may seem because the feed required for fish is much more expensive than that required for steers. The feed for fish should be around 35 percent protein, but a steer's ration can be around a third of that amount. Protein is usually the most costly portion of any animal's ration.

Another advantage fish have is in the percent of edible food obtained from the fish. A steer will yield only around 35 to 40 percent of the body weight in edible meat, but a catfish will yield about 55 percent, and a trout may yield as much as 85 percent! Some research indicates that fish may be more healthy to eat than red meat. Fish is lower in cholesterol and has fewer calories. Most health-conscious diets contain regular servings of fish.

There are some disadvantages to growing fish for profit. To become established in aquaculture requires the outlay of a lot of money to construct ponds or other structures. Ponds have to be constructed to exacting specifications in order to catch and hold enough water for the production of fish (Figure 16–5). The enterprise is also expensive to conduct because it is labor intensive. This means that a lot of labor is required from the time the fish are stocked until they are harvested. The water has to be constantly monitored to make sure that it is of sufficient quality for the fish to remain alive and healthy.

Figure 16–5 *Ponds have to be constructed to exacting specifications in order to hold enough water for the production of fish.* Courtesy of James Strawser, The University of Georgia.

Aquaculture is a high-risk operation (Figure 16–6). All of the fish in a pond can die suddenly if the quality of the water falls below acceptable levels. Perhaps one of the biggest disadvantages of aquaculture is marketing. In areas where the operations have been most successful, sound marketing plans and operations have been established before fish were produced. Because of the highly perishable nature of the product, aqua-culture operations have to be located near the processing plant so the fish can be delivered alive.

Figure 16–6 *Aquaculture is a high-risk operation. Fish can die suddenly. Courtesy of George Lewis, The University of Georgia.*

FOOD FISH

Fish are grown for food all over the world. The countries in Asia are the largest producers and consumers of fish in the world (Figure 16–7). China, Japan, Indonesia, Thailand, and Korea all produce substantial amounts of fish. The European countries follow Asia, with North America ranking third.

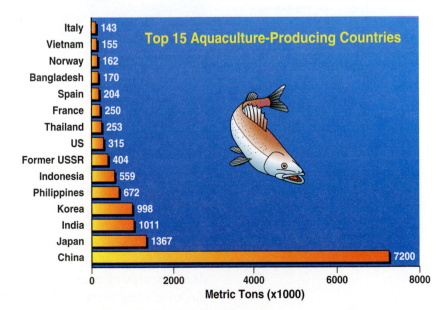

Figure 16–7 *The countries of Asia are the world's largest producers of fish.*

As humans began to discover ways to grow their own fish, they had to determine which species would be adaptable to culture. Most likely, many different species were tried just as were land animals. Early humans must have tried to domesticate antelope and gazelles just as they tried to domesticate pigs and cattle. They discovered that some animals were adaptable to domestication, and some were not. The same was probably true with fish, and only certain species were acceptable for domestication. The fish had to be acceptable for eating. Not all species of fish have a taste that is acceptable to humans. Also, the fish needed to have enough flesh to be worth the effort to raise. They had to withstand the rigors of captivity. This meant that they had to be hardy enough to live in confined areas with limited oxygen supplies. The fish also had to reproduce efficiently in order to provide enough young for stocking in large quantities.

Just as in the land animals people domesticated, species of fish were found that matched these characteristics. The most popularly grown fish in the world is **carp** (Figure 16–8). These species are hardy fish that can survive in crowded conditions

Figure 16–8 *The world's most popularly grown fish is the carp.* Courtesy of George Lewis, The University of Georgia.

with relatively low amounts of oxygen. They have been cultured for thousands of years and were probably one of the first species of fish to be grown in captivity. These fish have large coarse scales covering their bodies and range in color from silver to gold. Carp have been introduced in the United States in the wild but are not generally considered to be a very desirable food fish.

Closely rivaling carp in popularity worldwide is the tilapia. These fish are native to North Africa and the Mediterranean area and have been in production since the time of the ancient pyramids (Figure 16–9). Although more than a hundred species of tilapia exist, only about 10 are used in aquaculture. They are popular because of their rapid reproduction

Figure 16–9 *Tilapia have been produced since ancient times.* Courtesy of Gary Burrle, The University of Georgia.

rate, their ability to thrive in waters up to 90°F, their resistance to disease, and their ability to live in crowded conditions. These abilities are due, at least in part, to their capability to live in waters containing very low levels of oxygen.

Tilapia resemble our native bream species and have a mild flavored, medium-textured flesh. The taste of tilapia is quite acceptable to Americans and is gaining in popularity. One of the major drawbacks to the production of tilapia in the United States is the fear that they might escape into the wild. Some scientists and environmentalists fear that the Tilapia would overcome native populations of fish in lakes and streams because of their prolific reproductive capacities. Proponents of tilapia argue that the fish cannot thrive in water less than 65°F and will die in water less than 50°F. In most parts of the country, water temperature falls below this level during the winter months. Despite the controversy, tilapia are gaining in importance in the aquaculture industry in the United States.

Catfish Production

Several species of fish are grown in the United States, with trout, salmon, striped bass, and **catfish** being the most commonly produced. By far the most important of these is the catfish (Figure 16–10). Catfish are different from other species of fish. Instead of having scales, they are covered with a smooth skin. They have broad, flat heads that have long barbels (whiskers) on their upper lip that serve as sensory organs.

Figure 16–10 *The catfish is the most commonly grown food fish in this country.* Courtesy of George Lewis, The University of Georgia.

Many species of catfish can be found all across the United States. The species that is grown commercially is the channel catfish (*Ictalurus punctatus*). These fish grow rapidly, are very hardy, and can stand low levels of oxygen. They are classified as warm-water fish and grow best when the water temperature is about 85°F.

The characteristics of catfish make them well suited for production in shallow ponds in the southern parts of the United States. The state of Mississippi leads all others in the production of catfish. In fact, the fertile delta area of southern Mississippi where cotton was once king, now receives more income from catfish production than any other crop (Figure 16–11).

Figure 16–11 *Catfish are the top crop in the Mississippi Delta. They are raised in commercial ponds such as these.* Courtesy of James Strawser, The University of Georgia.

In the early years of catfish production, brood fish that supplied the eggs for production were captured from the wild. These fish served as the foundation for the selective breeding of catfish that were more suitable for culturing. Like most other agricultural food animals, they were selected for their growth efficiency, meatiness, and reproductive capacity. After several generations of selective breeding, catfish that are grown in modern operations are far superior to those in the wild.

The brood fish are generally from 3 to 10 pounds in weight (Figure 16–12). The weight of the fish seems to be the crucial factor rather than age in the production of eggs. In the wild, the male and female select a nesting place to lay the eggs. This may be a submerged hollow log or a small cavern in the bank. The commercial producer supplies the fish with a box or other type of nest for spawning (Figure 16–13). The female deposits from 2,000 to 4,000 eggs (called roe) per pound of

Figure 16–12 *Brood fish weigh from 3 to 10 pounds. The one on the right is a female, and the one on the left is a male.* Courtesy of James Strawser, The University of Georgia.

her body weight. This means that a 10-pound female may deposit as many as 40,000 eggs! After the female lays the eggs, the male deposits sperm into the egg mass, and fertilization occurs. In the wild the male cares for the nest and keeps predators away throughout the incubation period.

In commercial operations producers remove the egg (roe) masses, called clutches, and place them in hatcheries for incubation (Figure 16–14). This takes place in long tanks that have paddles moving in them to keep the eggs turned. When the eggs hatch, the new fish are referred to as "sac fry" because the have a small sac attached to them from the egg (Figure 16–15). This provides nourishment for the young fish until they are large enough to eat on their own. When the sac is dropped, the fry are placed into long trough-like tanks where they are fed and cared for until they are large enough to be placed in larger tanks or ponds.

Figure 16–13 *The producer supplies the catfish with a box or nest for spawning.* Courtesy of George Lewis, The University of Georgia.

Figure 16–14 *Egg masses are called clutches. Courtesy of James Strawser, The University of Georgia.*

Figure 16–15 *Newly hatched catfish are called sac fry. Courtesy of James Strawser, The University of Georgia.*

The most common way to produce catfish is in ponds that are around 4 to 6 feet deep. These specially constructed ponds are designed for maximum production and easy, efficient harvesting. Small catfish that are stocked in ponds are referred to as fingerlings (Figure 16–16). This size fish may vary from 2 to 6 inches in length. The stocking rate varies

Figure 16–16 *Young fish that are stocked in ponds are called fingerlings. Courtesy of George Lewis, The University of Georgia.*

with the capabilities and desires of the producer and may vary from a few hundred per acre to several thousand per acre. A well-managed pond may produce as much as 4,000 to 6,000 pounds of fish per acre. As with any livestock operation, the more intense the operation, the greater the management required and risk involved. There are more problems maintaining adequate oxygen levels for a pond that is heavily stocked. In addition, problems are encountered from the large amount of waste material excreted from the fish. Both these problems will be discussed later in the chapter.

An alternative to rearing the fish in open ponds is the use of cages (Figure 16–17). The cages are usually about 4 feet in diameter and about 4 to 6 feet deep. The advantages of raising the fish in cages are that the fish are easy to see, they are protected from predators, they are easy to treat for disease, parasites or other problems, and they are easily harvested. The disadvantages are that they require more intense management and they are more susceptible to disease and parasites. These problems are common to all animals that are raised in close confinement.

Figure 16–17 *Fish may be raised in cages. This method allows for easier management.* Courtesy of Ray Herren, The University of Georgia.

The grow-out period usually lasts from March, when the fingerlings are placed in the ponds or cages, to October, when they are harvested. This is the period when the water is the warmest and the fish are more active. Because they eat more during this period, growth is also greater during this period. Although the catfish can live during the winter, they are inactive, eat little, and grow very slowly.

Catfish are fed on a well-balanced diet of pelleted food that is designed to float on the water (Figure 16–18). This

Figure 16–18 *Catfish are fed pelleted feed that floats.* Courtesy of George Lewis, The University of Georgia.

BIO BRIEF

Oceanographic Instruments Monitor Fishpond Algae

Carl Jeffers manages Top Cat Fishery in Portland, Arkansas, a fast-growing company that sells 400,000 pounds of catfish weekly. The Catfish Farmers of America named him Arkansas' Catfish Farmer of the Year. But, he wants to do more.

"If we could find an economical way to control off-flavors and the blue-green algae that produce them, it would be a tremendous boon to the industry, no question," he says.

Many producers would agree. Currently, a major obstacle to catfish sales is the off-flavor compounds produced by microscopic algae. These natural algae byproducts are absorbed into the fishes' meat, giving them a muddy taste. The fish are still safe to eat, but they just don't taste good. It's estimated that during the late summer and fall, as much as 80 percent of U.S. catfish loses some flavor quality. As a result, many are unmarketable, causing an extra $5.8 to $12 million in overhead costs. That's why three times a week, Jeffers takes fish from harvest-ready ponds for flavor tesing by processors.

Catfish ponds are an ideal habitat for algae, which are often beneficial to the farmer. They provide oxygen for the fish and help stabilize pond temperatures, but it's not possible to predict when the algae will produce the off-flavor compounds. Recently, Agricultural Research Service scientists David F. Millie and Chris P. Dionigi joined forces with Oscar M. Schofield, a professor of oceanography at Rutgers University, to develop ways to monitor algae in catfish ponds. The studies focus on the relationship between the algae present, their health, and the presence of off-flavors.

In a series of experiments at Stoneville, Mississippi, the research team explored using oceanographic instruments to see whether they might provide an efficient means to monitor algae in catfish ponds. These instruments, which measure the optical properties of water, might be used to estimate how much and what kinds of algae are present in a pond.

Algae require light for growth, but some absorbed light is re-emitted as fluorescence

causes the fish to come to the surface to feed and allows the producer a chance to observe the fish. The fish reach a market weight of 1 to 2 pounds about 18 months from spawning. This size is preferred by consumers.

The fish are harvested by means of a **seine** (net) that is dragged across the pond by human or mechanical power. The fish are taken from the net and loaded into special tank trucks that deliver the fish live to the processing plant. The delivery

Scientists are divising new ways to monitor algae in fish ponds.

in a process that can provide basic biological information about algae cells. Researchers are looking at an instrument called the SAfire (Spectral Absorption and Fluorescence Instrument), which can measure the color of light that algae fluoresce. Scientists have long speculated this information could be used to assess the kinds of algae in a pond—particularly the blue-green algae associated with off-flavors. A second instrument, the pulse amplitude modulated fluorometer, measures how much light is fluoresced by algae cells, which allows researchers to estimate how healthy the algae are.

"What we hope to do is differentiate algae by their fluorescence properties and determine whether there is a connection between algae health and the production of off-flavor compounds," says Millie, a microbiologist at ARS's Southern Regional Research Center in New Orleans, Louisiana. "While initial results were promising, we need to finish analyzing all the results," says Millie's colleague, plant physiologist Dionigi. "The instruments were designed for clear, blue ocean water and might not provide reliable information in murky catfish ponds."

Source: Agricultural Research Magazine.

of live fish is essential to ensure that the fish are fresh and palatable when they reach the consumer (Figure 16–19).

The Culture of Cool-Water Food Fish

Other fish that are grown in the United States are salmon, trout, and striped bass. These are generally termed cool-water fish because they do best in cooler water than that suited for

Figure 16–19 *Fish must be alive when they reach the market.* Courtesy of James Strawser, The University of Georgia.

warm-water fish. For instance, trout thrive in water that is below 65°F and would die in the warmer temperatures where catfish thrive.

Salmon and trout are raised in the northern parts of the United States where the weather is cooler, and the water temperature is not as high as in the South. The largest portion of the trout produced comes from the Snake River Region of Idaho. These fish are raised mostly in long concrete structures called **raceways** that provide a constant flow of clean, cool water. The oxygen level is also enhanced by the free-moving water (Figure 16–20).

Figure 16–20 *Salmon and trout are raised in long raceways that provide constant movements of water.* Courtesy of The University of Georgia.

Figure 16–21 *Eggs are harvested from the female by gently applying pressure to the fish's abdomen.*
Courtesy of University of North Carolina Agricultural Research Services.

Figure 16–22 *The sperm is collected and mixed with the eggs.*
Courtesy of University of North Carolina Agricultural Research Services.

The spawning of trout, salmon, and bass is often done artificially. In this process, called **milking**, a technician injects the female with hormones to accelerate the ovulation process. After the proper time lapse, the technician gently applies pressure to the lower abdomen of the fish and causes the eggs to ooze out (Figure 16–21). A similar process is applied to the male, and sperm is collected. The sperm is applied to the eggs for fertilization, and the eggs are placed in hatching jars where the oxygen level, water motion, and temperature are carefully controlled (Figure 16–22). After hatching, the fry are placed in tanks where they are fed until they are large enough to stock the raceways.

The Production of Nonfood Fish

A rapidly growing segment of the aqua-culture industry is the production of aquarium fish (Figure 16–23). Hundreds of species of fish are raised to go into home aquariums. These fish are native to the tropical areas and vary greatly in size, shape, and color. Because some of these fish can be quite valuable, the producers can afford to spend more time and money in the production. They are generally raised in a closed-loop system that consists of a tank, a filtration system, and an aeration system. One type of fish, the koi, is particularly valuable (Figure 16–24). The fish are native to Japan and are a variety of carp. These fish have beautiful varying patterns of gold,

Figure 16–23 A rapidly growing segment of aquaculture is the production of tropical fish. *Courtesy of James Strawser, The University of Georgia.*

white, and black colors and are used in landscape ponds. Depending on the color pattern and pedigree, individual fish can cost into the thousands of dollars. In fact, breeding fish have been known to sell for more than a million dollars.

Another type of nonfood fish that is produced is the grass carp. These fish eat aquatic vegetation and are raised to help control weeds in fish ponds (Figure 16–25). They grow to large sizes and when stocked in adequate numbers can be quite effective in controlling aquatic weeds. These fish represent a welcome alternative to chemical weed control in lakes

Figure 16–24 Koi are raised for landscaping ponds. *Courtesy of James Strawser, The University of Georgia.*

Figure 16–25 *A common problem in ponds is weeds.* Courtesy of Educational Television, Auburn University.

and ponds. Grass carp are usually sold as sterile fish so they cannot reproduce and take over ponds (Figure 16–26). Some states restrict the release of grass carp that are not sterile.

Figure 16–26 *Sterile grass carp are released to eat weeds in ponds. They can be very effective.* Courtesy of George Lewis, The University of Georgia.

WATER QUALITY

In a sense fish are no different than other animals in their requirements. They must have the proper environment in which to live. This includes a food supply, water, oxygen, and a clean place in which to live. Animals such as humans get all of these requirements from the land. We obtain both our food and our oxygen from our environment, and fish are no different in this respect. The difference is that the environment of fish is enclosed in water while ours is encased in air. The quality of the environment (the water) determines whether or not they live and how well they grow.

Oxygen Levels

Land animals take air into their lungs where the oxygen is separated and taken into the bloodstream. Fish live entirely in the water and have to obtain oxygen from the water. Instead of lungs, fish are equipped with **gills** that take the oxygen from the water and put it into the bloodstream (Figure 16–27). Although a major component of water is oxygen, the gills of fish do not break the water down into the component elements of hydrogen and oxygen. The usable oxygen comes from oxygen that has been dissolved in the water. In most fish, the water is taken in through the mouth of the fish and

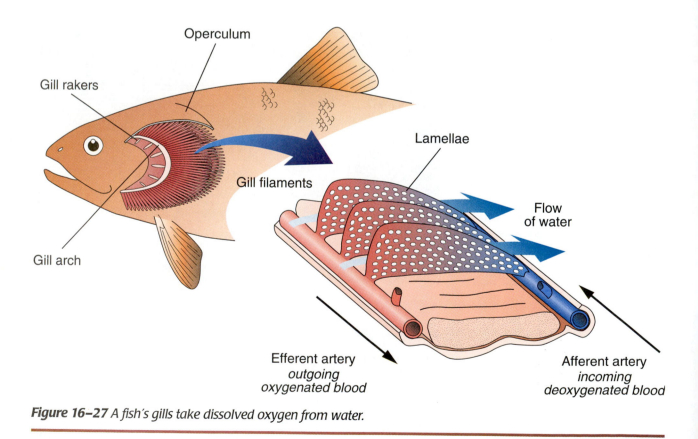

Operculum

Gill rakers

Gill arch

Gill filaments

Lamellae

Flow of water

Efferent artery
outgoing oxygenated blood

Afferent artery
incoming deoxygenated blood

Figure 16–27 A fish's gills take dissolved oxygen from water.

is forced out through the gills. You may have noticed in watching fish that their mouths are constantly opening and closing. The mouth acts like a valve that lets water in and pushes it out again.

The gills are composed of rows of curved structures called the gill arch. The gill arch is made up of fine, feather-shaped filaments (also known as primary lamellae) that run across the length of the gill arch. These filaments contain very thin, flat plates called secondary lamellae. The secondary lamella contain blood vessels that absorb the oxygen as the water passes through the gills. The blood flows through the vessels in the opposite direction to the water that flows through the gills. This countercurrent provides a more efficient means of removing the dissolved oxygen from the water. The surface area of the gill filaments can be from 10 to 60 times as great (depending on the species) as the total surface area of the fish. The gills of a fish are very efficient. They can, under ideal conditions, remove more than 80 percent of the dissolved oxygen from the water. In contrast, human lungs can extract only about 25 percent of the oxygen that is in the air.

The greatest limiting factor in the number of fish that can be grown in a particular body of water is the dissolved oxygen level of the water. This oxygen gets into the water in several ways. The wind blowing over the water puts oxygen into it, and the ripples caused by movement of the water over stones or waterfalls can also add oxygen (Figure 16–28).

Figure 16–28 *The wind can place oxygen in the water.* *Courtesy of James Strawser, The University of Georgia.*

However, the greatest amount of dissolved oxygen added to the water comes about through the photosynthesis of aquatic plants. Remember from Chapter 6 that as plants undergo photosynthesis, oxygen is released. If the plants live in the water, the oxygen released is dissolved in the water. Aquatic plant life can vary from the tiny one-celled phytoplankton and algae to the large leafy aquatic weeds such as duckweed (Figure 16–29). To a large degree, the fish owe their lives to the plants in the water that give off the oxygen. However, too

Figure 16–29 Phytoplankton can also put oxygen in the water.

thick of a plant population in the water can have a detrimental effect on the amount of oxygen in the water. If the water is too full of plant life, the sun may be shaded from the lower levels of the water. If this happens, that part of the water will not have enough sunlight to support the photosynthesis of plants in that region. Without this process, the plants cannot live and provide oxygen. It is essential that the proper balance of plant life be maintained for the adequate supply of oxygen.

Warm-water fish such as catfish normally need about 4 to 5 parts per million (ppm) of dissolved oxygen in the water. Cool-water fish may require much higher levels. Although fish can survive for short periods at levels around 0.5 ppm, when the levels get to around 1 to 2 ppm for several hours, fish begin to die. Because the depletion of oxygen can take place very suddenly, oxygen levels are monitored constantly in fish ponds.

Several factors can cause the depletion of oxygen from the water. Lakes and ponds literally teem with life. Aquatic animal life includes crustaceans such as crayfish, water insects, and bacteria. All of these compete for the dissolved oxygen in the water. By far the greatest competitors are the bacteria in the water. When any organic material begins to decay, bacteria are instrumental in the process. Waste from the fish, dead vegetation, and dead aquatic animals all provide a rich food supply for bacteria. Given the warm wet conditions of the pond and the food supply of waste material, bacteria multiply extremely fast. Because these bacteria (called aerobic bacteria) need oxygen to live, a significant increase in the level of bacteria in a pond can cause the oxygen level to fall dramatically.

Another phenomenon that causes problems is **thermoclines** in the water. Ideally, water in a pond is constantly moving. Warm water is less dense than cold water, and the warmer the water becomes, the less dense it becomes. As the water temperature increases, it may become so light that it will not mix with the cooler water beneath it. The cooler water is more dense, so it lies at the bottom of the pond. If the water is deep enough, the sun will not penetrate to the deeper, cooler regions of the pond. When this occurs, the pond is said to be stratified. Three distinct layers of water develop (Figure 16–30). The top layer, called the epilimnion, contains water that is circulated by the wind and contains dissolved oxygen from the wind action. The second layer, called the thermocline, is the portion that the sun does not penetrate very well and is as much as 10°F cooler than the epilimnion. The bottom layer is known as the hypolimnion. This layer is the coolest, densest water and is not circulated by the wind. In this layer is the decaying organic matter that has sunk to the bottom of the lake or pond. Bacteria can multiply and use up

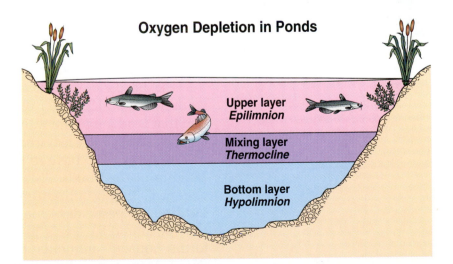

Figure 16–30 *Ponds can develop thermoclines where layers of water have little dissolved oxygen.*

all of the oxygen, and if the water is not recirculated, the oxygen is not replenished. Fish will not enter this layer because of the lack of oxygen. Problems for producers occur when a windstorm brings this layer to the surface. The dissolved oxygen content of the top layer is diluted with the water from the bottom layers, and there may not be enough oxygen for the fish to survive.

Oxygen may be depleted by hot summer nights. Remember that photosynthesis only occurs in the daylight when the plants can get energy from the sun. At night the fish depend on the oxygen that was given off during the day. Warm water holds less dissolved oxygen than cool water, so the hotter the water temperature the lower the oxygen content. So on very hot nights when the water is warm and there is no photosynthesis and no wind to circulate the water, the fish can use up their reserve of oxygen.

Producers are well aware that hot still nights promote conditions that cause oxygen depletion and monitor the level of oxygen in the pond throughout the night. They use a device called an **oxygen meter** to determine the oxygen level in the water (Figure 16–31). The device consists of a probe that is lowered into the water and a meter that registers the amount of dissolved oxygen in the water. The meter sends a charge of electricity from its batteries out to the probe in the water. If oxygen is present, the electrical current flows between two cathodes. The higher the content of oxygen, the stronger the current. The meter interprets the flow of the electrical current as the amount of dissolved oxygen in the water.

If the level of the oxygen is lower than the ideal, the producer operates **aerators** that are placed in the pond (Figure 16–32). There are several types of aerators ranging from permanently mounted electrically driven models to homemade devices that operate from the power takeoff (PTO) shaft of a tractor. All the devices serve the same purpose—throwing water into the air to increase the oxygen level. The higher the water is thrown and the smaller the particles of water, the more oxygen that is added to the water. By using these devices, producers may save all the fish from dying in a pond overnight.

Producers also have to be concerned about the chemical buildup in the ponds. Fish may be very sensitive to certain chemicals that occur naturally or that seep into a pond from fertilizers or pesticides. One of the most common problems is that of ammonia levels in ponds. Ammonia comes in two forms: ionized ammonia (NH_4^+) and unionized ammonia (NH_3). Ammonia in the unionized form is toxic to fish and can

Figure 16–31 *The oxygen content of ponds is carefully checked using an oxygen meter.*
Courtesy of George Lewis, The University of Georgia.

Figure 16–32 *Mechanical aerators are used when the oxygen levels become too low. The machine throws water into the air.*
Courtesy of The University of Georgia.

kill them if the concentration is high enough. This problem occurs in ponds that have very dense populations of fish. Producers who grow more than 5,000 pounds of catfish per acre can have problems with the buildup of ammonia in their ponds.

This chemical comes from the several sources. The largest single source is fish feed. As the excess feed is broken down, ammonia is released. Another very critical source is the waste excreted by the fish. This material contains a high concentration of ammonia, and the excreta from thousands of fish can be substantial. Ammonia is also added to the water by the decay of dead plant materials such as algae.

A proper amount of healthy algae is the best means of dissipating the levels of ammonia. If the algae suddenly die off, the levels of ammonia can reach dangerous levels. When this happens, the producer may suspend feeding for a short period, flush the pond with fresh water, and aerate the water. With the removal of the ammonia and the introduction of fresh aerated water, the algae can again grow.

Runoff from crop lands can also cause problems. Runoff is the water that flows along the ground after a rain. Rainwater usually goes to the lowest point, and ponds are commonly constructed in areas to take advantage of the runoff. This can cause a lot of soil to enter the lake from cultivated fields. The pond then becomes muddy, and the quality of the water is lowered (Figure 16–33). The suspended soil particles prevent the sun from penetrating the lower levels of the water. If the sunlight does not reach the plant life, photosynthesis cannot occur, and the oxygen level is lowered.

Figure 16–33 *Muddy water is of low quality because it blocks sunlight and impairs photosynthesis.*

If crops are sprayed with pesticides and the pesticides are carried into the pond in runoff, fish kills can occur. Remember from the section on the respiration of fish that a high volume of water passes through the gills of fish in order for them to absorb oxygen from the water. As oxygen is absorbed, chemical contaminants can also be absorbed. The contaminants absorbed by the gills go into the bloodstream and are in the fat tissues of the fish. If the levels of contaminants are high enough, the fish may die. If the fish lives and is used for food, the contaminants stored in the fish can be passed on to humans. Producers are very careful that chemical pesticides used on crops do not enter the ponds of fish being raised for food. Fish that are sold to the public are closely monitored to prevent a contaminated food supply.

SUMMARY

The future looks bright for the aqua-culture industry. The demand for fish and other aquatic products will increase, and through the efforts of scientific research and cultural practices, this demand will be met.

CHAPTER 16 REVIEW

Student Learning Activities

1. Visit the grocery store and compare the cost of ocean fish to that of farm-raised catfish. List all of the types of food that were grown in the water.

2. Gather information on a type of fish or other aquatic animal. Analyze the potential of the fish or animal for aquaculture. Be sure to list all the advantages and disadvantages. You might want to contact game and wildlife officials in your area. Report to the class.

Define the Following Terms

1. carp
2. catfish
3. seine
4. raceways
5. milking

6. gills
7. thermoclines
8. oxygen meter
9. aerators
10. runoff

True/False

1. Aquaculture is one of the fastest growing industries in agriculture.
2. There is no limit to the amount of fish that can be grown.
3. Most fish is lower in cholesterol and has fewer calories than red meat.
4. Carp is not generally considered to be a very desirable food fish.
5. The age of the fish, not the weight, seems to be the crucial factor in egg production.
6. Oxygen may be depleted by hot summer nights.
7. Warm water holds more dissolved oxygen than cool water.
8. Grass carp are usually sold as sterile fish so they cannot take over ponds.
9. Gills of fish break down water into the component elements of hydrogen and oxygen.
10. The mouths of fish are constantly opening and closing.

Fill in the Blanks

1. The most important phase of the aqua-culture industry is the production of _____.

2. Groups such as the _____ promoted the rearing of fish to replenish waterways in the wild.

3. The countries in _____ are the largest producers and consumers of fish in the world.

4. North America ranks _____ in the production of fish.

5. The most commonly produced species in the United States is _____ .

6. The most common way to produce catfish is in _____.

7. An alternative to rearing fish in an open pond is the use of _____.

8. A rapidly growing segment of aquaculture is the production of _____ fish.

9. A closed loop system consists of a _____, _____, and a _____.

10. The four most common species of fish grown in the United States are _____, _____, _____, and _____.

Multiple Choice

1. Fish food should have
   A. 35% protein.
   B. 50% protein.
   C. 75% protein.

2. The state leading in catfish production is
   A. Missouri.
   B. Mississippi.
   C. Louisiana.

3. The greatest competitors for oxygen in the water are
   A. bugs.
   B. plants.
   C. bacteria.

4. One of the most common problems producers face is
   A. algae in ponds.
   B. ammonia in ponds.
   C. insects in ponds.

5. The largest single source of ammonia is
   A. fish food.
   B. plants.
   C. chemicals.

6. The best means of dissipating levels of ammonia is
   A. insects.
   B. algae.
   C. snails.

7. Grass carp can help control
   A. oxygen depletion.
   B. insects.
   C. aquatic weeds.

8. The bottom layer of the water is known as
   A. hypolimnion.
   B. thermocline.
   C. epilimnion.

9. The top layer of water is circulated by
   A. decaying organic matter.
   B. bacteria.
   C. winds.

10. Fish will not enter the bottom layer of water due to a lack of
    A. food.
    B. oxygen.
    C. warmth.

Discussion

1. What advantages do fish have over other agricultural animals?
2. What are the disadvantages of growing fish for profit?
3. Why is the tilapia popular as a food fish?
4. What are the advantages and disadvantages of cages?
5. What is pelleted feed?
6. List some cool-water fish.
7. Give two examples of nonfood fish.
8. Explain how the gills of fish take in oxygen.
9. How is oxygen dissolved in water?
10. How can pesticides affect fish?

AGRICULTURE AND THE ENVIRONMENT

KEY TERMS

environment
hydrologic cycle
groundwater
pollution
nonpoint
water solubility
suspended solution
half-life
surfactant
erosion

STUDENT OBJECTIVES

After studying this chapter, you should be able to:

✦ Explain the relationship between agriculture and the environment.

✦ Explain how the hydrological cycle is related to water storage.

✦ Distinguish between point and nonpoint sources of pollution.

✦ List the ways agriculture can potentially pollute the environment.

✦ Discuss how the use of modern pesticides is less environmentally hazardous than previously used chemicals.

✦ Contrast the hazards and control measures associated with animal waste.

✦ Explain how soil erosion causes pollution.

✦ Defend the preservation of wetlands.

✦ Explain the dangers of water table depletion and conservation methods.

Almost six billion people are in the world, and during your lifetime this number will probably double (Figure 17–1). This statistic has a profound impact in at least two areas, agriculture and the **environment**. Because the planet is not getting any larger, the amount of space for each human

Figure 17–1 *During your lifetime the population of the earth will probably double.* Courtesy of Photo Disc, Adalberto Rios Szalay/Sexto Sol.

Figure 17–2 *When European settlers arrived, forests of huge trees had to be cleared.*

will get smaller. The environment is all the external conditions that affect an organism or community and influence its development or existence. For example, the surrounding air, light, moisture, temperature, wind, soil, and organisms are all parts of the environment. A limited amount of all these factors exists, and as the world's population increases, the amount available to individual humans will shrink.

There will be increased demand for more food and clothing for the growing population. If the population doubles, it stands to reason that double the amount of food and double the amount of fiber will be needed to feed, clothe, and shelter the people. All food and a large portion of the fiber for clothing and housing comes from the growing of plants and animals. As more of these essential materials are needed, the greater the pressure on the agricultural sector to produce them. This increased need will probably have to be met through more intensive efforts because the available land suitable for the production of food and fiber is shrinking rather than growing. This is because of the need for housing, factories, and other growth that takes productive land out of use. As the intensity of production increases, steps must be taken to protect the environment.

Since humans first began to grow their own food either as crops or in herds of animals, the environment has changed. Trees and other plants had to be removed to provide room for cultivated crops or grazing for animals. This caused an interruption of the natural succession of plants in the area. As people settled in one particular area, the nature of the environment changed. As the dominant animal species of the area, humans affected the way other animals lived and the way the plant life grew. The larger the number of humans in a confined area, the greater the effect on the environment. The history of the United States is a good example. Even before European settlers arrived, humans had changed the environment. Huge tracts of land grew tall grass instead of trees. This was due at least in part to the Native Americans who periodically burned the plains to make better grazing for the buffalo.

As pointed out in Chapter 15, when European settlers arrived, they began to clear away the huge forests and plant crops (Figure 17–2). The environment of the continent was dramatically changed. By the turn of the 20th century, much of the native wildlife had disappeared from the areas that were settled. Because a majority of the land had been cleared for farming, the wildlife simply lost their habitat. At one time, animals such as the buffalo, white-tailed deer, and wild turkey were rare and almost extinct. Some species, such as the passenger pigeon and the Carolina parakeet, became extinct through a combination of loss of habitat and over-hunting. According to the U.S. Fish and Wildlife Service, more than 500 species, subspecies, and varieties of plants and animals in this country have become extinct since the Pilgrims landed at Plymouth Rock in 1620. However, all of the extinctions were not due to the intervention of humans. Some scientists believe that all species of animals will eventually become extinct as others take their place. The point is that humans have been the cause of the destruction of at least some plant or animal life.

There are strong voices on each side of the issue of providing habitat for wildlife, especially for those species listed as endangered. Some people say we should save every species of plant and animal life we can. They argue that all life has value and a specific role in the ecology of an area, and that once these species are lost, they can never be regained. They point out that uses may be found for plants and animals that were never dreamed of. For example, a lowly fungus that was considered undesirable was used to develop the life-saving drug penicillin. Wildlife advocates feel that agriculture or any other type of encroachment should not be allowed to interfere with the habitat of wildlife. People with opposing views feel that a strong agriculture is essential to the survival of humans and that it should be given priority regardless of the consequences to wildlife.

Fortunately, compromises between the two viewpoints have been worked out. Most agriculturists care about the environment and wildlife habitats and may be among the strongest advocates for achieving a balance between agricultural production and wildlife. Areas where wildlife had previously been almost eliminated now have substantial populations. A striking example is that of the white-tailed deer (Figure 17–3). At one time the loss of habitat and strong hunting pressure had almost wiped out the deer. Today, through a variety of measures, there are more white-tailed deer in the United States than before Europeans arrived. In fact, in some areas the deer are overpopulated and are causing problems with crop production. Because of this, the controversy between proponents of wildlife and agricultural interests continue (Figure 17–4). Other areas are affected also. In the West, sheep herders have problems with coyotes preying on lambs. In remote areas, these predators can cause severe damage to

Figure 17–3 The white-tailed deer in this country has made a tremendous comeback. *Courtesy of PhotoDisc.*

a herd of sheep. Wildlife advocates contend that it is only natural for the coyotes to prey on the sheep and that they should not be destroyed because of it.

Figure 17–4 Wildlife can be considered a nuisance when they destroy crops. *Courtesy of Tom Dodge, Successful Farming.*

WATER POLLUTION

The pollution of our water in the United States has developed into an extremely serious problem. In a sense, all water on earth is connected. In a cycle known as the **hydrologic cycle**, water is constantly cycled through the process of evaporation, condensation, and precipitation (Figure 17–5). Precipitation in

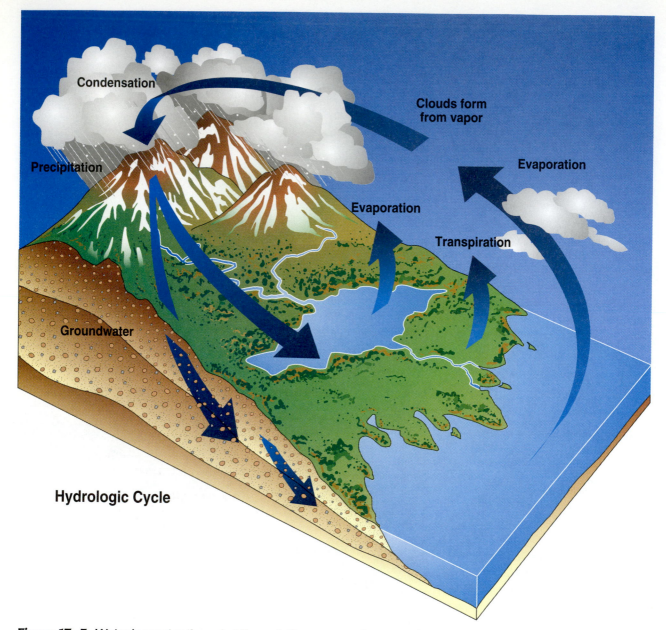

Figure 17–5 *Water is constantly cycled through the process of evaporation, condensation, and precipitation.*

the form of rainfall or melting snow and sleet eventually winds up (through a process called runoff) in storage in either surface water or **groundwater**. Water collects in surface storage (surface water) in streams that empty into larger streams that empty into lakes or the oceans. By far the largest storage of surface water is in the oceans. Unfortunately, ocean water contains salts that render it unfit for drinking, irrigation, and most manufacturing needs. The remaining water, called freshwater, sustains the life of most of the Earth's plants and animals. Groundwater (also known as phreatic water) is water that is under the surface of the ground and makes up the water table that supplies wells. This water is stored in rock, sand, or gravel formations called aquifers. These formations are permeable (allow water to pass through) and are saturated with water. As rainwater falls it hits the ground and is either absorbed into the ground or begins to run off the surface of the ground. As runoff water goes across the ground, some is absorbed into the ground, and the rest winds up in surface water storage such as lakes and streams. The water that seeps into the ground continues through the ground until it reaches the saturation point or the point beyond which it can travel no further. Groundwater is not stored in large underground lakes or oceans except in the rare instances of underground caverns. Rather it accumulates and is stored in the porous rock and gravel far below the ground surface. Wells are drilled through the ground into the layers of the porous rock, sand, or gravel and water is pumped from this area. According to the U.S. Environmental Protection Agency (EPA), groundwater accounts for about 96 percent of the world's total freshwater resources.

Water **pollution** occurs in both surface water and in groundwater. Pollution is the presence of substances in water, air, or soil that impair usefulness or render it offensive to humans. Pollution became so bad that at one time many large streams and lakes in the United States could not support fish that were fit for human consumption. Some bodies of water were so polluted that nothing would live in them. In 1970, former President Nixon signed into law the National Environmental Policy Act, which led to the establishment of the Environmental Protection Agency (EPA). This agency was charged by Congress to "protect the nation's land, air, and water systems." Policies, laws, and regulations have been and continue to be established to prevent pollution and to clean up pollution that has already occurred. In the years the agency has been in operation, a lot of progress has been made in cleaning up the air and water. Many bodies of water that were once considered dead now have been brought back to life by cleaning up the water and stopping pollution.

Pollution comes about as a result of contaminants entering the water in two main ways: point and **nonpoint** sources. A point source is from a specific place. For instance, a factory that dumps waste in a river is a point source of pollution because the contaminants can be traced to a single point. The solution to this problem is relatively easy because the source of the pollution can be identified and the problem corrected. In this area the EPA has made tremendous strides in halting stream pollution. Industries have been required by law to stop dumping or allowing contaminants to enter waterways (Figure 17–6). Water used in industry and returned to streams or lakes is carefully monitored to ensure that it contains no polluting substances.

Figure 17–6 *Industry has in the past been sources of point pollution. Most have now stopped polluting.* Courtesy of James Strawser, The University of Georgia.

The other type of pollution, nonpoint, is more difficult to deal with. Nonpoint means that the pollution comes not from a single point but from a wide area that is difficult to identify. The pollution is usually from a number of sources that collectively affect the environment. Agriculture is the largest origin of nonpoint pollution in the United States (Figure 17–7). This is due in part to the enormous size of the agricultural industry, in which millions of acres of land are in production. Each year this land receives treatment in the form of plowing, cultivating, storage of animal wastes, and the application of pesticides. As a result, agricultural pollution comes from several sources: fertilizers, pesticides, animal wastes, and soil erosion (Figure 17–8).

Pesticides

Each year about 570 million pounds of pesticides are used to control insects, weeds, and fungi (Figure 17–9). The vast majority of the crops produced in this country use chemical pesticides as a major component of the production plan. Most

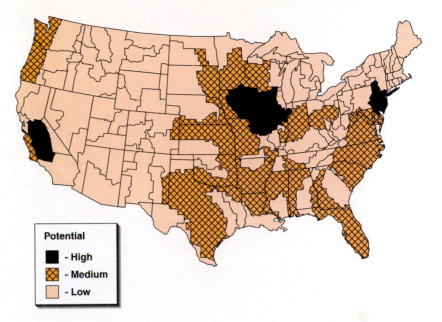

Figure 17–7 Agriculture is the largest source of nonpoint pollution. *Courtesy of the National Council for Agricultural Education.*

Potential
- High
- Medium
- Low

of these are applied directly to crops either over the top or in the soil. Rainwater dissolves pesticides and provides the means for them to enter the ground. This creates the potential for a buildup of the pesticides in both surface and groundwater. At one time it was thought that pesticides would not be able to reach groundwater because the chemicals would have to travel such a long distance through the soil. However, in 1980 pesticide residues were discovered in groundwater, and that theory was dismissed. Since that time, the EPA has conducted several extensive studies to determine the extent of pesticide pollution in groundwater. The EPA has concluded that fewer than 1 percent of rural domestic and community water system wells contain unacceptable levels of pesticides. Where dangerous levels of pesticide contamination occur, it is usually the result of accidental pesticide spills rather than normal application.

Certain areas of the country are more susceptible to pesticide residues in the groundwater than others (Figure 17–10). These differences are referred to as the vulnerability to contamination by pesticides. Several factors can cause a particular site to be more vulnerable than another site.

One large factor is called the *hydrological vulnerability* and is based on such geological factors as soil depth and texture. Obviously, the deeper the groundwater is below the surface, the farther the pesticides have to travel. Because of this, water from a deep well is less likely to be contaminated than water from a shallow well. Soil texture is also a large factor.

Nonpoint Source Pollution Effects

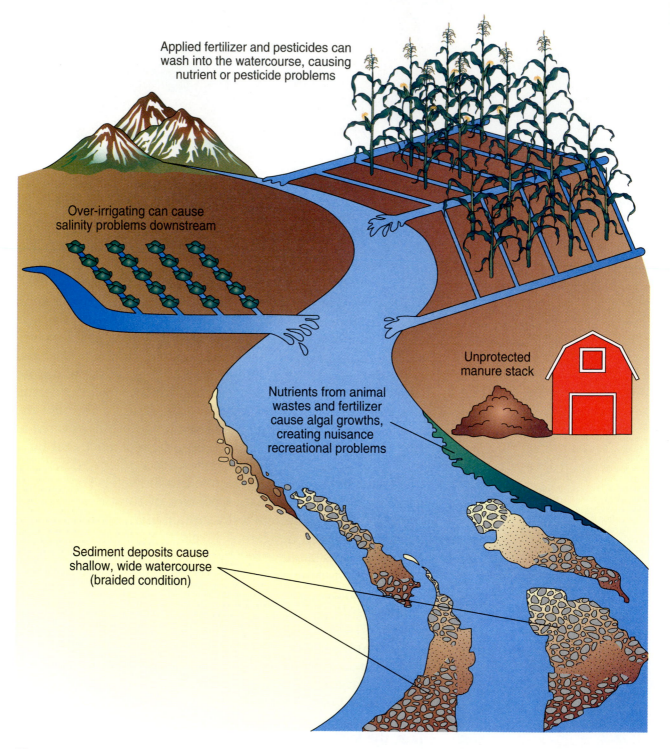

Applied fertilizer and pesticides can wash into the watercourse, causing nutrient or pesticide problems

Over-irrigating can cause salinity problems downstream

Unprotected manure stack

Nutrients from animal wastes and fertilizer cause algal growths, creating nuisance recreational problems

Sediment deposits cause shallow, wide watercourse (braided condition)

Figure 17–8 *Agricultural pollution comes from several sources.*

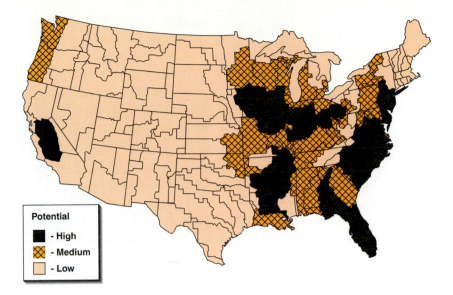

Figure 17–9 *Each year about 570 million pounds of pesticides are used in the United States. Courtesy of the National Council for Agricultural Education.*

Potential
- ■ – High
- ⬛ – Medium
- ⬜ – Low

Coarse, sandy soils allow water to pass through more rapidly than heavier clay soils. The more clay content a soil has, the slower pesticide-contaminated water moves. Also, pesticide molecules adhere more readily to clay than they do to sand.

Other factors that make a site more vulnerable to contamination are the practices used in applying the pesticides. Logically, the larger the amounts of pesticides that are applied, the greater the risk (Figure 17–11). However, other factors, such as the characteristics of the pesticide, contribute to the risk. Several of these factors are discussed here:

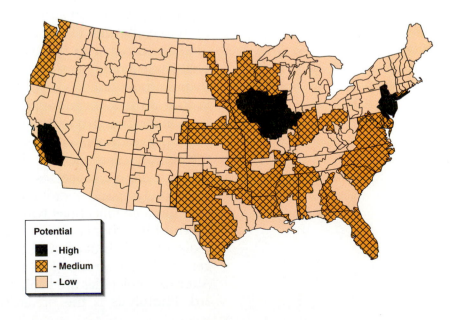

Figure 17–10 *Some areas are more susceptible to water pollution. Courtesy of the National Council for Agricultural Education.*

Potential
- ■ – High
- ⬛ – Medium
- ⬜ – Low

Figure 17–11 *Low-volume sprayers can help reduce the amount of pesticides used.*
Courtesy of P. Sumner.

Water Solubility. How easily a chemical pesticide dissolves in water, its **water solubility**, determines to a large degree how much of the chemical is transported through the soil. The more readily a pesticide goes into solution with water, the more likely the chemical is to be transported through the soil. Pesticides must be able to be used in water to allow them to be applied. Most go into suspension in water rather than being dissolved. A **suspended solution** means that the particles separate from the water if they are allowed to sit for a while. This is why most spray equipment recalculates and agitates the pesticide solution.

Half-Life of the Chemical. Remember from Chapter 14 that modern pesticides do not last very long in the soil. They begin to break down almost as soon as they are applied, and therefore, they are less of a threat to the environment. This is in sharp contrast to the older (now banned) pesticides that could last many years in the soil. For example, chlorinated hydrocarbons such as chlordane, toxaphene, and DDT could last for 20 years or more in the soil. A pesticide with a **half-life** longer than two to three weeks may have the potential to cause groundwater contamination. A chemical's half-life is the time it takes for half of the chemical to be eliminated through natural processes. These processes include hydrolysis, photolysis, and microbial transformation. Hydrolysis is the reaction a chemical has with water that results in the breakdown of the chemical molecule. Bonds between parts of the molecules are broken, and a hydrogen (H^+) ion and an oxygen hydrogen group (HO^-) become attached to the parts. After the molecules are broken down, they may not be a hazard. Photolysis is the breakdown of a substance from the energy of the sun. Absorbed light energy can cause the breakdown of some compounds that will render them non-hazardous. Microbes in the soil also can help dissipate chemical compounds in the pesticides. Modern pesticides are designed to be receptive to break down by these forces to make them less residual. Some pesticides resist these processes more than others, so the type of pesticide has a bearing on how readily groundwater is contaminated.

Modern production methods lessen the impact of pesticides on ground and surface water as well as other aspects of the environment. Choosing pesticides that are least residual and using only the minimal amount necessary helps the problem. The use of integrated pest management (IPM) greatly reduces the amount of pesticide needed (see Chapter 14).

Modern application methods also can be effective in reducing the potential for contamination. The size of the droplets in the spray affects how much of the spray sticks on the plant. Generally, the smaller the drops, the better the spray can be retained on the plant, and the amount contacting the soil is minimized. Also, the use of a **surfactant** in the spray solution helps the spray to adhere to the plant. Most plants have a waxlike coating on the exterior that causes water to "bead up" (Figure 17–12). A surfactant is a detergent-like substance that is added to a spray mixture to cause the liquid to disperse across the plant surface better. Electrostatic sprayers create a negative electrical charge that is attracted by a positive charge on the plant (Figure 17–13). This causes the spray droplets to go directly to the plant surface and stick, resulting in a lower volume of pesticide and less contact with the ground.

Figure 17–12 *Plants have a waxylike substance that causes water to "bead up." A surfactant helps the spray to adhere. Courtesy of USDA-ARS.*

Figure 17–13 *Electrostatic sprayers help get more spray on the plant and less on the ground.* Courtesy of Ed Law, The University of Georgia.

Fertilizers and Animal Wastes

Water is also polluted by commercial fertilizers and animal wastes (Figure 17–14). These materials both contain high amounts of nitrates, which are easily dissolved in water. Rainfall in areas with high concentrations of commercial fertilizers or large amounts of animal manure can carry a concentration of nitrates in the runoff. When these materials reach streams and lakes, high levels of nitrates can accumulate. Remember from Chapter 16 that water with a high level of nitrates can be detrimental to fish and other aquatic wildlife.

Figure 17–14 *Water is polluted through the use of commercial fertilizers and animal wastes.* Courtesy of John Deere & Company.

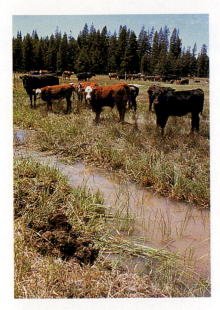

Figure 17–15 *Manure that is dropped in a stream can carry nitrates and bacteria for miles.* Courtesy of USDA-ARS.

The Clean Water Acts of 1970 and 1990 contain rules and regulations aimed at preventing pollution from agricultural sources. A major part of the legislation deals with regulating animal wastes that might get into surface or groundwater. Agricultural animals should have no contact with running streams of water. Manure dropped directly into a stream can carry nitrates and bacteria miles from the source (Figure 17–15).

Animals raised in confinement produce a lot of manure in a small area (Figure 17–16). This material is generally used on fields as a fertilizer. However, because such enormous amounts of the material is produced in a large operation, it must often be stockpiled until it can be spread on the fields. Large piles of animal waste can be the source of high levels of bacteria and nitrates in runoff water. The *E. coli* bacteria in animal feces can cause several diseases in humans and also may infect other agricultural animals. If these bacteria get into a water supply that is to be used for human consumption, disease may occur.

Two practices can be used to help control the problem. The first is to control runoff from rainwater. Diversion ditches are cut into the area surrounding the livestock facility to prevent runoff containing waste materials from getting into streams. The drainage ditches divert the runoff into holding areas where the risk from runoff is at a minimum. Also, the rainwater runoff from the roof of the facilities can be collected by a gutter system. This prevents the water from running across the barnyard and adding to the amount of runoff.

Figure 17–16 *Animals in confinement operations produce a lot of waste.* Couresty of Progressive Farmer.

BIO BRIEF

Managing Poultry Manure Nutrients

Scientists in the ARS Poultry Production and Product Safety Research Unit at Fayetteville, Arkansas, are focusing on soil properties and slope, the seasonal dynamics of surface runoff, and the effects of adding aluminum sulfate to poultry litter to optimize manure application.

"We're searching for ways to reduce nutrient runoff from poultry litter into lakes and streams," says soil chemist Philip A. Moore, Jr. "The first step is determining where runoff does and does not occur, so we can find better ways to manage it."

More than 7.5 billion broiler chickens raised in the United States each year produce up to 7 million tons of poultry litter. The litter is a mixture of chicken manure, feathers, spilled food, and bedding material. Many farmers use it as an inexpensive fertilizer for cropland because the manure contains nitrogen and phosphorus, two important fertilizer ingredients. However, water that runs off fields fertilized with poultry litter may carry excess nutrients to nearby waterways, hurting water quality and aquatic life.

Soil scientist Thomas J, Sauer is looking at ways to manage litter to minimize this runoff. "Many things can influence nutrient runoff after animal manure is applied, including weather, physical and chemical properties of the soil, and land use," he says.

In a recent field study, Sauer found that trends in the measured soil's chemical and physical properties were related to its slope and plant cover—for example, pasture versus forest. If farmers take these things into consideration, they can reduce nutrient runoff. For example, silty alluvial soils found near rivers are potentially important in capturing runoff from upland pastures before it reaches the river.

Sauer says additional research is needed to verify the role that areas along streams and rivers play in retaining nutrients delivered from uphill pastures. He also says that farmers developing best-management practices for poultry litter application should pay close attention to areas with low water infiltration rates and to upland pasture soils' ability to retain applied nutrients.

At a site in Arkansas, Sauer and Moore studied two small watersheds within a tall fescue pasture. They installed automated equipment to measure the rate and composition of runoff. They also installed sensors to measure precipitation and soil moisture and to continuously record evaporation.

"Some important trends are already apparent," reports Sauer. "Throughout the

The second practice is to use a lagoon as a holding tank for the manure. A lagoon is a pond of water into which the waste is directed. Here the material is decomposed by anaerobic bacteria (Figure 17–19). These bacteria live beneath the surface of

Scientists are researching ways to reduce nutrients in runoff water.
Courtesy of USDA-ARS.

"Preliminary figures from our site in Arkansas suggest that information on soil properties and slope could be used to develop and improve strategies for applying animal waste to land. This type of approach could supplement existing methods for assessing nutrient runoff.

"Poultry litter," Sauer says, "can be a valuable nutrient and organic matter resource that can help sustain and revitalize a grower's soil, if soil and landscape position information are used in designing land application programs."

summer, evaporation dominated the water balance. The tall fescue grass showed strong drought tolerance and withdrew water from deep down, after the surface layers became very dry," he says. "By summer's end, it took several significant rainfalls to replenish soil moisture in the root zone before groundwater movement began."

"We hope to relate the occurrence, timing, and volume of surface runoff to storm characteristics, height of the grass, and surface soil moisture content," says Sauer.

Farmers may be able to reduce runoff simply by planting the right grass. Moore is interested in seeing whether different grasses affect runoff. LeAnn Davis, a graduate student at the University of Arkansas who works with Moore, found that less water and phosphorus runoff occurred with tall fescue grass compared to eastern gamagrass, switchgrass, bermudagrass, or bluestem grass. They're not sure at this point in the study why differences exist in the effectiveness among the grasses.

Source: Agricultural Research Magazine.

the water and thrive without the presence of oxygen. Anaerobic bacteria break down the waste material and convert the nitrogen into a gaseous state. At the proper stage of decomposition, the sludge is pumped from the lagoon and spread on the fields.

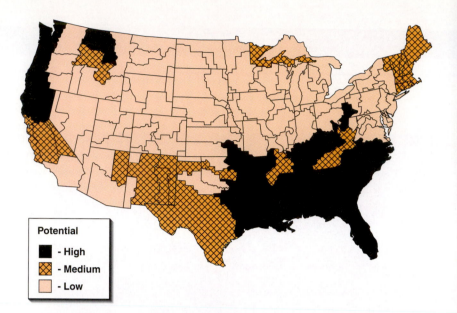

Figure 17–17 *Some parts of the country have a higher potential for pollution by animal wastes.* Courtesy of the National Council for Agricultural Education.

Potential
- **High**
- **Medium**
- **Low**

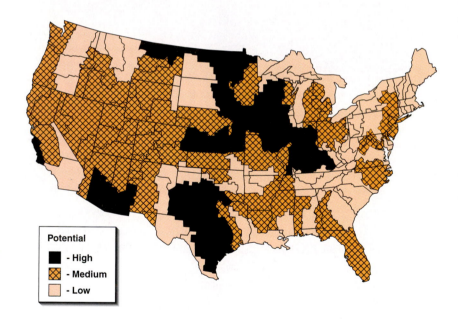

Figure 17–18 *Nutrient levels applied to crops are higher in some parts of the country.* Courtesy of the National Council for Agricultural Education.

Potential
- **High**
- **Medium**
- **Low**

Soil Erosion

Our largest environmental problem stemming from agriculture is soil **erosion** (Figure 17–20). Erosion is the wearing away of the soil by wind or water. This problem is twofold—the loss of topsoil and the pollution of surface water by the displaced soil. Remember from Chapter 2 that one of the methods of building up soil is through the transportation of soil from one place to the next. Although this may be helpful in building new soils in another area, the area from which the

Figure 17–19 In a lagoon, animal waste is decomposed by anaerobic bacteria. *Courtesy of James Strawser, The University of Georgia.*

soil was removed was damaged. This lowers productivity because the soil lost is the topsoil that contains the organic material that holds plant nutrients (see Chapter 2). Also, the thinner the topsoil, the less water-holding capacity of the soil. This makes plants more susceptible to drought. Soil that is lost is slow to recover. The Soil Conservation Service estimates that it takes about 30 years to form topsoil from the subsoil parent material.

Research indicates that the sediments from soil cause around $6 billion in damage each year. Almost all improved land in the United States has a system of drainage ditches to carry off excess water. Eroded soil clogs these waterways and has to be periodically removed (Figure 17–21). Similarly, navigable

Figure 17–20 Our largest environmental problem stemming from agriculture is soil erosion. *Courtesy of James Strawser, The University of Georgia.*

Figure 17–21 *Eroded soil clogs waterways and has to be periodically removed. Courtesy of USDA.*

waterways can become so clogged that they have to be dredged out in order for the vessels to pass through. This is a tremendously expensive process that takes a lot of time and resources.

Problems with sediments filling waterways and reservoirs is by no means a new problem (Figure 17–22). Almost from the beginning of recorded history, we have evidence of soil sediments creating problems for whole civilizations. For example, the ancient Sabeans who occupied the eastern portion of Yemen around 1000 B.C. were a wealthy society based

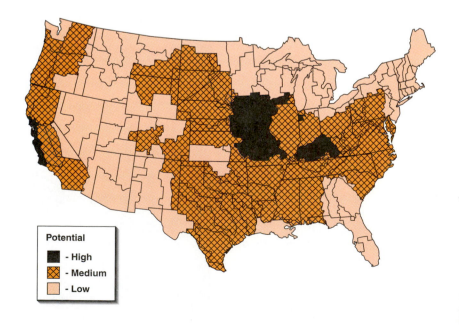

Figure 17–22 *Sediments in some parts of the country can be high. Courtesy of the National Council for Agricultural Education.*

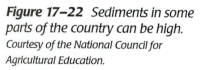

Potential
- ■ - High
- ▨ - Medium
- ▢ - Low

on agriculture. This wealth came from a complicated irrigation system that used water from a large reservoir created by building a dam across a relatively small stream. The water literally made the desert bloom as crops were grown that fed the population and provided a source of income. As a result of the sediments that moved in from the eroding of the soil, the reservoir became filled, and the water flowed over the dam. This eventually caused the dam to burst, and the reservoir was destroyed. As a result, the wealth of the nation was lost, and the whole civilization faded away. History is filled with such examples of irrigation systems that brought about prosperity and was lost due to sedimentation.

Another serious problem caused by sediment pollution is the loss of fish and other aquatic life in lakes and the streams. Some species cannot tolerate the muddy water and die from lack of oxygen or lack of clean water. As soil is washed into the water, nutrients are also brought into the streams. As a result, aquatic plant life grows at a more rapid rate, clogging the waterways. When the plants die, they fall into or remain in the water. Remember from Chapter 16 that as organic matter decays, oxygen levels can be depleted.

In Chapter 2 physical and chemical bonding of soil particles were discussed. As water droplets hit the soil during a rainfall, these bonds are broken and allow the soil to be more easily moved. As water gathers on the surface, the point is reached where the force of gravity pulls the water and the dissolved soil particles downhill. The steeper and the longer the slope, the more force the water has as it moves as runoff. The force of the running water moves additional particles that may or may not be in suspension with the water. In areas where the rainfall can be very intense, huge amounts of soil can be removed in a short time if the ground is not protected (Figure 17–23).

The agricultural industry has developed methods of protecting the soil from erosion. Intensive efforts began back in the 1930s with the creation of the Soil Conservation Service (SCS). The mission of this agency was to help prevent the loss of soil and to stop pollution caused by the washing away of the soil. The SCS has made tremendous strides in the control of erosion (Figure 17–24). Through conservation plans and demonstration projects, we have learned a lot about controlling erosion. Early efforts were aimed at teaching farmers to terrace fields and to plow across slopes instead of up and down the hill. Today, relatively little land that is steeply sloping is used for row crops.

Figure 17–23 *High amounts of rainfall can cause severe gully erosion. Courtesy of Progressive Farmer.*

Figure 17–24 *The Soil Conservation Service has developed several practices that have slowed soil erosion. Terracing is a good example.* Courtesy of USDA Natural Resources Conservation Service, Iowa.

Modern efforts are aimed at keeping as much vegetation on the ground as possible. Plants hold the soil together by creating a support system for the soil particles. Roots penetrate the soil and hold it in place. Many crops are now grown with a minimum of plowing. This not only allows more plant roots and plant cover to hold the soil but also slows the flow of water over the land. In some areas, residues of harvest are now left until spring, and the new crop is planted in the stubble (Figure 17–25).

New varieties of grasses and other plant covers are also being developed. Bare hills are planted in grasses that grow quickly, have an expansive root system, and do well on eroded soils.

Figure 17–25 *The use of no-till planting allows stubble from the last crop to aid in holding the soil.* Courtesy of James Strawser, The University of Georgia.

Wetlands

For many years, we have drained and used areas we thought to be useless swamps. These areas, known as wetlands, were once thought to serve no good purpose and were drained to be productive (Figure 17–26). It is estimated that more than half of the wetlands of the United States have been destroyed for construction or for agricultural purposes.

Figure 17–26 Wetlands were once thought to be wasteland. We now know better. Courtesy of James Strawser, The University of Georgia.

Scientists now know that wetlands provide several essential roles in the ecosystem of an area. These swampy areas filter water before it reaches open areas that store surface water. Earlier in this chapter, problems with chemicals, sediments, and other pollutants in runoff water were discussed. Wetlands catch runoff water before it reaches open, deeper bodies of surface water. The vegetation removes nutrients and other pollutants from the water, and as a result the open areas of surface water are cleaner. Wetlands also act as a sponge or storage area that collects and absorbs flood water before it can damage crops, homes, or other property.

Wetlands are critical to the life cycle of many plants and animals (Figure 17–27). They provide water for drinking, food, or shelter for a variety of animals. Some plant species will only grow in areas where the ground remains saturated for most of the growing season. If wetlands are destroyed, many of these plants may be lost forever.

The EPA has established regulations that protect wetlands by restricting their use. Incentives are offered to landowners to promote the preservation of wetland areas. Permits must be obtained from the government when wetlands are to be

Figure 17–27 Wetlands provide habitat for wildlife. Courtesy of Progressive Farmer.

disturbed. Through these measures, valuable wetlands will be protected, and in turn the whole environment will be enhanced.

DEPLETION OF WATER RESERVES

Another concern that involves both agriculture and the environment is the depletion of the nation's reserves of water. Agriculture is the nation's largest user of water, accounting for more than 80 percent of the nation's total water use (Figure 17–28). The production of crops takes a tremendous amount of water each year. Millions of acres of land are under cultivation that could not support crops by relying on rainfall. Even in areas of relatively high rainfall, producers use irrigation to supplement rainfall when precipitation is not adequate. As discussed earlier, underground aquifers are supplied with water from rainfall and other precipitation. The exception is a type of groundwater called fossil water that was stored in the ground during geologic time. This source in nonrenewable. Aquifers, both renewable and nonrenewable, are tapped as a source for water to supply irrigation systems during the growing seasons. Particularly in the western parts of the country, more water is being taken from the underground storage areas than is being replaced. This can have

Figure 17–28 Agriculture accounts for more than 80 percent of the nation's water use. Courtesy of James Strawser, The University of Georgia.

serious implications for future irrigation water, drinking water, and other uses for water.

Fortunately, in recent years the increase in water use has slowed. Several practices have led to better water conservation. For example, at one time, irrigation water was distributed by high-pressure pumps that shot water into the air. This method lost a lot of water to evaporation and runoff. Even the overhead sprinklers that were used wasted a lot of water. Today, overhead sprinklers use low-pressure, low-volume pumps that allow the water to be distributed much more slowly, and thus allow more of the water to enter the ground. Also, ground moisture meters and infrared photos tell producers when the water is needed (Figure 17–29). This monitoring helps alleviate unnecessary watering.

Figure 17–29 *Infrared photos can show areas where plants need water. Courtesy of James Strawser, The University of Georgia.*

SUMMARY

Both agriculture and the environment will continue to be important issues. We must feed and clothe the world and, at the same time, keep our planet and our local environment healthy and pleasant. Scientific inquiry and careful cooperation and planning will create new and better ways to produce food and to protect the environment.

CHAPTER 17 REVIEW

Student Learning Activities

1. Examine your community for point and nonpoint sources of pollution. Develop ways to stop the pollution. Also develop a plan to clean up the pollution that exists. Discuss the different solutions in class.

2. Photograph several places in your area where soil erosion is occurring. Devise a plan for controlling the erosion.

3. Examine a wetland area. List the different plants and animals that live there and determine which would be lost if the wetland was lost.

Define the Following Terms

1. environment
2. hydrologic cycle
3. groundwater
4. pollution
5. nonpoint
6. water solubility
7. suspended solution
8. half-life
9. surfactant
10. erosion

True/False

1. There are almost 6 billion people in the world.
2. Ocean water is best for irrigation, drinking, and most manufacturing needs.
3. Water pollution occurs in both surface and groundwater.
4. The more clay content a soil has, the slower pesticide-contaminated water moves.
5. Modern pesticides last a long time in the soil.
6. Agricultural animals should have no contact with running streams.
7. Our largest environmental problem stemming from agriculture is soil erosion.
8. Agriculture is the nation's largest user of water.
9. Fossil water is renewable.
10. Groundwater is usually stored in large, underground lakes or oceans.

Fill in the Blanks

1. Population has a profound impact on _____ and the _____.
2. The space that humans live in is called the _____.

3. More than _____ species, subspecies, and varieties of plants and animals in this country have become extinct since 1620.

4. The two major water storage types are _____ and _____.

5. Water that is stored in rock or gravel formations is called _____ .

6. _____ is the breakdown of a substance from the energy of the sun.

7. _____ in the soil can help dissipate or break down chemical compounds in some pesticides.

8. _____ hold the soil in place.

9. _____ were thought to be wastelands that served no purpose.

10. Fertilizers and animal wastes contain high amounts of _____.

Multiple Choice

1. The space that humans live in is called the
   A. environment.
   B. atmosphere.
   C. world.

2. When European settlers arrived, a majority of the land was cleared for
   A. building homes.
   B. making fires.
   C. farming.

3. Water is constantly cycled through the process of the
   A. hydrologic cycle.
   B. water cycle.
   C. both A and B.

4. The largest storage of surface water is in
   A. oceans.
   B. clouds.
   C. lakes.

5. The water that sustains the life of most of the earth's plants and animals is
   A. groundwater.
   B. freshwater.
   C. lakes.

6. Agricultural pollution comes from
   A. fertilizers.
   B. soil erosion.
   C. both A and B.

7. The hydrological vulnerability factor depends on
   A. soil taste.
   B. soil texture.
   C. both A and B.

8. Sprayed pesticide solution can adhere to the plant better when using
   A. wax.
   B. surfactant.
   C. fertilizer.

9. Many crops are now grown with a bare minimum of
   A. soil.
   B. plowing.
   C. none of the above.

10. Our largest environmental problem stemming from agriculture is
    A. pollution.
    B. runoff.
    C. soil erosion.

Discussion

1. What effect does population growth have on agriculture?
2. Explain the differences between point and nonpoint sources of pollution.
3. How can levels of nitrates and bacteria be controlled in runoff water?
4. What was the purpose of the Soil Conservation Service?
5. Why are wetlands essential?
6. What are two ways that the EPA is using to help preserve wetlands?
7. Why is ocean water unfit for drinking?
8. List the sources of agricultural pollution.
9. What is the hydrological vulnerability factor based on?
10. Should agricultural animals have contact with running streams? Why or why not?

A SAFE FOOD SUPPLY

STUDENT OBJECTIVES

After studying this chapter, you should be able to:

✦ Analyze the risks involved in the use of pesticides in food production.

✦ Explain safeguards to prevent harmful pesticide residues on food.

✦ Evaluate the risk of using hormones and antibiotics on animals raised for food.

✦ Describe the safeguards used in the meatpacking industry to ensure a wholesome product.

✦ Discuss the use of chemical preservatives in food.

✦ Distinguish between saturated and unsaturated fats.

✦ Describe the use the body makes of cholesterol.

✦ Describe the government regulations regarding the labeling of food.

Americans enjoy an abundant, diverse, and relatively inexpensive food supply. Just go into any supermarket and observe the variety of fresh and processed fruits, vegetables, dairy products, meats, seafood, and just about any other type of food you can imagine. Never in the history of the world has any civilization had access to such a broad array of foods (Figure 18–1). In sharp contrast to times in the past, most people in this country are not involved in the production and processing of the foods they eat. As our society becomes more affluent, we become more conscious of the safety of the food we eat (Figure 18–2). In the past few years, the media has reported research that seems to indicate that foods, or components of the food we eat, are either unhealthy

Figure 18–1 *Never in the history of the world has any civilization had access to such a broad array of food. Courtesy of Dow/Elanco.*

or hazardous to our health. This has caused public concern over the safety of food. It is easy to forget that properly conducted and reported scientific research is responsible for the abundant, safe supply of food that we enjoy. However, inconclusive or poorly reported research can often mislead the public. The majority of these concerns can be grouped into three categories: **pesticide** and chemical residues in the foods, questions regarding the healthiness of the food itself, and the cleanliness of the food.

Figure 18–2 *In the past few years, Americans have been increasingly concerned over the food they consume. Courtesy of James Strawser, The University of Georgia.*

PESTICIDE AND CHEMICAL RESIDUES

In 1989, the Natural Resources Defense Council published a report entitled "Intolerable Risk: Pesticides in our Children's Food." Along with this report, the media began a series of stories on the chemical daminozide, which was called by its trade name, Alar. This chemical is a growth regulator used on apples to prevent them from falling off the tree too early. This allows the fruit to have a better shape and color and also prolongs storage life.

According to the reports, the chemical, when fed to laboratory rats, proved to be a **carcinogen** (a substance that causes cancer). As a result, near panic ensued among consumers. Apples and apple products were removed from sale, dumped, and the fruit was banned from school cafeterias. The apple industry suffered tremendous losses, with apple sales alone falling by more than $100 million—and that does not include income lost from support services of the apple industry (Figure 18–3). Many apple producers reached the brink of bankruptcy.

This example points to the awareness of the public as to the safety of food. Although the measures taken would have been justified if there was a real safety problem, a closer look proves that the scare had no real justification.

Figure 18–3 A scare over a chemical led to the loss of more than $100 million to the apple industry. *Courtesy of California Agriculture, University of California.*

Figure 18–4 *A person would have to eat the equivalent of 28.000 pounds of apples a day for 10 years to equal the amount of Alar fed to laboratory rats before they develop tumors.* Courtesy of DOW/Elanco.

Figure 18–5 *A chemical must pass more than 120 tests before it is released for use.* Courtesy of DOW/Elanco.

Because cancer causes more than one-fifth of all the deaths in the United States, people are concerned about this disease. Although scientists still do not fully understand the process that causes cancer, they know that there are substances that can trigger its onset. To be classified as a carcinogen, a substance must cause a significant increase in the rate of cancer in laboratory animals over that occurring in a nontreated control group. In the Alar scare, rats had been given enormously high amounts of the substance before they developed tumors. At much lower rates, the chemical had no effect at all. According to the American Council on Science and Health, a person would have to eat the equivalent of 28,000 pounds of apples per day for 10 years to equal the amount of Alar fed to laboratory animals before they developed tumors (Figure 18–4). In spite of this fact, the chemical was pulled off the market.

Chemical companies are constantly developing new chemicals to be used as pesticides, preservatives, or for other uses. Chemicals used in the production, harvesting, transporting, storing, and marketing of foods are closely regulated. This process begins with the testing of chemicals that are to be used in association with food. Before it can be released for use, each chemical must be approved by the Food and Drug Administration, the Environmental Protection Agency, and the regulatory agencies of the individual states where the chemical is to be used. In all, the chemical must pass more

Figure 18–6 *Plants produce natural toxins to combat insect infestation. Courtesy of USDA-ARS.*

Figure 18–7 *Producers cannot grow the quality of fruit we are accustomed to without chemical pesticides. Courtesy of USDA-ARS.*

than 120 separate tests over a period of 8 to 10 years (Figure 18–5). On average, only one in 20,000 chemicals are approved for use by agricultural producers. The National Cancer Institute has concluded that there is no scientific evidence that pesticide residues at the allowable levels on fruits and vegetables cause cancer. In fact, most of our foods contain no detectable traces of pesticides.

Because pesticides are poison, can even minute trace amounts be harmful to humans? The answer lies in two points. First, pesticides are used to kill insects, weeds, and plant disease organisms. Modern insecticides, fungicides, and herbicides have a low level of toxicity to humans, and even at their full strength would take a considerable amount to be lethal. In fact, many of the pesticides used have very little toxicity to humans. The allowable levels in food are measured in parts per billion or parts per trillion. It takes very sophisticated tests and equipment to detect levels this low. The second point is that almost all substances can be toxic or hazardous at some level. For example, salt is a necessary component of our diets. We simply cannot function without it, but if we eat the amount of salt that can be held in two hands, we would die from salt poisoning. Similarly, vitamins are essential for good health, but an overdose of vitamins can cause severe and permanent damage to our bodies.

A recent trend is to eat organically grown foods to avoid even any traces of pesticide residue. Though people may feel safer eating foods that were grown in this manner, research shows that there is no real advantage in organically grown food. In fact, there are several disadvantages. One of the primary reasons for organically grown food is to avoid the toxins from the trace amounts of insecticides that could possibly be on regularly produced fruits and vegetables. Plants have defense mechanisms that produce natural toxins to help ward off insects (Figure 18–6). If attacked by pests, the plants release these substances because of the stress caused by the insects or disease. These natural toxins can be many times the amount produced by plants that are protected by pesticides.

Therefore, fruits and vegetables that are grown organically are more expensive and are of much poorer quality. Produce simply cannot be grown that can have the quality we are accustomed to without the use of chemical pesticides (Figure 18–7). Natural methods of controlling pests can be effective, but when used alone cannot offer the protection that will ensure that the food does not contain insect damage. This damage on fruits and vegetables provides a natural harbor for

Figure 18–8 *The aim of the USDA Food Safety and Inspection Service (FSIS) is to ensure that the most wholesome meat reaches the consumer.* Courtesy of Progressive Farmer.

bacteria and other organisms that cause the food to spoil. Most scientists agree that the greatest danger we have from our food supply is food spoilage and the ingestion of bacteria, not the presence of pesticides.

MEAT INSPECTION

Essentially all meat that is sold to the public must be inspected by the USDA Food Safety and Inspection Service (FSIS). The aim is to ensure that the very best, most wholesome meat reaches the consumer (Figure 18–8). Although all meat must be inspected, it can be sold without being graded (Figure 18–9). However, most meat sold retail is graded. Quality grading refers to the eating quality of the meat. It is determined by the amount of intermuscular fat content and the age of the animal at slaughter. Yield grade refers to the amount of lean retail cuts the carcass will yield. Meat inspection guarantees that the meat will be safe, wholesome, and accurately labeled. Meat inspection includes several phases. First, animals that are to be slaughtered must be inspected while they are alive in a process called ante-mortem inspection ("ante" means before and "mortem" means death). As the animals are brought in prior to slaughter, a government inspector examines them (Figure 18–10). Animals that are down, disabled,

Figure 18–9 *The USDA grades meat sold to consumers.* Courtesy of National Meat Board.

Figure 18–10 All meat sold to the public must be government inspected. *Courtesy of James Strawser, The University of Georgia.*

diseased, or dead are condemned as unsafe for human consumption. Animals that the inspector thinks may have a problem are set aside for further examination. When the inspector suspects that the animal is not in good health, it is moved to a well-lighted area for further inspection. These animals are examined thoroughly and if found to be ill, are tagged as condemned and are not allowed to be slaughtered for human consumption.

After the animals are slaughtered they must again undergo inspection. Slaughterhouses are required to provide adequate lighting (50 foot-candles) for the inspector to see and thoroughly inspect the carcass. In cattle, sheep, and hogs, the head, lungs, heart, spleen, and liver are inspected for signs of disease, parasites, or other problems that might render the meat less than wholesome. The internal and external cavity of slaughtered poultry must be examined as well as the air sacs, kidneys, sex organs, heart, liver, and spleen. Carcasses that do not pass inspection are condemned and not allowed to be used for human consumption. These condemned carcasses undergo a process called **rendering** where they are placed under enough severe heat to kill any organism that could cause problems. The rendered meat is then used as a by-product other than for human consumption. These products may be pet food, lubricants, animal feed, soap, or other uses. Condemned carcasses usually represent less than one percent of carcasses inspected. Producers, buyers, and packers all try to avoid sending animals to slaughter that will not

Figure 18–11 *Processing is stopped periodically for cleaning the equipment. Courtesy of James Strawser, The University of Georgia.*

pass inspection. Those that do get by are discovered by the federal inspectors at the processing plant.

Each slaughter plant is constantly inspected to ensure that the facilities are clean and free from any matter that might harbor bacteria or other disease-causing pathogens. Meat is a perfect medium for the growth of bacteria and other microbes, so precautions must be taken to prevent their growth and spread. Each day all equipment and facilities are cleaned and sanitized before processing begins (Figure 18–11). In plants that run around the clock, the processing is stopped periodically for cleaning.

Along the processing line, all carcasses must be thoroughly cleaned to remove debris from the slaughter process—materials such as blood, loose tissue particles, or any foreign particles. If the meat becomes contaminated by materials, such as fecal matter, that portion of the carcass must be trimmed.

Even with the best measures, meat will contain bacteria or other harmful microbes. For this reason it is essential that all meat be thoroughly cooked to destroy these pathogens. Almost all of the sickness caused from eating meat can be attributed to improper cooking procedures.

HORMONE AND ANTIBIOTIC RESIDUES

An area of consumer concern is the use of synthetic growth and reproductive hormones in beef and dairy animals. These products were introduced into the livestock industry more than 20 years ago, and almost from the beginning there has been controversy over their use. Because humans also use hormones to regulate bodily functions, there is concern among some people that hormone residues in meat can cause harmful effects in humans.

The use of growth hormones in animals raised for meat (mainly beef animals) redirects energy from the production of fat to the production of lean. This means that more weight can be gained by the animal at a lower cost because lean is more efficiently produced than fat. Also, the product is healthier because of the lower fat content. The hormones are placed just under the skin in the animal's ear and are slowly

Table 1 Daily Human Estrogen Production (Nanograms)

Human	Amount
Female child, before puberty	54,000
Male child, before puberty	41,000
Nonpregnant woman	480,000
Pregnant woman	20,000,000
Adult male	136,000

Table 2 shows the estrogen level in beef from nonimplanted and implanted steers.

Table 2 Estrogen Levels in Beef

Beef Source	Nanograms per gram of muscle	Nanograms per 3 oz. of muscle
Steer, implanted	0.022	1.9
Steer, non-implanted	0.015	1.3

Table 3 shows estrogen levels in various food products.

Table 3 Estrogen Levels in Foods

Food	Nanograms per gram of food	Nanograms per 3 oz. of food
Beef from implanted steer	0.022	1.9
Wheat germ	40	3,400
Soybean oil	20,000	1,680,000
Milk	0.13	11

Figure 18–12 *The amount of estrogen from a pound of beed from an implanted steer is inconsequential.*

released throughout the animal's growth period. When the animal is slaughtered, the ears are removed and discarded to prevent residues from accumulating in edible meats.

The hormones used are naturally occurring anabolic compounds that stimulate growth. The implants contain estradiol or zeranol. As a result of using these compounds, the levels of the hormone estrogen are increased in beef. Beef from animals that have not been implanted contains about 6 nanograms of estrogen compacted to 7 to 11 nanograms of estrogen per pound in beef from implanted animals. The amount varies according to the type of implant used. This may seem like quite an increase, but a nanogram is equal to one billionth of a gram (Figure 18–12). Also, this is a small amount compared to the amount of naturally occurring estrogen in other foods. For example, one egg contains 1,750 nanograms; a glass of milk contains 75,000 nanograms; and a serving of peas contains 400 nanograms. In addition, the human body produces estrogen every day. An adult male produces about 136,000 nanograms per day, and an adult female produces about 480,000 nanograms per day. In perspective, the estrogen from a pound of beef from an implanted steer is inconsequential. To ensure that the levels of hormone residue are kept to an extremely low level, the FDA closely monitors hormone levels in beef.

Livestock and poultry producers periodically give their animals antibiotic drugs to cure and prevent diseases. Because humans rely on the same types of drugs to kill the same disease organisms, concern has been raised that the pathogens (disease-causing organisms) may build up immunity to the drugs as a result of residues of the drugs in meat. According to the National Academy of Science, no data have been found that implicates the use of antibiotics in animals used for food as a health risk to humans.

Another concern for humans has been the use of a relatively new treatment of dairy cattle using **bovine somatotropin (BST)**. This naturally occurring hormone stimulates cows to produce more milk. A lot of misinformation has circulated among the public about the risks of the treatment (Figure 18–13). There has been widespread concern that the milk from cows treated with BST will be a health risk. Exhaustive studies have found no evidence of risks to humans. In fact, milk from cows treated with BST is almost identical to milk from nontreated cows. BST is a naturally occurring hormone that is in all milk, whether the cow is

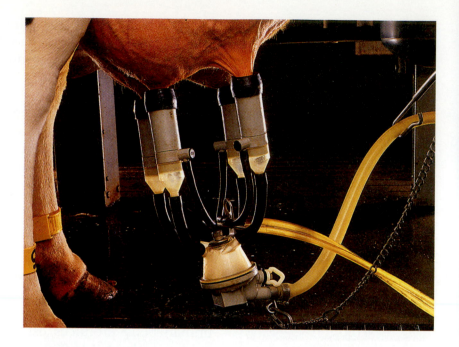

Figure 18–13 *A lot of misinformation has been circulated about milk from cows treated with BST. Courtesy of James Strawser, The University of Georgia.*

treated with BST or not. BST is a protein, is easily digested (including the very minute amount from milk of BST-treated cows), and is not passed into the systems of humans.

The FDA closely monitors meats, dairy products, and all foods from animals for residues of pesticides, drugs, and hormones. If levels are detected that go beyond safe tolerances, the product is condemned at the producer's expense. This gives producers incentives to adhere to guidelines and **withdrawal periods**. A withdrawal period is the time between when the substance was administered to the animal until the animal can be slaughtered or the milk used.

PRESERVATIVES

Many people are concerned over adding **preservatives** such as nitrites, citric acid, sodium benzoate, and phosphoric acid to foods. The purpose of these additives is to prevent the formation of toxic substances associated with the spoilage of food. Foods such as processed meats, margarine, egg products, and mayonnaise are susceptible to the growth of bacteria that can cause food poisoning. The *Clostridium botulinum* and the *Staphylococcus aureus* bacteria produce waste materials that are extremely toxic. Additives are included in these foods to prevent or retard bacterial growth (Figure 18–14). The likelihood of

Figure 18–14 *Additives are put in processed meat to prevent spoilage. Courtesy of James Strawser, The University of Georgia.*

any health problems arising from chemical preservatives added to foods is insignificant compared to the food poisoning that can occur in untreated food.

FAT CONTENT IN FOOD

In recent years a major concern about the food supply has been fat content. Studies have shown a relationship between diets high in fat and health problems such as obesity, heart disease, cancer, and hypertension. Particular concern has been expressed over eating red meat, not only because of the relatively high amount of fat, but also because of a substance called **cholesterol** that is associated with heart disease. Both fat and cholesterol are necessary components of human diets, but excessive amounts of either can cause problems (Figure 18–15).

Fats are derived from both plant and animal sources and are divided into two broad classifications, **saturated fat** and **unsaturated fat**. A fat molecule is made up of one molecule of glycerol (a three-carbon alcohol that contains three hydroxyl groups) bonded to three fatty acids. If the fatty acids in the molecule contain one or more double bonds, it is considered to be unsaturated. If there are two or three double bonds, the fat is said to be polyunsaturated. If there are no double bonds

BIO BRIEF

Safer Salad Is "In the Bag"

Bagged salads are one of the most popular items in the fresh produce section of supermarkets today. The major reason: Salads are healthy foods. They help consumers meet the recommended quota of five servings each day of fruits and vegetables to maintain good health. Sales of packaged lettuce in the United States were more than $1.2 billion in 1997.

From a food safety perspective, salads are considered by some to be among the safest foods. However, some segments of our population often exclude salads and other uncooked fruits and vegetables from their diets. Because of the high levels of microbial agents found on fresh-cut produce, salads are often not recommended for the young, old, pregnant, or immuno-compromised. These people can't risk exposure to microorganisms that, for the general population, are normally considered nonpathogenic.

Even though commercial food processors use chlorine to control microbes on fresh-cut lettuce, the treatment doesn't eliminate all the organisms that can be present, such as *Shigella* and *E. coli* 0157:H7. Although *E. coli* is primarily found on meat, it has recently shown up in apple juice, sprouts, and lettuce. Outbreaks of food poisoning from *Shigella* on iceberg lettuce have occurred in Sweden, England, and Wales.

Robert D. Hagenmaier, an Agricultural Research Service chemist at the U.S. Citrus and Subtropical Products Laboratory in Winter Haven, Florida, has found a way to reduce these and other pathogenic and nonpathogenic microorganisms. He combines an ionizing irradiation treatment with the chlorine wash. Technician Kelly Alger assists with the research.

Ionizing radiation passes through food in the form of radiant energy, without leaving any residue. It does not make food radioactive. Although the U.S. Food and Drug Administration has approved up to 1 kilogray (kGY) of ionizing irradiation for fresh produce, Hagenmaier uses much less. In lab experiments, he found that irradiation significantly reduced the microbial and yeast populations on cut iceberg lettuce.

in the molecule's fatty acids, it is a saturated fat. Unsaturated fats are in a liquid state at room temperatures and include fats found in soy oil, corn oil, and cottonseed oil (Figure 18–16). Fats that are in a solid state at room temperature are saturated and include pork fat (lard), beef fat (tallow), and palm oil.

Saturated fats are associated with cholesterol. This is an essential compound found in all cells of higher animals where it plays an important role in the structure of cell membranes. Cholesterol is converted by the adrenal glands into hormones

A combination of ionizing irradiation treatment and chlorine wash reduces microorganisms on fresh salads. Courtesy of USDA-ARS.

Eight days after zapping chlorine-washed lettuce with only 0.2 kGy of irradiation, microbial counts were 290 colony-forming units (CFU), and yeast counts were 60 CFU.

Control samples that had not been irradiated showed microbial counts of 220,000 CFU and yeast counts of 1,400 CFU.

"Low levels of irradiation were used to minimize changes in the texture or appearance of the lettuce," Hagenmaier says.

Irradiated lettuce had about the same shelf life as untreated samples. Normal shelf life claimed by manufacturers for retail sales of salads is between 14 and 16 days from the packaging date. Hagenmaier also irradiated chlorine-washed, shredded carrots in modified-atmosphere packaging. Nine days after irradiation, on the expiration date, the microbial count was 1,300 compared to 87,000 for nonirradiated, chlorinated controls, he says, and texture and appearance were unchanged.

"This research could help fresh-cut salads to be included in diets of people who otherwise couldn't enjoy them because of a potential microbiological health risk," Hagenmaier says.

Source: Agricultural Research Magazine.

such as cortisone that regulate glucose and the body's mineral balance. It is also critical in the development of the sheath that protects nerves.

Because cholesterol is critical to the well-being of the human body, the liver synthesizes enough to make up for any amount lacking in the diet. Problems may arise when the intake of cholesterol far exceeds the amount needed. Cholesterol from the diet is carried from the intestines to the liver by means of large lipoprotein molecules in the bloodstream. The liver then

Figure 18–15 *Concern has been expressed over eating red meat, however, both fat and cholesterol are necessary components of human diets.* Courtesy of James Strawser, The University of Georgia.

secretes a substance called very low-density lipoprotein (VLDL) into the blood. This substance contains cholesterol and other compounds. VDVL is then transported to fat–adipose tissues where it is converted to low-density lipoprotein (LDL). LDL transports cholesterol to the body tissues. This cholesterol is what is sometimes deposited on the inside of artery walls. In some individuals the excess cholesterol causes clogging of arteries, which may in turn lead to heart and circulatory

Figure 18–16 *Unsaturated fats are liquid at room temperature.* Courtesy of James Strawser, The University of Georgia.

problems. High-density lipoprotein (HDL) helps to transport cholesterol back to the liver, where it is broken down for excretion from the body. In this process some LDL cholesterol is removed from the arteries. Recent research indicates that cholesterol problems are also related to the genetic makeup of the individual and to the amount of exercise the person gets.

Cholesterol levels are relatively high in meats from ruminant animals such as cattle and sheep. Eating red meats from these and other animals is safe as long as the amounts are moderate and are part of a balanced diet (Figure 18–17). In fact, the intake of all fats should be limited to 30 percent of the daily intake of calories. This, combined with a balanced diet and plenty of exercise, should allow people to enjoy eating red meat and gain the benefits of its nutritive value.

Food Guide Pyramid
A Guide to Daily Food Choices

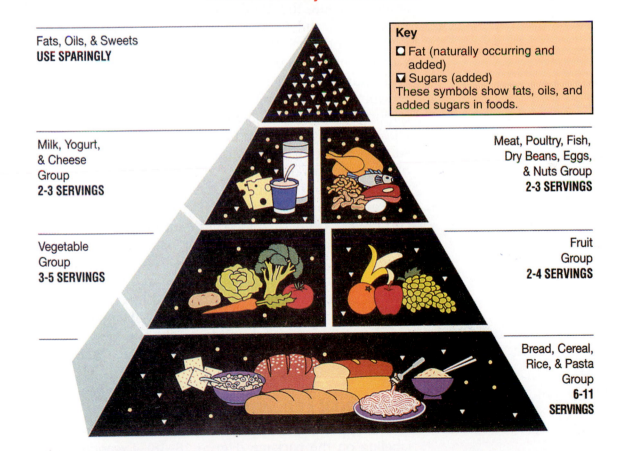

Fats, Oils, & Sweets
USE SPARINGLY

Key
☐ Fat (naturally occurring and added)
☑ Sugars (added)
These symbols show fats, oils, and added sugars in foods.

Milk, Yogurt, & Cheese Group
2-3 SERVINGS

Meat, Poultry, Fish, Dry Beans, Eggs, & Nuts Group
2-3 SERVINGS

Vegetable Group
3-5 SERVINGS

Fruit Group
2-4 SERVINGS

Bread, Cereal, Rice, & Pasta Group
6-11 SERVINGS

Figure 18–17 The food pyramid is designed to help plan a healthy diet. *Courtesy of USDA.*

LABELING OF FOODS

According to legend, President Theodore Roosevelt was eating sausage one morning while he was reading a novel by Upton Sinclair entitled *The Jungle*. Sinclair described in graphic terms some of the unsanitary practices of the meat processing industry. He wrote of how rodents sometimes fell into the hoppers of meat that were being ground for sausages. The President was so upset by the allegations that he began investigations into the practices of the food industry. His efforts culminated in the passage of the Federal Food and Drug Act and the Federal Meat Inspection Act. These laws were the first steps in the regulation of the food industry as to the safety and quality of foods.

These acts began a series of legislation aimed at assuring that foods sold to the public would be healthy and wholesome. In 1938, the 1906 act was replaced with the Federal Food, Drug, and Cosmetic Act. Among other provisions, it required the label of every processed and packaged food to contain the name of the food, its net weight, and the name and address of the manufacturer or distributor. For many products, a list of ingredients was required. One of the most important parts of the law was to prohibit statements in food labeling that were misleading. Prior to that time, food processors could put just about anything they wanted on the labels of foods. Claims of purity, quality, and nutritional value were often either misleading or entirely false.

Laws regarding the processing and labeling of foods have continually been upgraded and improved since that time. This effort culminated with the passage of the Nutrition Labeling and Education Act of 1990. This effort was aimed at mandatory nutrition labeling for most foods, standardized serving sizes, and uniform use of health claims. By 1994, all food processors, with the exception of small businesses, were required to label foods under the new regulations. These regulations have several components that have to be on the labels.

Nutritional Labeling

About 90 percent of all processed foods must have nutritional labeling on the package (Figure 18–18). Exceptions are such foods as coffee and tea, certain spices, and flavorings that contain no significant amounts of nutrients. Also, ready-to-eat food

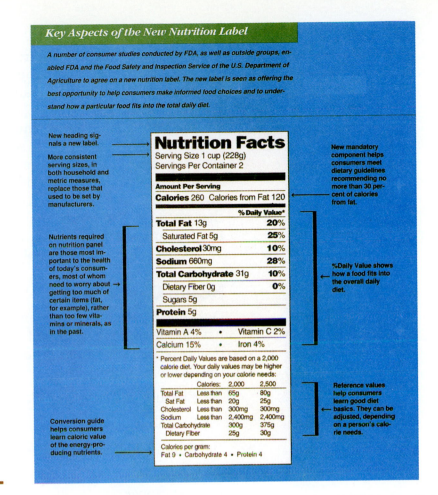

Key Aspects of the New Nutrition Label

A number of consumer studies conducted by FDA, as well as outside groups, enabled FDA and the Food Safety and Inspection Service of the U.S. Department of Agriculture to agree on a new nutrition label. The new label is seen as offering the best opportunity to help consumers make informed food choices and to understand how a particular food fits into the total daily diet.

New heading signals a new label.

More consistent serving sizes, in both household and metric measures, replace those that used to be set by manufacturers.

Nutrients required on nutrition panel are those most important to the health of today's consumers, most of whom need to worry about → getting too much of certain items (fat, for example), rather than too few vitamins or minerals, as in the past.

Conversion guide helps consumers learn caloric value of the energy-producing nutrients.

New mandatory component helps consumers meet dietary guidelines recommending no more than 30 percent of calories from fat.

%Daily Value shows how a food fits into the overall daily diet.

Reference values help consumers learn good diet basics. They can be adjusted, depending on a person's calorie needs.

Nutrition Facts

Serving Size 1 cup (228g)
Servings Per Container 2

Amount Per Serving

Calories 260   Calories from Fat 120

	% Daily Value*
Total Fat 13g	**20%**
Saturated Fat 5g	**25%**
Cholesterol 30mg	**10%**
Sodium 660mg	**28%**
Total Carbohydrate 31g	**10%**
Dietary Fiber 0g	**0%**
Sugars 5g	
Protein 5g	

Vitamin A 4%	•	Vitamin C 2%
Calcium 15%	•	Iron 4%

* Percent Daily Values are based on a 2,000 calorie diet. Your daily values may be higher or lower depending on your calorie needs:

	Calories:	2,000	2,500
Total Fat	Less than	65g	80g
Sat Fat	Less than	20g	25g
Cholesterol	Less than	300mg	300mg
Sodium	Less than	2,400mg	2,400mg
Total Carbohydrate		300g	375g
Dietary Fiber		25g	30g

Calories per gram:
Fat 9 • Carbohydrate 4 • Protein 4

Figure 18–18 About 90 percent of all processed food must have nutritional information on the package. Courtesy of USDA.

prepared on the site such as deli and bakery items, restaurant foods, and foods processed by small businesses are exempt.

The nutritive content of the food is expressed in terms of **Daily Value (DV)**. The Daily Value of a food refers to the percent of the nutrient in the food that is recommended based on a 2,000-calorie diet. For example, the Daily Value for fat based on a 2,000-calorie diet is 65 grams. A food that has 13 grams of fat per serving would state on the label that the percent Daily Value for fat is 20 percent. These recommendations (based on a 2,000-calorie diet) for energy producing nutrients are always calculated as follows:

- Fat based on 30 percent of the calories
- Saturated fat based on 10 percent of calories
- Carbohydrates based on 60 percent of calories
- Protein based on 10 percent of calories
- Fiber based on 11.5 grams per 1,000 calories.

Content Claims

As Americans became more health conscious and new research discoveries gave insight into the foods that are most healthy for humans, food manufacturers began appealing to diet awareness (Figure 18–19). Many new products came on the market claiming to be "lite," "diet," "low calorie," "high in fiber," or "low in fat." The law sets standards for such terms so that any used to describe the nutrient content of a food will mean the same on every product on which it is used. There are 11 terms that have been defined in terms of the DV for that nutrient. The terms are: free, low, lean, extra lean, high, good source, reduced, less, light, fewer, and more. Figure 18-20 gives some examples of how these terms may be used.

Health Claims

Only under certain circumstances will a product be labeled as having an ingredient linked with the risk of a disease or health-related condition. The FDA allows statements about the relationship between

- Calcium and osteoporosis
- Fat and cancer
- Saturated fat and cholesterol and coronary heart disease
- Fiber-containing grain products, fruits, and vegetables and cancer
- Fruits, vegetables, and grain products that contain fiber, particularly soluble fiber, and coronary heart disease
- Sodium and hypertension
- Fruits and vegetables and cancer

Ingredient Labeling

All ingredients in all processed, packaged foods, including foods such as bread, macaroni, and mayonnaise, which were previously exempt, must be included on the label. For example, prior to 1994, the term "jelly" was all that was required on the labels of jelly jars. The FDA had specified that to be termed jelly, a product must be 45 percent fruit and 55 percent sugar. Now, the ingredients must be specified on the label so consumers will know what the components of jelly are.

One of the targets of the regulations was fruit juice drinks (Figure 18–21). Previously, processors were allowed to label a

Sugar	Fat
Sugar free: less than 0.5 grams (g) per serving.	*Fat free:* less than 0.5 g of fat per serving
No added sugar, Without added sugar, No sugar added:	*Saturated fat free:* less than 0.5 g per serving, and the level of trans fatty acids does not exceed 1 percent of total fat
• No sugars added during processing or packing, including ingredients that contain sugars (for example, fruit juices, applesauce, or dried fruit).	*Low fat:* 3 g or less per serving, and if the serving is 30 g or less or 2 tablespoons or less, per 50 g of the food
• Processing does not increase the sugar content above the amount naturally present in the ingredients. (A functionally insignificant increase in sugars is acceptable from processes used for purposes other than increasing sugar content).	*Low saturated fat:* 1 g or less per serving and not more than 15 percent of calories from saturated fatty acids
• The food that it resembles and for which it substitutes normally contains added sugars.	*Reduced or Less fat:* at least 25 percent less per serving than reference food
• If the food doesn't meet the requirements for a low- or reduced-calorie food, the product bears a statement that the food is not low-calorie or calorie-reduced and directs consumers' attention to the nutrition panel for further information on sugars and calorie content.	*Reduced or Less saturated fat:* at least 25 percent less per serving than reference food
Reduced sugar: at least 25 percent less sugar per serving than reference food	

Calories	Cholesterol
Calorie free: fewer than 5 calories per serving	*Cholesterol free:* less than 2 milligrams (mg) of cholesterol and 2g or less of saturated fat per serving
Low calorie: 40 calories or less per serving and if the serving is 30 g or less or 2 tablespoons or less, per 50 g of the food	*Low cholesterol:* 20 mg or less and 2 g or less of saturated fat per serving and, if the serving is 30 g or less or 2 tablespoons or less, per 50 g of the food
Reduced or Fewer calories: at least 25 percent fewer calories per serving than reference food	*Reduced or Less cholesterol:* at least 25 percent less and 2 g or less of saturated fat per serving than reference food

Sodium	Fiber
Sodium free: less than 5 mg per serving	*High fiber:* 5 g or more per serving. (Foods making high-fiber claims must meet the definition for low fat, or the level of total fat must appear next to the high-fiber claim.)
Low sodium: 140 mg or less per serving and, if the serving is 30 g or less or 2 tablespoons or less, per 50 g of the food	
Very low sodium: 35 mg or less per serving and, if the serving is 30 g or less or 2 tablespoons or less, per 50 g of the food	*Good source of fiber:* 2.5 g to 4.9 g per serving
Reduced or Less sodium: at least 25 percent less per serving than reference food	*More or Added fiber:* at least 2.5 g more per serving than reference food

Figure 18–19 *The meanings of some descriptive words for specific nutrients. Courtesy of FDA.*

Claim Specifics

	Nutrient/Food-Disease Link	Typical Foods	Requirements	Sample Claim
Calcium and osteoporosis:	Low calcium intake is one risk factor for osteoporosis, a condition of lowered bone mass, or density. Lifelong adequate calcium intake helps maintain bone health by increasing as much as genetically possible the amount of bone formed in the teens and early adult life and by helping to slow the rate of bone loss that occurs later in life.	Low-fat and skim milks, yogurts, tofu, calcium-fortified citrus drinks, and some calcium supplements.	Food or supplement must be "high" in calcium; must not contain more phosphorus than calcium. Claims must cite other risk factors; state the need for regular exercise and a healthful diet; explain that adequate calcium early in life helps reduce fracture risk later by increasing as much as genetically possible a person's peak bone mass; and indicate that those at greatest risk of developing osteoporosis later in life are white and Asian teenage and young adult women, who are in their bone-forming years. Claims for products with more than 400 mg of calcium per day must state that a daily intake over 2,000 mg offers no added known benefit to bone health.	"Regular exercise and a healthy diet with enough calcium helps teen and young adult white and Asian women maintain good bone health and may reduce their high risk of osteoporosis later in life."
Sodium and hypertension (high blood pressure):	Hypertension is a risk factor for coronary heart disease and stroke deaths. The most common source of sodium is table salt. Diets low in sodium may help lower blood pressure and related risks in many people. Guidelines recommend daily sodium intakes of not more than 2,400 mg. Typical U.S. intakes are 3,000 to 6,000 mg.	Unsalted tuna, salmon, fruits and vegetables, and low-fat milks, low-fat yogurts, cottage cheeses, sherbets, ice milk, cereal, flour, and pastas (not egg pastas).	Foods must meet criteria for "low sodium." Claims must use "sodium" and "high blood pressure" in discussing the nutrient-disease link.	"Diets low in sodium may reduce the risk of high blood pressure, a disease associated with many factors."
Dietary fat and cancer:	Diets high in fat increase the risk of some types of cancer, such as cancers of the breast, colon and prostate. While scientists don't know how total fat intake affects cancer development, low-fat diets reduce the risk. Experts recommend that Americans consume 30 percent or less of daily calories as fat. Typical U.S. intakes are 37 percent.	Fruits, vegetables, reduced-fat milk products, cereals, pastas, flours, and sherbets.	Foods must meet criteria for "low fat." Fish and game meats must meet criteria for "extra lean." Claims may not mention specific types of fats and must use "total fat" or "fat" and "some types of cancer" or "some cancers" in discussing the nutrient-disease link.	"Development of cancer depends on many factors. A diet low in total fat may reduce the risk of some cancers."

Figure 18–20 Food labels must be specific about the terms used. *Courtesy of FDA.*

Claim Specifics

	Nutrient/Food-Disease Link	Typical Foods	Requirements	Sample Claim
Dietary saturated fat and cholesterol and risk of coronary heart disease:	Diets high in saturated fat and cholesterol increase total and low-density (bad) blood cholesterol levels and, thus, the risk of coronary heart disease. Diets low in saturated fat and cholesterol decrease the risk. Guidelines recommend that American diets contain less than 10 percent of calories from saturated fat and less than 300 mg cholesterol daily. The average American adult diet has 13 percent saturated fat and 300 to 400 mg cholesterol a day.	Fruits, vegetables, skim and low-fat milks, cereals, whole-grain products, and pastas (not egg pastas).	Foods must meet criteria for "low saturated fat," "low cholesterol", and "low fat." Fish and game meats must meet criteria for "extra lean." Claims must use "saturated fat and cholesterol" and "coronary heart disease" or "heart disease" in discussing the nutrient-disease link.	"While many factors affect heart disease, diets low in saturated fat and cholesterol may reduce the risk of this disease."
Fiber-containing grain products, fruits, and vegetables and cancer:	Diets low in fat and rich in fiber-containing grain products, fruits, and vegetables may reduce the risk of some types of cancer. The exact role of total dietary fiber, fiber components, and other nutrients and substances in these foods is not fully understood.	Whole-grain breads and cereals, fruits, and vegetables.	Foods must meet criteria for "low fat" and, without fortification, be a "good source" of dietary fiber. Claims must not specify types of fiber and must use "fiber," "dietary fiber," or "total dietary fiber" and "some types of cancer" or "some cancers" in discussing the nutrient-disease link.	"Low-fat diets rich in fiber-containing grain products, fruits, and vegetables may reduce the risk of some types of cancer, a disease associated with many factors."
Fruits, vegetables, and grain products that contain fiber, particularly soluble fiber, and risk of coronary heart disease:	Diets low in saturated fat and cholesterol and rich in fruits, vegetables, and grain products that contain some fiber, particularly soluble fiber, may reduce the risk of coronary heart disease. (It is impossible to adequately distinguish the effects of fiber, including soluble fiber, from those of other food components.)	Fruits, vegetables, and whole-grain breads and cereals.	Foods must meet criteria for "low saturated fat," "low fat," and "low cholesterol." They must contain, without fortification, at least 0.6 g of soluble fiber per reference amount, and the soluble fiber content must be listed. Claims must use "fiber," "dietary fiber," "some types of dietary fiber," "some dietary fibers," or "some fibers" and "coronary heart disease" or "heart disease" in discussing the nutrient-disease link. The term "soluble fiber" may be added.	"Diets low in saturated fat and cholesterol and rich in fruits, vegetables, and grain products that contain some types of dietary fiber, particularly soluble fiber, may reduce the risk of heart disease, a disease associated with many factors."
Fruits and vegetables and cancer:	Diets low in fat and rich in fruits and vegetables may reduce the risk of some cancers. Fruits and vegetables are low-fat foods and may contain fiber or vitamin A (as beta-carotene) and vitamin C. (The effects of these vitamins cannot be adequately distinguished from those of other fruit or vegetable components.)	Fruits and vegetables.	Foods must meet criteria for "low fat" and, without fortification, be a "good source" of fiber, vitamin A, or vitamin C. Claims must characterize fruits and vegetables as foods that are low in fat and may contain dietary fiber, vitamin A, or vitamin C; characterize the food itself as a "good source" of one or more of these nutrients, which must be listed; refrain from specifying types of fatty acids; and use "total fat" or "fat," "some types of cancer" or "some cancers," and "fiber," "dietary fiber," or "total dietary fiber" in discussing the nutrient-disease link.	"Low-fat diets rich in fruits and vegetables (foods that are low in fat and may contain dietary fiber, vitamin A, or vitamin C) may reduce the risk of some types of cancer, a disease associated with many factors. Broccoli is high in vitamins A and C, and it is a good source of dietary fiber."

Figure 18–20 Food labels must be specific about the terms used. Courtesy of FDA.

Figure 18–21 *Fruit juice processors must label packages with the type and the approximate percentage of fruit juice contents.* Courtesy of James Strawser, The Univeristy of Georgia.

drink as a certain type containing real fruit juice even though there was little fruit juice in the drink. Now processors must either state that the drink is flavored by the juice or state the amount within 5 percent of the actual amount. If they claim that a drink is 100 percent fruit juice, all of the juice in the drink must be stated in terms of percents.

Preservatives, spices, colorings, and other additives must also be listed. Ingredients such as "hydrolyzed plant protein" must specify the source, such as corn or soybeans. This portion of the regulations is aimed at helping people avoid foods or substances they are allergic or sensitive to. This also helps people of different cultures avoid foods that are forbidden by their religion.

SUMMARY

Foods are safer now than at any time in history. Problems with foods are generally caused after the foods get to the home.

Improper cooking, spoilage, and mishandling of the foods by the consumer are by far the biggest causes of food safety problems. The food producing and distribution industry of this country is huge and so is the infrastructure that ensures that the food is sanitary and wholesome.

CHAPTER 18 REVIEW

Student Learning Activities

1. Make a list of all the types and quantities of food you consume in a week. Analyze the amount of nutrients and calories you consume each day. Compare your average intake to the Daily Values of each nutrient.

2. Analyze the labels of five different foods that claim to be "lite" or "low calorie." Are they comparable? Give the rationale for your answer. Report to the class.

Define the Following Terms

1. pesticide
2. carcinogen
3. rendering
4. bovine somatotropin (BST)
5. withdrawal periods
6. preservatives
7. cholesterol
8. saturated fat
9. unsaturated fat
10. Daily Value (DV)

True/False

1. There is scientific evidence that pesticides at the allowable tolerances on fruits and vegetables cause cancer.

2. Pesticides are used to kill insects, weeds, and plant disease organisms.

3. Most scientists agree that the greatest danger we have to our food supply is food spoilage and the ingestion of bacteria.

4. Meat cannot be sold without being graded.

5. Almost all of the sickness caused from eating meats can be attributed to poor meat inspection.

6. Cholesterol levels are highest in meats from ruminant animals such as cattle and sheep.

7. The intake of all fats should be limited to 30 percent of the daily intake of calories.

8. Coffee and tea are two of the foods that require nutritional labels.

9. To be termed jelly, a product must be 45 percent fruit and 55 percent sugar.

10. Growth hormones are used in animals to redirect energy from the production of fat to the production of lean meat.

Fill in the Blanks

1. Any substance that causes _____ is called a carcinogen.

2. If there are two or three double bonds in a fatty acid, it is considered to be _____.

3. A recent trend is to eat _____ _____ in order to avoid any traces of pesticide residue.

4. _____ are included in certain foods to prevent or retard microbial growth.

5. Concern has been expressed over eating red meat because of its relatively high _____ and _____ content.

6. The _____ synthesizes enough cholesterol to make up for the amount lacking in the diet.

7. The daily value for fat based on a 2,000-calorie diet is _____ grams.

8. Nitrites, citric acid, sodium benzoate, and phosphoric acid are all considered _____.

9. Meat is the perfect medium for the growth of _____ and other microbes, so caution must be taken to prevent their growth and spread.

10. All meat that is sold to the public must be inspected by the _____.

Multiple Choice

1. The purpose of preservatives is to
   A. prevent the growth of microbes.
   B. cure and prevent disease in animals.
   C. redirect energy from the production of fat to lean.

2. Diets high in fats can cause health problems such as
   A. obesity.
   B. cancer.
   C. hypertension.
   D. all of the above.

3. Cholesterol is associated with
   A. unsaturated fats.
   B. saturated fats.
   C. polyunsaturated fats.

4. Pesticides are used to kill
   A. mice.
   B. insects.
   C. weeds.
   D. all of the above.

5. Hormones are placed just under the skin in the animal's
   A. stomach.
   B. leg.
   C. ear.

6. The daily recommendation for saturated fat is based on
   A. 30 percent of daily caloric intake.
   B. 10 percent of daily caloric intake.
   C. 60 percent of daily caloric intake.

7. Excessive cholesterol may cause
   A. heart problems.
   B. circulatory problems.
   C. clogging of the arteries.
   D. all of the above.

8. Researchers have to deal with misleading information regarding
   A. the healthiness of food.
   B. pesticides and chemical residue in food.
   C. cleanliness of food.
   D. B and C.
   E. all of the above.

9. Plants have self-defense mechanisms that produce natural toxins to help ward off
   A. predatory insects.
   B. insects.
   C. harmful pesticides.

10. Fruits and vegetables that are grown organically are
    A. cheaper.
    B. more expensive.
    C. of much poorer quality.
    D. B and C.

Discussion

1. What happened to the sale of apples in 1989 and was it justified?

2. What is involved in the testing of chemicals used in association with food?

3. What is the meat inspection process?

4. What provisions were included in the 1938 Federal Food, Drug, and Cosmetic Act?

5. Name some of the relationships between ingredients and health-related conditions that the FDA allows on labels.

6. What are some of the biggest causes of food safety problems?

7. How has the FDA changed the way jellies and juices are classified?

8. What are some foods that are exempted from nutritional labeling?

9. How does the body use cholesterol?

10. What are the risks involved in the use of pesticides in food production?

THE SCIENCE OF FOOD PRESERVATION

STUDENT OBJECTIVES

After studying this chapter, you should be able to:

◆ Analyze the causes of food spoilage.

◆ Explain how bacteria cause food to spoil.

◆ Discuss how bacterial growth can be retarded.

◆ Explain how molds break down plant and animal tissues.

◆ Discuss chemical changes that cause food to spoil.

◆ Explain how drying, salting, fermentation, canning, and freezing preserve food.

◆ Analyze the advantages and disadvantages of irradiation as a means of preserving food.

When humans first began to grow their own food, many problems were solved, but others were created. Even though they had a ready food supply that could be harvested without having to travel long distances, the problem arose of how to preserve the food. As people began to learn how to produce large yields of food from plants and animals, they needed to find ways to preserve the food for future use. For example, if a large animal was killed, the people ate well, but if all of the meat was not consumed in a short time, most of it decayed and was wasted. Fruits that were harvested nourished them during the summer months but spoiled in a few days after the fruit was harvested.

451

People searched for ways to keep food from spoiling. The first ways were, by our standards, very primitive, but through trial and error, and later through scientific inquiry, the preserving of food became a reality. As civilizations prospered and more food was produced, new ways of keeping food until the next harvest were discovered and developed.

CAUSES OF FOOD SPOILAGE

When early humans began preserving foods, they had little understanding of the causes of spoilage. They knew that as soon as food was harvested or animals were slaughtered, the food began to spoil, but they did not know why. They first found methods that worked through accident or through trial and error, but only when scientific inquiry began to determine the specific causes of food spoilage were real advances made in food preservation.

There is no single cause of all food spoilage but rather a number of different causes. How rapid and how severe the spoilage is depends on the type of food and the conditions under which the food is stored (Figure 19–1). Most foods that spoil do so because of a combination of causes.

Figure 19–1 How rapidly and how severely food spoils depends on the type of food and the conditions under which the food is stored. *Courtesy of James Strawser, The University of Georgia.*

Microbes

Extremely small organisms known as **microbes** are a major cause of food spoilage. One of the basic schemes of nature is that all living material decays and is broken down after it dies. Because all food is derived from material that once was living, this holds true for food. Not until after the 1860s, when Louis Pasteur discovered the presence of microbes, did the science of food preservation make great strides. Of course it was always known that decay took place, but no one understood why or how. Pasteur established that living organisms caused food to spoil. His research brought about the process known as **pasteurization** that renders the microorganisms (microbes) in milk harmless.

The major means of breaking down organic matter is through the action of microbes. Given the proper moisture, temperature, and other conditions, microbes can rapidly decompose organic matter. Several types of microbes are responsible for decomposition, and all of them can cause food spoilage.

Bacteria. **Bacteria** are microbes that are responsible for much of the decay and spoilage of the plant and animal tissue in foods (Figure 19–2). When this spoilage process is fully developed, foods are no longer fit to eat because of the smell and taste. Bacteria can also make the food dangerous to eat. Illnesses caused by food-borne bacteria are separated into two categories: food poisoning and food infection. The most serious of the two is food poisoning, which comes about as a result of toxins secreted as waste materials by the bacteria. Even though the bacteria themselves may not cause harm, the powerful poisons they excrete can be fatal. Even after the bacteria are no longer living in the food, the toxins can remain.

Figure 19–2 Bacteria growth may cause the tops of cans to bulge outward. *Courtesy of James Strawser, The University of Georgia.*

As mentioned in Chapter 18, the bacterium *Clostridium botulinum* produces a toxic waste material that is one of the most potent poisons known. This toxin is so strong that a half ounce of the pure substance would be enough to cause the deaths of millions of people. The condition, known as botulism, was at one time considered to be almost always fatal. Now, however, some people who contract this type of food poisoning can be saved by modern anti-toxin drugs.

This bacterium is a rod-shaped organism found in the soil, water, and in the intestines of humans and animals. It forms a heat-resistant spore that can be difficult to destroy. Spores are resting bodies in which the bacteria live when conditions are not favorable for the bacteria to reproduce. If these organisms are present in food under the proper temperature, moisture, and pH, they will germinate from the spore and begin to grow. When the bacteria are well into the growth process, they have the capability of producing the toxin. Foods that are relatively high in pH such as meats and most vegetables provide a good growing medium for the bacteria. Most fruits are too acid to allow the growth of *Clostridium botulinum*. These microbes are **anaerobes**, which means that they can only reproduce in the absence of oxygen. Microbes that reproduce only in the presence of oxygen are called aerobes.

The other bacterium that causes food poisoning, *Staphylococcus aureus*, is a round organism that congregates in grape-like clusters. It occurs naturally on the skin and nasal passages of humans and is transmitted by workers' hands and through sneezing during food preparation. Although they grow best at around body temperature (98°F), they can survive at temperatures of 115°F. Even though the microbe is easily killed by heat, the toxin secreted by the bacteria survives after the bacteria die.

The key to preventing this poisoning is to not allow food to remain for a long period in an unpreserved state at temperatures between 40°F and 140°F. Refrigeration or heating retards the growth and reproduction of these microbes (Figure 19–3).

Food infection results from eating foods that are infected with bacteria that continue to grow and reproduce in the intestines. The bacteria responsible for most food infections are the *Salmonellae*. More than 1,200 types of these bacteria exist, and all are potentially dangerous to humans. The prime source of this infection is from meats, and because the organism lives in the intestines of animals, the bacteria can be

Figure 19–3 *Refrigeration of foods causes the retardation of food-spoiling microbes.* Courtesy of James Strawser, The University of Georgia.

Figure 19–4 *Molds can also produce harmful toxins.* Courtesy of James Strawser, The University of Georgia.

spread at slaughter. Fortunately, these organisms are easily killed with heat. Properly cooked meat poses no danger.

Molds. **Molds** are microbes that also aid in the breakdown of plant and animal tissues. Like some forms of bacteria, molds can also produce harmful toxins that can remain after the mold is removed (Figure 19–4). Toxins from molds are called mycotoxins and may cause illness when consumed by humans. One of the most potent is a toxin called aflatoxin that is produced by the mold *Aspergillus flavus.* This mycotoxin is responsible for the deaths and illnesses of humans and animals all around the world. Although many molds are harmful, many have also been used to produce life-saving drugs such as penicillin.

Molds appear as fine, hairy filaments on the surfaces of foods such as bread. Many of the filaments grouped together give the mold a fuzzy or cottony appearance. At the end of the filaments, spores form, which are the dormant stage of the mold. These spores can be carried to other places such as other foods, and new mold growth can occur. In the spore state, molds are resistant to heat and are much more difficult to destroy.

Because molds need oxygen to live, they are generally found only on the surface of foods. Also, they are mesophilic microbes, which means they grow well at room temperature. The optimal temperature for most mold growth is 77°F to 86°F, although some grow very well at temperatures from 95°F to 100°F, and others may thrive at temperatures around the freezing point of water (32°F).

Yeasts. **Yeasts** are single-celled fungi of the family *Saccharomycetaceae.* These microbes are used extensively in the food industry because the enzymes they produce cause the fermentation of carbohydrates. One type, *Saccharomyces cerevisiae,* commonly known as baker's yeast, is used in baking bread and other products in order to cause the dough to rise (Figure 19–5). Other types are used in the **fermentation** process. This will be discussed later in the chapter.

Although many yeasts are beneficial, they may also cause food spoilage by growing and reproducing on the food. Both molds and yeasts use proteins and carbohydrates in the food for their own growth, and this can cause problems if the molds and yeasts become abundant. Optimal conditions for the growth of yeasts are similar to those optimal for the growth of molds. However, some yeasts can grow in anaerobic (without oxygen) conditions.

Figure 19–5 Yeasts are used in making bread. The rising of the dough is caused by the fermentation of carbohydrates brought about through the action of yeast. Courtesy of James Strawser, The University of Georgia.

Chemical Causes of Food Spoilage

Throughout the life of plants and animals, chemical processes are constantly taking place that aid in the life functions. When the plant or animal is harvested to use for food, certain chemical reactions continue to take place. Chemical reactions are generally aided or caused by protein molecules that serve as catalysts. These catalysts, called enzymes, can cause sugars to turn into fats or may cause proteins or fats to change in makeup or function. An example is bananas that are bought green from the grocery store (Figure 19–6). When they are removed from storage and placed on the shelf in the kitchen, they begin changing from green to yellow, then to a splotched brown and yellow, and finally to brown and black. As the banana changes color, the taste and odor changes from undesirable (green) to desirable (yellow) to undesirable again (brown to black). All these changes are caused by enzymes that trigger chemical reactions in the fruit.

Bacteria also secrete enzymes. For the bacteria to obtain the nutrients from the food they are living in, the cell walls of the food must be broken down. When this happens, the cells that make up the food tissue are destroyed, and this leads to the spoilage of the food.

Up to a point, certain enzymatic changes are desirable in that they trigger the creation of sugars in the ripening process. But if the process continues, enzymes can cause food to become inedible. Enzymes will continue to work in the food as long as they remain viable. In the preservation process, steps are taken to stop the actions of the enzymes. This topic will be dealt with later in the chapter.

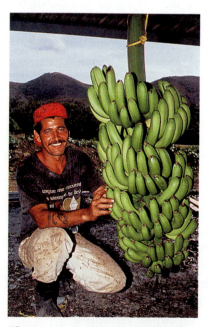

Figure 19–6 Bananas are picked green. Through the action of enzymes they ripen. Courtesy of USDA-ARS.

Rancidity. Another type of chemical food spoilage is the chemical processes that take place in fats (Figure 19–7). Fats are made of a series of molecules known as fatty acids. These acids are susceptible to being broken down into simpler molecules. Remember from Chapter 18, that unsaturated fats (usually found in fats from plants) differ from saturated fats (animal derived) in that they have one or more double bonds in their chemical makeup. The double bonding makes the

Figure 19–7 Such foods as butter, lard, and meat that contain fat turn rancid in a short time if left at room temperature. Courtesy of James Strawser, The University of Georgia.

unsaturated fats more prone to an undesirable chemical reaction with oxygen. The more double bonds the fatty acid has, the more susceptible the fat is to break down in the presence of oxygen. This process is called oxidative rancidity. This breakdown of the fatty acids causes the fat to have a bad odor and taste and is what causes french fries to taste stale when they have been sitting exposed to the air. The oil the potatoes were fried in undergoes oxidative rancidity.

Hydrolytic rancidity comes about as a result of reactions of the fatty acids with water molecules. When water molecules react with other molecules, the reaction is called **hydrolysis**. The rate of this reaction of fatty acids in the presence water molecules alone is not great. However the enzymes in the food and the presence of microbes such as bacteria, molds and yeasts hasten the process by acting as catalysts.

Animal fat is much more susceptible to rancidity than fat from plant sources. Such foods as butter, lard, and meat containing fat turn rancid in a short time if left at room temperature. The storage of foods containing fats at a cold temperature is essential to delaying rancidity.

FOOD PRESERVATION

As techniques become more sophisticated, the quality of preserved food becomes better. Not only must a food be preserved so that it can be stored, it must also retain much of its initial flavor, odor, appearance, and nutritive value. In addition, the food must be convenient for consumers to purchase, be of high quality, and easy and quick to prepare. Frozen dinners that are ready to take from the freezer and place in the microwave are in demand. Meats that are fully cooked and only need a minimum of preparation are often preferred to meat that needs to be cooked. Instant breakfasts, lunches, and dinners are sought by our highly mobile society (Figure 19–8). Not only must food meet all these measures, it must also retain its nutritive value and be healthy to eat. Today, a wide variety of methods are used that apply scientific principles to ensure that the conditions mentioned here are met.

Figure 19–8 Instant breakfast lunches, and dinners are sought by our highly mobile society. *Courtesy of USDA-ARS, K-3550-13.*

Fresh Food

There is no substitute for the taste of fresh food (Figure 19–9). Although a cooked apple may taste good, there seems to be no substitute for the taste and enjoyment of a fresh apple. This is true in regard to nutrition as well. Generally, the longer foods are stored or cooked, the less the nutritional value. Much research and development has gone into the preservation of fresh foods, and although these foods are still considered perishable, a lot can be done to preserve the freshness.

Figure 19–9 There is no substitute for the taste of fresh fruit. Courtesy of James Strawser, The University of Georgia.

The process begins when the produce is first picked (Figure 19–10). Pickers are careful not to damage fruits and vegetables as they are harvested. Bruises damage the food because of the rupturing of tissues within the fruit, and tears in the skin allow bacteria to enter and grow. This damage to the fruit hastens the spoilage of the food. After the fruits are washed and cleaned, they are often coated with a thin layer of **paraffin** to prevent damage to the skin and to help seal the surface of the fruit (Figure 19–11).

Figure 19–10 The process of preserving fresh fruits and vegetables begins with the harvest. Courtesy of James Strawser, The University of Georgia.

As fruits begin to ripen, several chemical processes take place. Most fruits contain high levels of starch that converts to sugar in the ripening process. Organic acids such as malic acid begin to decrease, and the pH level of the fruit increases. The fruit softens because of a substance called pectin that begins to break down as the fruit ripens. All these processes continue to ripen the fruit until the fruit is no longer palatable for human consumption. Modern techniques delay or control these processes until the fruit is sold for consumption.

Light, atmospheric gases, and temperature all play important roles in the ripening of fruit. In a process called controlled-atmosphere storage, the respiration of the fruit cells is controlled. The fruit is placed in dark chambers where the temperature is held at a constant temperature of between 32°F and 55°F, depending on the type of fruit. Oxygen and carbon dioxide levels are decreased and are replaced with the inert gas nitrogen. With the proper balance of oxygen, carbon dioxide, and nitrogen, the correct temperature, and the absence of light, fruits such as apples can be kept fresh for many months (Figure 19–12).

Figure 19–11 *Fruits are washed and coated to preserve freshness.* Courtesy of USDA-ARS.

Figure 19–12 *With the proper storage, fresh fruit can be kept fresh for weeks or even months.* Courtesy of James Strawser, The University of Georgia.

Drying

Perhaps the first method of preserving food for an extended period was drying. People probably learned this method by observing that as seed pods ripened, they began to dry. Beans, peas, kernels of grain, and other seeds could be kept through the winter if they were allowed to mature and to dry. They probably reasoned that if this were true for seed, it must be true for other foods. Meats, vegetables, milk, fruits, and other

Figure 19–13 *Grapes are preserved by drying. The dried grapes are sold as raisins. Courtesy of USDA-ARS.*

foods all have been successfully preserved by dehydration. The principle is that microbes and enzymes must have a certain amount of water to live and reproduce. If the water is removed from food, the food can be preserved. Fruits such as grapes (raisins), apples, peaches, plums (prunes), and other fruits have enough sugar content that water can be bound up with the sugar in the fruit to make the dried fruit palatable, flavorful and good to eat (Figure 19–13).

Meat and fish have been dried over fires for many years for preservation. In addition to the heat from the fire driving out moisture, the smoke gave an additional coating of particles that helped limit the action of enzymes. Some meats and fish are still preserved by drying. On the counters of many stores are jars containing individually wrapped sticks of jerky, which are meat products preserved by drying and smoking (Figure 19–14). Modern products such as fruits, powdered milk, instant potatoes, and instant coffee are examples of foods we commonly dry.

The most modern technique of drying is freeze-drying (Figure 19–15). Foods that have a low fat content such as fruits and vegetables can be dried by this process. Water boils at 212°F at sea level. At higher altitudes, lower temperatures are required to boil water. This is because atmospheric pressure is less at higher altitudes. By using a chamber that withdraws all the air and creates a vacuum, the boiling point of water can be lowered considerably. In the process of freeze-drying, the product is quickly frozen and placed in a chamber where a vacuum is drawn. Using modern techniques, the chamber removes so much air that the atmospheric pressure

Figure 19–14 *Meats are preserved by smoking and drying. These meats are known as jerky. Courtesy of James Strawser, The University of Georgia.*

Figure 19–15 *Freeze-drying is accomplished through the use of low temperatures in a vacuum.*
Courtesy of James Strawser, The University of Georgia.

in the chamber is lowered so that water boils well below the freezing point. This allows the water molecules to go almost directly from ice to vapor. The shape, color, and flavor of the product is much better preserved than by conventional drying.

Salting

Another old method of preserving food is the use of large amounts of salt. Meat was preserved using this procedure long before most other methods were used. Remember from Chapter 3 that osmosis causes moisture from a low concentration of salts to pass to an area of high concentration of salts. By covering meats with salts, the water in microbes passes to the salt, and the microbes dry out and die. Similarly, sugar can be used as a preservative because the same principle prevents microbes from growing in foods with a high concentration of sugar. This is why foods such as jelly, which are high in sugar content, resist spoilage.

The problem with salting is removing the salt from the meat or other food being preserved. A high concentration of salt in the food can lead to health problems. Also the palatability of the meat preserved only by using salt was not very good. About the only foods preserved in salt today are those such as fat back and salt pork that are considered specialty items.

Fermentation and Pickling

Pickling and fermentation have been used for hundreds of years and are still used extensively. The scientific principle is that microbes cannot multiply if the pH of a substance is below a certain level. For example, *Clostridium botulinum* will not grow and reproduce at a pH lower than 4.6, *Salmonellae* at less than 4.8, and *Staphylococcus aureus* at a pH lower than 4.2. Solutions high in salt and vinegar give the advantages of the salt solution dehydrating the microbes and the vinegar lowering the pH to a level intolerable to the microbes.

Fermentation involves the conversion of complex compounds such as sugar into simpler compounds such as alcohol, acetic acid, and lactic acid. This is accomplished through the use of certain yeasts and other microbes that secrete enzymes. These microbes are called starter cultures because they begin the process of fermentation. Wines, bread, cheese, sauerkraut, and pickles are all foods that are preserved by fermentation (Figure 19–16).

Figure 19–16 *Solutions containing high amounts of salt and vinegar preserve through pickling. These large vats contain cucumbers in brine during the pickle-making process.* Courtesy of James Strawser, The University of Georgia.

Canning

Remember from Chapter 1 the story of Nicolas Appert, who won a prize for preserving peas by heating wine bottles filled with peas and sealing the top with a cork. The method worked, although no one at the time (1795) had any idea why. Since that time the method has received widespread use. At one time the majority of households in the country preserved food grown in home gardens by canning. Today the canning industry has grown into a gigantic enterprise and has become so efficient that home canning is no longer economically feasible. Each year billions of cans of fruits, meats, vegetables, and other foods are preserved through canning, and in most instances the canned goods are cheaper than home canning (Figure 19–17).

Figure 19–17 *Each year billions of cans of food are preserved through the use of heat under pressure. Courtesy of FMC Corporation.*

After the food has been cleaned, peeled, shelled, or otherwise prepared, it is placed into containers made of metal or glass. Metal cans are made of or coated in metals that resist corrosion. Most home canning uses glass jars that have a sealing lid. The cans or jars filled with the food are heated until steam rises and pushes out the air. The tops are then sealed tightly enough to prevent the entrance of air or microbes.

Remember from the section on freeze-drying that water in a vacuum boils at far lower temperatures than at sea level atmospheric pressure. The inverse is also true. Water under higher than normal pressure boils at a higher temperature. To use this principle, cans are put into a device known as a steam retort that places the cans under steam pressure (Figure 19–18). This allows the contents of the cans to be heated to a high temperature without boiling the water within the food. If the water boiled in the sealed can, the steam would have no place to go and the can would eventually explode. By placing the cans and contents under pressure, high enough temperatures can be reached to kill any microbes in the food. This sterilizes the food, and the sealed can does not allow any new microbes to enter. As long as the can remains sealed, the contents remain sterile.

Figure 19–18 *Heat and pressure are applied to the cans in a device called a retort.* Courtesy of FMC Corporation.

The exact temperature and length of time the cans of food are held under pressure depend on the type of food. High-acid foods like tomatoes require less temperature and time than low-acid foods such as meat. Remember that most microbes are not tolerant of high-acid environments. Not only does high temperature kill the microbes, but the enzymes in the food are deactivated also. This prevents chemical changes in the food that take away taste and other food qualities.

Freezing

People have known for thousands of years that if food is kept at a low temperature, the rate of spoilage can be retarded. At one time, people cut ice blocks from frozen lakes and put the ice in insulated crates inside icehouses. If properly stored, the ice could be used during the warm months to keep food cold and retard spoilage. Even as late as the 1940s, iceboxes were used that cooled food with a large block of ice stored in the compartment with the food.

In the late 1800s, mechanical refrigeration was developed that allowed food to be kept cool for an extended period of time. This innovation had a large impact on the livestock industry. When fresh meat could be kept long enough to be shipped anywhere in the country, the demand for meat grew tremendously. Refrigerated trucks, rail cars, and ship compartments allow meat and all sorts of fruits and vegetables to be shipped all over the world.

Remember from earlier in this chapter that in order for microbes to grow, the temperature has to be in the correct range. Because microbes do not grow well in cool temperatures, the lower the temperature, the longer food can be preserved. As a general rule, for every 20°F lower in temperature, the shelf life of a particular food is increased by two to five times over what it would be at room temperature. At freezing, microbial action ceases (Figure 19–19). Not only is the cold temperature an advantage, but as the water in the food freezes, it ceases to be a medium for microbial and enzymatic processes. Most microbes and enzymes must have water in order to function and if the water is in solid form (ice), it is not available for the microbes and the enzymes to use.

Figure 19–19 *Most microbial action ceases at freezing.* Courtesy of James Strawser, The University of Georgia.

For food to be preserved at its best quality, freezing should be done as quickly as possible. This is because as the water freezes, ice crystals form within the cells of the food. The slower the freezing process, the larger the ice crystals. If the ice crystals become too large, the cell walls are ruptured and the fluid within the cells is lost when the food is thawed. Slow freezing causes foods to have a lot of liquid left in the containers when the food thaws. For this reason most foods are frozen as quickly as possible. Berries, peas, beans, and other foods are placed in a device that resembles a wind tunnel. This device subjects the food to a temperature of –40°F with air movement reaching 50 miles per hour. This method freezes the individual berries, or beans, to a hard gravel-like consistency that allows the consumer to pour out a part of the package contents without defrosting.

The instant freezing of small packages of food can be accomplished by submerging them in a liquid that has an extremely low temperature. Liquid nitrogen, liquid carbon dioxide, and liquid air all have temperatures around −100°F. When packages are dipped into this medium, they freeze almost instantly. This process, called cryogenic freezing, has limited use because of the expense involved.

Irradiation

One of the newest methods of preserving foods that may be used in the future is the use of radiation. This method subjects food to gamma rays produced by radioactive cobalt-60. As the gamma rays penetrate bacteria in the food, molecules within the bacteria are split, and the bacteria either die or are rendered nonfunctional. Food that is sealed in a vacuum package and subjected to the proper dose of gamma radiation will remain unspoiled for months at room temperature (Figure 19–20).

NON - IRRADIATED -   IRRADIATED - (0.2 M RAD)

STRAWBERRIES -
15 DAYS STORAGE 38°F (4°C)

Figure 19–20 *Irradiated foods have been treated with radiation to kill the microbes.* Courtesy of Eugene Wierbicki, USDA-ARS, Philidelphia, PA.

In 1997, the Food and Drug Administration (FDA) granted approval to use radiation to help preserve red meat. In 1990 the technology was approved for use on poultry to control salmonella and on pork to stop trichina, but the approval for beef was delayed. In fact irradiation is currently used in more than 40 countries to preserve food.

BIO BRIEF

Pest-Proofing Food Packaging

No one wants to open their breakfast cereal or pancake mix and find it infested with bugs. Even if these occurrences represent one in a million, they make a lasting impression. Manufacturers of food, feed, and other processed grain products want to avoid these incidents and provide consumers with high-quality products. That's why Agricultural Research Service's entomologist Michael A. Mullen at the U.S. Grain Marketing and Production Research Center in Manhattan, Kansas, has been working with food and feed manufacturers for more than 10 years to help design insect-proof packaging.

"Packaging should protect the commodity from the point of manufacture to the point of consumption," says Mullen. Nine times out of ten, an insect infestation isn't the manufacturer's fault. Often, insects get into packages during transportation or storage in a warehouse. Stored product insects are one of two types: either invaders or penetrators.

"The invaders look for opportunities to get inside food containers by searching for cracks, cervices, and holes," says Mullen. "The pentrators simply chew holes in the packages." Invaders include the red flour beetle, confused flour beetle, sawtoothed grain beetle, Indianmeal moth, and almond moth. Penetrators like the lesser grain borer, cigarette beetle, warehouse beetle, and rice moth can bore through one of more layers of packaging materials.

"There is no perfect package," says Mullen. "Packages are usually tailored to fit the product and designed to last throughout its shelf life. Often, this means that the package will have to provide this protection for more than a year."

Although packages can become infested anywhere along the marketing chain, they are most likely to become infested during long-term storage. Inside warehouses, insects start by attacking vulnerable packaging and later jump to sturdier material. Most stored-product insects are invaders, entering food and feed packages through seams and closures. They lay their eggs in the tight spaces formed when packages are folded. These spaces give the newly hatched larvae an ideal starting spot to invade. Dry pet foods are usually packaged in bags like these.

Seals and closures can often be improved by changing the type or pattern of sealant glue. A pattern that forms a barrier is usually the most insect resistance. Recently, paper bag manufacturers discovered, through working with Mullen, that closures on bag bottoms were prone to insect entry and needed reinforcement as much as top closures. Mullen is helping one bag manufacturer expand its customer base from nonfood agricultural products to food products.

Another packaging problems involves smell. Insects are attracted to packages that allow food odors to escape. Certain plastic film overwraps that fit tightly around a package can help prevent insects from smelling its contents. Interior plastic liners like those used in breakfast cereal boxes can be effective, blocking air from carrying aromas outside to hungry insects.

Dr. Mullen is researching ways to construct better food packaging. Courtesy of USDA-ARS.

Mullen helped to develop an odor neutralizer that can be incorporated into packaging materials. He devised a laboratory choice test, which allowed insects to choose between a food protected by the odor neutralizer or by only untreated packaging. They chose the food in untreated paper. "There's really no one thing that makes a package insect proof," says Mullen. "Each additional improvement added to the package design helps to keep insects out. All packages provide some protection against invasion, but tightening up the seals and adding a repellent adds even more," he says.

Mullen has developed scientifically proven methods to evaluate packaging materials against insects in the laboratory. He places 32 to 40 of each package type in an environmentally controlled room for about three months. These packages are exposed to five species of insects. Each month, the researchers examine packages for holes and flaws in seams and closures. Finally, they open the packages and examine the contents for insect infestations.

Manufacturers rely on these findings to improve future package designs or to conduct larger packaging studies. Test results have led to insect-resistant, pesticide-free packages for dry pet foods, raisins, baby cereals, pancake mixes, and breakfast cereals for domestic consumption and export. One company has reported a 75-percent reduction in consumer complaints from insect-related problems.

The food industry is facing increasing restrictions on pesticide use. Insect-resistant packaging can help reduce dependence on insecticidal treatments. This research helps assure consumers of insect-free food and protects manufacturers against the loss of goodwill arising from insect-infested packaging.

Source: Agricultural Research Magazine

Preserving meat by irradiation makes use of low levels of radiation to kill pathogens in food products. Additional preservation techniques, such as refrigeration, are required because the food is not permanently preserved. However, the rate of spoilage is greatly reduced, and dangerous bacteria such as salmonella and *E. coli* are killed. This makes meat much safer for the consumer.

The FDA has declared the process to be safe and effective. A symbol indicating that the product has been treated by irradiation is required on all packages. After treatment, the food is no more radioactive than your teeth are after a dental X-ray.

Although this process has been around for many years, currently there is little commercial utilization compared to other methods. There are at least two reasons why irradiation for food processing is not widely accepted.

First, people have a psychological aversion to eating anything that has been irradiated. Radiation in large doses can do irreparable harm to the human body. Just as molecules in the bacteria are split, radiation can split molecules within human cells with disastrous results, often without the person being aware that it is happening. This fear is unfounded with the use of irradiation for food preservation because when the source of the radiation is removed, the radiation is gone. However, people have an understandable fear of anything that has received radiation.

The second concern is better founded. As the radiation breaks down molecules in the bacteria, molecules in a small portion of the food are also split. This causes chemical changes in the food. Although research has provided no evidence that these chemical changes are harmful, it is impossible to test for all of the chemical changes. Even though these chemical changes are no more dramatic than those occurring in other food processing methods, the effects are difficult to prove or disprove.

SUMMARY

If all of these problems could be solved, food irradiation could prove to be an effective means of food preservation. Other possibilities also exist for this method. Foods such as potatoes could be prevented from sprouting, and insect infestation in stored foods such as grain could be eliminated. As with any aspect of the agricultural industry, the answers will be in the extensive use of the scientific method applied in research.

CHAPTER 19 REVIEW

Student Learning Activities

1. Take an inventory of all the food eaten by your family during a week. Determine how each of the items was preserved. List the reasons why a particular food preserved in a particular way was purchased. For example, why buy frozen beans as opposed to canned beans? Was it because of price, taste, or other factors?

2. Choose a method of food preservation and obtain more information on the history of how the process was developed. Report to the class.

Define the Following Terms

1. microbes
2. pasteurization
3. bacteria
4. anaerobes
5. molds
6. yeasts
7. fermentation
8. hydrolytic rancidity
9. hydrolysis
10. paraffin

True/False

1. There is no single cause of all food spoilage.
2. All food is derived from material that was once living.
3. When bacteria die, toxins can no longer remain in the food.
4. All molds are harmful.
5. The longer foods are stored or the longer they are cooked, the less nutritional value.
6. As a general rule, for every 20°F higher in temperature, the shelf life of a food is increased two to five times.
7. The bacteria responsible for most food infections are the *Salmonellae*.
8. Yeasts appear as the fine hairy filaments that grow on the surfaces of food.

9. Because molds need oxygen to live, they are generally found only on the surface of foods.

10. Enzymes are a type of catalyst that can cause sugars to turn into fats or may cause proteins or fats to change in makeup or function.

Fill in the Blanks

1. _____ _____ discovered the presence of microbes.

2. Illnesses caused by food-borne bacteria are separated into two categories: _____ and _____.

3. Foods that are relatively high in _____, such as meats and most vegetables, provide a good growing medium for bacteria.

4. Bacteria secrete poisonous waste products known as _____.

5. The more _____ the fatty acid has, the more susceptible the fat is to break down in the presence of oxygen.

6. _____, _____, and _____ all play important roles in the ripening of fruit.

7. _____ fat is more susceptible to rancidity than is fat from plant sources.

8. Most fruits contain high levels of _____ that convert _____ to sugar in the ripening process.

9. The most modern technique of drying is _____ drying.

10. Fruits are often coated with a thin layer of _____ to prevent damage to the skin and to help seal the surface of the fruit.

Multiple Choice

1. *Staphylococcus aureus* is a
   A. square-shaped organism.
   B. round-shaped organism.
   C. rectangular-shaped organism.

2. *Staphylococcus aureus* can be transmitted through
   A. workers' hands during the food preparation process.
   B. sneezing during the food preparation process.
   C. both A and B.

3. Examples of animal fats are
   A. butter and lard.
   B. apples and bananas.
   C. peas and carrots.

4. Microbes only grow in
   A. hot temperatures.
   B. cool temperatures.
   C. a wide range of temperatures.

5. The most popular choice for the industrial canning of food is
   A. metal cans.
   B. glass jars.
   C. plastic jars.

6. Foods resist spoilage better when they are high in
   A. fat content.
   B. sugar content.
   C. protein content.

7. The first method of preserving food was probably
   A. salting.
   B. drying.
   C. irradiation.

8. Molds are generally found
   A. deep within foods.
   B. only on breads.
   C. on the surface of foods.

9. Filaments grouped together in a mold give it an appearance of
   A. cotton.
   B. salt.
   C. grass.

10. The changing color of bananas is a
    A. process that is not natural.
    B. result of good preservation.
    C. chemical process.

Discussion

1. What causes bananas to change color?
2. How do bacteria obtain nutrients from the food they live in?
3. Give an example of oxidative rancidity.
4. Can bruises to fruit be harmful? Why or why not?
5. What is controlled atmosphere storage?

6. What is the main problem with salting?

7. Give the three disadvantages in using irradiation.

8. How can bacterial growth be retarded?

9. Why are cans of food put into a steam retort?

10. What factors are considered in choosing a food preservation method?

CHAPTER 20

THE SCIENCE OF FIBER PRODUCTION

KEY TERMS

cotton gin
boll
cellulose
keratin
yolk
scouring
sericulture
metamorphosis
retting
scrutching

STUDENT OBJECTIVES

After studying this chapter, you should be able to:

✦ Explain why natural fibers are superior to synthetic fibers for making clothing.

✦ Discuss the origins of cotton and its importance to the development of the United States.

✦ Discuss how cotton fibers are formed.

✦ Explain the chemical makeup of wool fibers.

✦ Explain how the properties of wool make it a good clothing material.

✦ Explain how the life cycle and metamorphosis of the silk moth is used in the production of silk.

✦ Discuss how the phloem of the flax plant is used to make clothing.

Agriculture is involved in the production of materials used for food, clothing, and shelter. We produce a wide variety of plants and animals for food and vast amounts of wood used in building shelters. A large part of agriculture is involved with the production of fiber to make clothing and other items such as rugs, tapestry, and cloth goods (Figure 20–1).

Much of the cloth produced today (such as rayon, nylon, and polyester) is manufactured from petroleum. In a process that was developed in the first half of the 20th century, petroleum is processed into long fibers that are used to make cloth. This cloth is relatively inexpensive and very durable. For

Figure 20–1 *A large part of agriculture is involved with the production of fiber used to make clothing. Courtesy of National Cotton Council of America.*

these reasons, agriculturally produced fiber declined in popularity for several years. However, today, natural fibers are rebounding for use as a clothing material. Human-made fibers may be more wrinkle resistant and durable, but they cannot match fiber produced by agriculture for comfort. Also, a lot of cloth produced is a combination (called a blend) of artificial and natural fibers that gives the flexibility and durability of synthetics and the comfort of natural fibers (Figure 20–2). Also, the use of natural fibers makes us less dependent on imported petroleum.

Figure 20–2 *Some cloth is a blend of natural and artificial fibers. Courtesy of James Strawser, The University of Georgia.*

Although several kinds of natural fibers exist, the four main types of fibers grown provide millions of people with materials that can be made into clothing and other types of cloth and fabrics. These fibers are cotton, wool, silk, and linen. As with most aspects of agriculture, scientific principles are used in the growth and utilization of these fibers.

COTTON

Cotton was probably first used in the Nile Valley in ancient Egypt. The climate and the fertile soil was ideal for the growth of the plant. In North America, humans have used cotton for at least 7,000 years, and its cultivation goes back hundreds of years before the European settlers arrived. Much of the history of the United States was developed around the growing of cotton. Although cotton was used as a clothing material for hundreds of years, it was not until around 1800 that cotton production made a major impact on agriculture in this country. The problem centered around the type of cotton that was grown. *Gossypium barbadense* was the species of cotton grown because of the long length of the fibers (called staple) and the ease with which the loosely bound seeds could be removed from the lint. This type of cotton, called sea island cotton, had a major drawback—it grew only along the coastal areas of the South and the Caribbean. Another type of cotton, *Gossypium hirsutum,* had shorter fibers but could be grown in areas away from the southern seacoast. The problem with this species was that the seeds were attached to the fibers so tightly that it was almost impossible to remove them. Then in 1793, an inventor named Eli Whitney constructed a **cotton gin** ("gin" was short for engine) that would remove the cotton fibers from the seeds (Figure 20–3). Because the fibers could be removed from cotton that could be grown all across the southern portion of the country, a whole new crop and industry was started. The short staple or upland cotton could be grown from Virginia to what is now California. Settlers pushed west through the South into fertile regions that had an ideal climate for growing cotton. A single family could clear several acres of land, plant, cultivate, and harvest cotton, and make a fairly good living. By 1860, the textile mills of Europe ran day and night processing cotton from the southern United States.

Figure 20–3 *(A) This is a version of the cotton gin invented by Eli Whitney. (B) This modern gin on the right evolved from Whitney's machine.* Courtesy of National Cotton Council of America.

Figure 20–4 *Cotton flowers are pale yellow, then turn red.* Courtesy of James Strawser, The University of Georgia.

Cotton is a shrublike perennial plant (although it is cultivated as an annual plant) that requires a long growing season and warm temperatures. About midseason the plant blooms from a bud that appears on the stem. As it opens, the flower is a pale yellow that gradually turns pink and then red (Figure 20–4). The blooms are complete flowers in that they have both pollen-bearing stamens and an ovary. The blossoms can self-pollinate, although cross-pollination does occur through pollen brought from other plants by bees and other insects. After the flower is fertilized the seed pod, known as a **boll**, forms. This structure is shaped like a small football and contains the fiber and the seed. The seed grows within the boll, and the fibers form in the epidermal layer on the seeds. Each fiber is actually a single long plant cell that contains cytoplasm. The outer layer of the cell is the cell wall, composed of **cellulose**. Cellulose is composed of molecules containing carbon, hydrogen, and oxygen arranged in hydroxyl groups (Figure 20–5). These groups react with dyes and other chemicals to give cotton fabric its finish.

As the fiber cell matures, layers of cellulose form at the rate of about one a day for about three weeks. This growth of the seeds and the fibers causes the boll to enlarge until it bursts open, and the fibers are displayed. Within each boll are from 20 to 50 seeds, each with thousands of fiber cells attached to the surface (Figure 20–6). When the boll opens, the cytoplasm or cell fluid dries from the tubelike fiber cell. As it dries, the fiber becomes narrower in diameter until it reaches a diameter of from .0005 to .0009 inch.

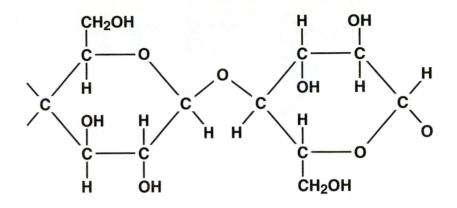

Figure 20–5 *Cellulose is composed of molecules containing carbon, hydrogen, and oxygen.*

As the fiber dries it begins to convolute or twist as the central canal of the fiber cell collapses. The fiber takes on a twisted ribbon appearance when viewed under a microscope. The outside of the fiber is composed of a waxy substance known as the cuticle. Within this layer is a primary wall, a secondary wall, and a central core or lumen (Figure 20–7). The secondary wall is made of layers of cellulose deposited as the fiber grew. The layers are composed of bundles of cellulose chains called fibrils. The fibrils are arranged spirally, and at certain points reverse the direction of the spirals. These reversals are what causes the twists or convolutions of the fiber. Short-staple cotton will have around 200 twists or convolutions per inch, and long-staple cotton may have around 300. This is a very important characteristic because it allows the fibers to lock together in the spinning process to create long threads of yarn that are woven into cloth.

Figure 20–6 *The fiber cells are attached to the seeds. Courtesy of National Cotton Council of America.*

BIO BRIEF

A Close Up Look at White-speck Neps in Cotton

The scanning electron microscope (SEM) was developed in 1942 and has been commercially available since the early 1960s. It uses electrons to scan a sample's surface and form images, in much the same way a television does. This microscope allows researchers to look at fiber samples three-dimensionally, providing valuable information about morphological structure.

Wilton R. Goynes used an SEM to confirm that small, undyeable clumps of cotton fibers—known in the textile industry as white-speck neps—are the result of underdeveloped cotton. Goynes is a chemist in the Cotton Fiber Quality Research Unit at ARS's Southern Regional Research Center (SRRC) in New Orleans, Louisiana. His finding proved what researchers had suspected since the 1940s. Under an SEM, white-speck neps appeared as mats of ribbon-like material, giving researchers proof of their origin.

Cotton fibers are usually made up of a thin primary cell wall and a thicker secondary one. It's the secondary wall that gives the fiber a rounded, tube-like shape and makes for easy dying. Without that secondary cell wall, the fibers look flat—like the ribbons that are revealed by the SEM.

Cotton-grading systems such as HYI (high-volume instrumentation) can track neps, but not all neps produce white specks. Technically, neps are tangles of fiber. A tangle of mature fiber can still take up dye. It's just the tangles of very immature fibers that become undyable white-speck neps.

"The white-speck problem actually comes from the field, when plants or growing conditions don't allow the fibers to mature properly," says Goynes. "Neps can sneak up on mills: The money is spent to dye the fabric, and it comes out spattered with white specks where the dye didn't take."

Bales of high-quality cotton can be blended with lower grades to use more of the cotton crop. However, if a bad growing season produces bales with a large number of undeveloped fiber clumps, blending cannot solve the problem.

Now Goynes' verification that the white-speck neps are really the result of underdeveloped fibers allows him and other researchers to focus on solutions. Some will involve improving conditions in the field.

Others will be directed toward detecting immature white-speck neps before they reach the dye bath. For example, because the clumps of undyeable flat fibers reflect light differently from mature cotton, Goynes says there may be a way to use special lighting to detect these immature fibers before money is wasted on trying to dye them.

Goynes and Blanchard explored whether enzymes such as cellulase can modify undeveloped fibers to improve fabric dyeability. Even with cotton varieties that were prone to white-speck neps, Blanchard saw reductions of 33 percent.

It's not enough that fabric dyes well. Consumers also want lasting color for their clothes. Detergent makers now add enzymes to reduce piling, but there is a risk of fading the color and weakening the fabric.

Scanning electron microscopes are used to closely examine cotton fibers. Courtesy of Photo Disc, S. Solum/ Photo Link.

The electron microscope lets researchers see fiber wear long before consumers can, so industry can choose treatments that keep clothing looking good and lasting longer. It also helps chemists like Blanchard prove that experimental enzyme treatments are effective in controlling neps. And, Blanchard found, some dyes may work better with enzymes than others.

"There are various dyes for cotton, including direct and reactive classes," he says. "Since reactive dyes chemically bind to fabrics, their colors stay true with enzyme detergents. Some direct dyes, which are just positioned within the fiber structure, may fade after several washes. It also appears that some dyes—both direct and reactive—actually limit enzyme damage," Blanchard adds.

Bel-Berger, a textile engineer in the SRRC's Cotton Fiber Quality Research Unit who also worked on the award-winning paper, has found that mechanical processing can play a role in white-speck nep control. Cotton mills clean and straighten fibers using a process known as carding, in which the fiber is run through a large drumlike roller with combing wires. Most mills use two carding cylinders to perform tandem carding.

Bel-Berger's surprising find was that when cotton has lots of underdeveloped fiber neps, single carding is better than tandem carding, which tends to open and separate the white-speck neps, making the problems appear worse. Here image analysis of dyed fabrics showed tandem-carded white-speck neps to be larger and more numerous than single-carded ones and to result in a higher percentage of white on the dyed fabrics. Bel-Berger is now working with industry collaborators to confirm her results. If the findings prove true, mill operators can pre-test their cotton and process it accordingly.

Source: Agricultural Research Magazine.

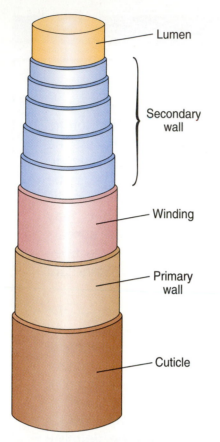

- Lumen
- Secondary wall
- Winding
- Primary wall
- Cuticle

Figure 20–7 *A cotton fiber is composed of a cuticle, primary wall, secondary wall, and lumen.*

The dried fibers are from 2 inches in length depending on the species and variety of the cotton plant. Most of the cotton grown in the United States is around 11/8 inches in staple length. The longer stapled cotton is produced in tropical areas such as the Nile River Valley in Egypt.

Cotton is picked using large machines with rotating fingerlike projections on opposite rotating drums to remove the lint from the bolls (Figure 20–8). This ingenious invention replaced long strenuous hours of manual labor that were once required to handpick the cotton. The lint is removed from the picker and is either stored in the field in long, loaf-shaped modules that hold 10 to 12 bales of cotton or it is placed in trailers and taken directly to the gin. A bale is around 1,200 pounds of cotton before the seeds are removed and around 500 pounds after the seeds are removed.

At the gin, the seeds are removed, and the cotton is cleaned of trash and foreign material. The seeds are a secondary industry. They are sent to a processor where they are pressed to remove the oil. The oil is used as a cooking oil or in a variety of products from margarine to salad dressing. The cake that is left after the oil is pressed out is ground into a meal that is a valuable source of protein. Cottonseed meal is fed to cattle and other ruminants, but cannot be used by pigs and certain other simple-stomach animals because of a toxic substance in the seeds called gossypol. Ruminants break gossypol down in the digestive process.

Figure 20–8 *Cotton is picked with a mechanical harvester that uses rotating fingerlike projections on opposite rotating drums.*
Courtesy of James Strawser, The University of Georgia.

Once the seeds are separated and the cotton is cleaned, the lint is pressed into 500-pound bales, graded, and stored or sent to the mill (Figure 20–9). At the mill, the bales are broken open, and the lint from several bales is mixed together to ensure uniformity. The lint is then sent to a carding machine that separates and aligns the fibers. In order to be made into yarn, the fibers all have to be laid parallel in a thin web (Figure 20–10). The fibers are then drawn through a funnel-like device that forms the fibers into ropelike strands called slivers. Spinning machines draw the slivers into threads by twisting the fibers and stretching them out more thinly. This process continues until the proper thickness or count is achieved. The thread or yarn is then wound tightly around tubes called bobbins and is then ready to make fabric.

Figure 20–9 *After ginning, cotton is stored in 500-pound bales.* Courtesy of National Cotton Council of America.

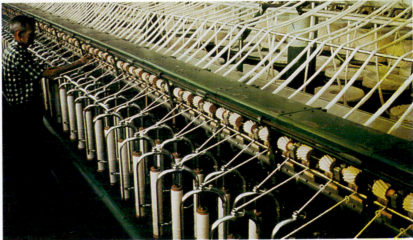

Figure 20–10 *Cotton fibers are spun into yarn that is in turn woven into cloth.* Courtesy of National Cotton Council of America.

Figure 20–11 *People discovered that cloth could be made from animal hair without killing the animal.* Courtesy of American Sheep Industry Association.

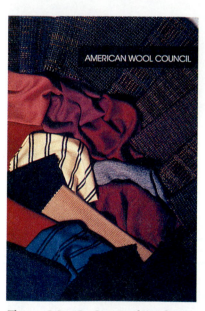

Figure 20–12 *Some of the finest fabric in the world is made of wool.* Courtesy of American Sheep Industry Association.

WOOL

Wool has perhaps been used longer than any other fiber by humans to make clothing. Its use probably began when sheep skins were used for clothing. It may have been accidentally discovered that wool fibers from the pelt would hold together when twisted, and thread could be made. The discovery led to the realization that the animal did not have to be killed to obtain clothing and that the wool could be harvested as the animal grew it (Figure 20–11). This most likely began a selective breeding process through which our modern breeds of fine wool sheep were developed.

Wool has been used for clothing since before recorded time and to this day, many people think that there is no finer fabric than wool (Figure 20–12). Wool is durable, warm in the winter and can be worn in the summer because it absorbs moisture. Some of the finest cloth in the world is made from wool because it lends itself to such a wide variety of applications—from fine garments to beautiful and durable tapestries and rugs.

Wool is the hair from sheep, although the hair of certain other animals such as angora goats and alpacas is also classified as wool. Wool is composed of a type of protein called **keratin** that is also found in tissues such as horns, hooves, and fingernails. It consists of carbon, hydrogen, oxygen, nitrogen, and sulfur. The wool fibers grow as hairs that originate from small cavities or follicles in the skin of the animal. The hair grows in groups called follicular bundles that normally consist of three primary follicles and a varying number of secondary follicles (Figure 20–13). The primary follicles produce the coarse outer hair, and the secondary follicles produce the finer inner hairs. The primary hairs help keep water away from the skin of the animal, and the inner hair provides insulation that keeps the animal warm.

The hairs grow by adding cells that contribute to the structure of the hair. This structure is made up of two distinct layers—the cuticle on the outside and the cortex on the inside. As cells are added to the cuticle layer, they are arranged in an overlapping manner that resembles the scales on a pine cone or the shingles on the roof of a house. This scaly texture allows the fibers to be matted together in a process called felting that bonds the wool into a solid mass under high pressure (Figure 20–14). The resulting felt is used in the manufacture of rigid items such as hats. Also, the

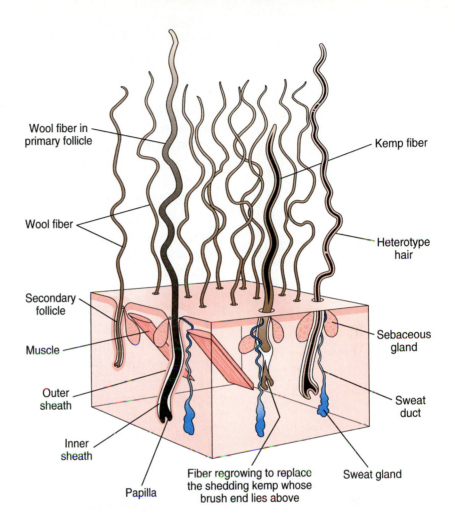

Figure 20–13 *Wool grows out of the skin in follicular bundles.*

Labels:
Wool fiber in primary follicle
Kemp fiber
Wool fiber
Heterotype hair
Secondary follicle
Muscle
Sebaceous gland
Outer sheath
Sweat duct
Inner sheath
Papilla
Fiber regrowing to replace the shedding kemp whose brush end lies above
Sweat gland

Close-up of Wool Fiber

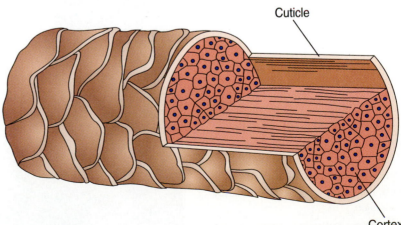

Cuticle
Cortex

Figure 20–14 *The fiber of the wool is composed of two distinct layers, the cuticle on the outside and the cortex on the inside.*

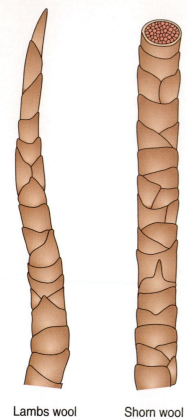

Lambs wool   Shorn wool

Figure 20–15 *Wool fibers have a scaly texture that helps them to mat together.*

scaling on the fibers (Figure 20–15) helps hold the fibers together when they are spun into yarn.

The keratin in the cells of the wool fibers consists of helical or coiled chains of amino acids that are cross-linked to each other by disulfide bonds. Because these chains of amino acids are shaped like coils, the fibers spring back when they are stretched. This characteristic is known as crimp and gives the wool a wavy appearance (Figure 20–16). Generally the more crimp, the more desirable the wool.

As the wool fiber grows from the epidermal layer of the animal's skin, the follicle secretes a grease-like substance that helps keep the wool pliant and helps it shed water. The term for this substance, which is visible on the fleece of a live animal, is **yolk**. The yolk is removed when the wool is processed and is refined into a product called lanolin that is used in skin care products such as soaps, lotions, and hand creams (Figure 20–17).

After the wool has been shorn from the sheep, it is bundled together and marketed. This wool is called grease wool because it has not yet gone through the process of cleaning or **scouring**. In this process all of the grease and foreign material is removed from the wool, and the wool is graded. Next the wool is blended with other types of wool fibers to achieve a particular type of fabric. After blending, the fibers are untangled and laid parallel to each other in a process known as carding. This is done to prepare the fibers for the spinning process. As in the spinning of cotton, the fibers are twisted together to form thread or yarn that is woven into cloth.

Figure 20–16 *The fibers spring back when they are stretched. This is called crimp and is the result of chains of amino acids shaped like coils.* Courtesy of American Sheep Industry Association.

Figure 20–17 *The yolk is visible on the fleece of the animal. Lanolin is produced from yolk.* Courtesy of American Sheep Industry Association.

SILK

Silk has been used for thousands of years and has been said to be the most prized of all fabrics. Silk fibers are very fine, ranging from .00059 to .00118 inch in diameter. The long length of the filaments helps give the fabric a high degree of strength. Silk is the strongest of all the fabrics made from natural fibers and tends to hold its shape well (Figure 20–18). It is also very receptive to dyeing and produces beautiful colors in the finished material. Cloths made from silk are comfortable and absorb moisture better than any of the other natural fibers.

Figure 20–18 *Silk is the strongest of all of the natural fibers and produces beautiful cloth.* Courtesy of James Strawser, The University of Georgia.

Figure 20–19 *Silk is made by the* Bombyx mori *moth. Courtesy of Bob Tompkins, Canadian Forestry Service.*

Silk probably originated in ancient China around 2600 B.C. The Chinese kept the process of silk production a closely guarded secret for hundreds of years. When the secret finally got out, silk was produced in many parts of Asia. Many fortunes were made in producing and marketing silk. Trade routes were established all over the ancient world to bring silk to buyers.

This fabric originates as a fiber produced by an insect. Although more than 30,000 species of spiders and 113,000 species in the insect order *Lepidoptera* make silk, most of the silk is produced by the moth *Bombyx mori* (Figure 20–19). Thailand, China, and Japan are the leading countries in **sericulture**, the culture of the silk worm.

Sericulture begins when the adult moth lays eggs on specially prepared paper (Figure 20–20). A female lays around 700 eggs, each of which is no larger than a pin head. When these eggs begin to hatch, they are laid on bamboo frames that are covered in mulberry leaves. When the tiny larvae hatch, they feed on the leaves and may devour 30,000 times their weight at hatching and may increase to 10,000 times their weight at hatching. During a growing period of about 35 days, the larva molts (sheds its skin) four times and grows to a length of about 3 inches with a diameter of about 1/2 inch.

At the end of the growth period, a straw structure is placed on the bamboo frame and the larva attaches itself to the straw (Figure 20–21). It then begins to spin a cocoon that covers its body. The silk material for the cocoon is secreted from two glands inside the larva's mouth and is forced through two tiny openings called spinnerets in the head of the larva. The two strands of silk are covered with a gummy, water-soluble substance called sericin. The cocoon is completed in two to three days and may contain as much as a mile of silk filament.

When the cocoon is finished and the larva is completely covered, it begins the next stage of **metamorphosis** and converts to a pupa. The pupa is then killed with heat, and the silk is collected by gathering several filaments together and winding them onto a reel. Each cocoon usually produces about 100 yards of the combined filaments.

If the pupa is allowed to live in the cocoon, it again begins metamorphosis, changes into the adult stage, and emerges in a period of about 8 days. Those that grow to the adult stage are used for breeding and laying eggs.

The silk as it is taken from the cocoon is called raw silk and has to be boiled in a solution of soap and water to remove

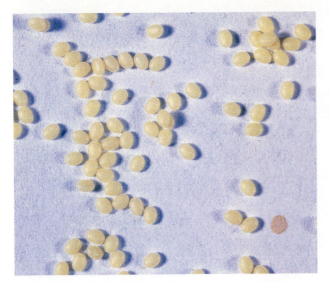

Figure 20–20 Sericulture begins when the adult moth lays eggs on specially prepared paper. *Courtesy of Bob Tompkins, Canadian Forestry Service.*

Figure 20–21 Larva attach to straw layed on bamboo frames. The larva spin silk to create cocoons. Each cocoon may contain a mile of silk filament. *Courtesy of Photo Disc, Tim Hall.*

the gummy coating. After boiling, the silk appears soft and lustrous and is then woven into cloth. It takes about 2,500 to 3,000 cocoons to produce a yard of silk cloth.

FLAX

The use of linen dates back to the Stone Age when humans still wore skins for clothing but used linen for making fishnets. Over the years, linen has been used for clothing in most countries of the world. Much of its popularity as a fabric for clothing was lost with the widespread availability of cotton that began in the early 1800s. Today it is still popular as a cloth for making tablecloths, napkins, and in recent years has regained some popularity as a clothing material (Figure 20–22).

Linen comes from fibers produced in a plant called flax (*Linum usitatissimum*) that grows in climates that have a lot of rainfall and temperatures that do not get very hot. Most of today's flax is produced in Europe and New Zealand. Plant stems contain two structures, the xylem and the phloem, that transport water and nutrients through the plant. Linen comes from the fibers (called bast fibers) that make up the phloem of the plant (Figure 20–23).

Figure 20–22 *Linen is used for clothing, tablecloths, and napkins.* Courtesy of James Strawser, The University of Georgia.

The plants grow to a height of about 3 feet and are harvested. The outer layer of woody material must be dissolved before the fibers can be removed. This is done either by soaking the stems in warm water where bacterial action decays the material or by a more modern method where the covering is dissolved using chemicals. This process is called **retting**.

After the outer layer is removed, the phloem fibers are passed through fluted rollers that break up the woody substances connected to the fibers. In a process called **scrutching**, the usable fibers are separated out. Then the fibers are combed out in much the same process used in carding cotton. This lays out the fibers parallel to each other so they can be spun into yarn.

Figure 20–23 *Linen comes from fibers that make up the phloem of the flax plant.* Courtesy of Flax Council of Canada.

SUMMARY

As petroleum becomes more scarce and prices rise, fabric made from agricultural fibers will increase in importance. Crude oil deposits will one day be exhausted, but with research, development, and management, we can have a constantly renewable supply of fiber from our agricultural industry.

CHAPTER 20 REVIEW

Student Learning Activities

1. Look through your closet and determine the fiber content of the clothes you wear. List the advantages and disadvantages of the types of fibers.

2. Research the origins and the manufacturing of synthetic fibers. Summarize your findings and develop an opinion as to whether or not synthetic fibers are superior to natural fibers. Organize your reasons and debate someone in your class who holds the opposite opinion.

Define the Following Terms

1. cotton gin
2. boll
3. cellulose
4. keratin
5. yolk
6. scouring
7. sericulture
8. metamorphosis
9. retting
10. scrutching

True/False

1. Fiber produced by agriculture is more comfortable than human-made fibers.
2. The North is the ideal place for growing cotton.
3. Flax is a shrublike perennial plant.
4. Wool can be worn in the summer and winter.
5. Cotton is the strongest of all natural fibers and tends to hold its shape well.
6. The Chinese kept the process of silk production a secret for hundreds of years.
7. The tiny silkworm larvae feed on mulberry leaves.

8. Most of today's flax is produced in Europe and New Zealand.

9. Plant stems contain two structures, the xylem and phloem, that transport water and nutrients through the plant.

10. Wool is the hair from sheep.

Fill in the Blanks

1. Agriculture is involved in the production of materials used for _____, _____, and _____.

2. Much of the cloth produced today is manufactured from _____.

3. The four main types of fiber are _____, _____, _____, and _____.

4. _____ invented the cotton gin.

5. _____ has perhaps been used longer than any other fiber by humans to make clothing.

6. _____ has been said to be the most prized of all fabrics.

7. _____ may be more wrinkle resistant than cotton.

8. _____ is a shrublike perennial plant.

9. Wool is composed of a type of protein called _____.

10. Silk probably originated in ancient _____.

Multiple Choice

1. Much of the cloth today is manufactured from
   A. petroleum.
   B. cotton.
   C. animals.

2. Cotton was probably first used in the Nile Valley in ancient
   A. America.
   B. China.
   C. Egypt.

3. Much of the history of the United States was developed around the growing of
   A. silk.
   B. cotton.
   C. linen.

4. A durable fabric that can be worn in the summer and winter because of its ability to insulate and absorb moisture is
   A. wool.
   B. linen.
   C. silk.

5. Felt from wool is used in the manufacturing of rigid items such as
   A. shoes.
   B. hats.
   C. jewelry.

6. Yolk removed when sheep wool is processed and refined into a product called
   A. soap.
   B. perfume.
   C. lanolin.

7. The fabric that originates from an insect is
   A. silk.
   B. linen.
   C. cotton.

8. Linen comes from the fibers in the plant called
   A. fern.
   B. flax.
   C. boll.

9. Cellulose is composed of molecules containing carbon, hydrogen, and oxygen arranged in
   A. carbon bonds.
   B. hydrogen bonds.
   C. hydroxyl groups.

10. Sericulture is the culture of the
    A. sheep.
    B. silkworm.
    C. mulberry bush.

Discussion

1. Why are natural fibers superior to synthetic fibers for making clothing?
2. Why is wool such a good clothing material?

3. Explain the life cycle of the silkworm moth and how it is used in silk production.

4. Give some uses of petroleum in making cloth.

5. What were the major drawbacks with cotton?

6. When is wool called grease wool and why?

7. What are some of the uses for the yolk from the fleece of a sheep?

8. Discuss the uses of linen.

9. How is the phloem of the flax plant involved in making clothing?

10. What elements make up keratin?

CHAPTER 21

PRODUCING ORGANICALLY GROWN PRODUCTS

Key Terms

organic
organic fertilizers
Organic Food
Production Act
supply and demand
commercial fertilizers
cover crop
composting
humus
soil amendments
fish slurry
crop rotation
border plants
predator insects
natural toxins

STUDENT OBJECTIVES

After studying this chapter, you should be able to:

◆ Discuss the meaning of the term organically raised product.

◆ Analyze the reasons why organic production is increasing.

◆ Define the rules governing the certification of organic products.

◆ Discuss the role of the USDA in labeling organically raised products.

◆ Describe how organic products are raised.

◆ Analyze the controversial issues surrounding organically raised products.

In the late 1960s and early 70s, a new direction in the production of food and fiber began when a new industry known as **organic** farming was born. The movement came about as a result of public awareness of the tremendous amount of commercial pesticides and fertilizers that were applied to crops (Figure 21–1). Before this time, scientists who analyzed very small amounts of substances could detect toxins or other materials in parts per hundred. However, with

Figure 21–1 *The organic agriculture movement began as the public became aware of the amount of pesticides used on crops.* Courtesy of USDA-ARS/Ken Hammond.

new research techniques developed during this period, analysis could be achieved in parts per million (ppm) and later in parts per billion (ppb). When any amount of pesticide was detected in food, public concern was raised. Studies were conducted using laboratory rats that were fed huge quantities of pesticides, and these tests revealed cancer or other health problems in the test animals. Also, farms were getting larger and were perceived as factories rather than farms.

The idea began that inputs such as fertilizers derived from natural rather than manufactured means would produce healthier produce that would in turn become better food.

Many definitions of organic farming exist, but most commonly, the term refers to the use of strictly organic or natural inputs into the production of food and fiber. Organic refers to materials that originated from living organisms and/or materials that are nonmanufactured or natural in origin. This definition is broad and doesn't fit all areas of the industry called "organic agriculture."

In fact, one of the major holdbacks for the organic agriculture industry has been the definition of organically produced products. Over the years much debate has occurred over just what qualifies as organically produced products. For example, if corn is produced using only natural means such as **organic fertilizers** and no chemical pesticides, most likely the corn would be considered to be organically produced. But, what if chemical pesticides, such as rotenone or pyrethrums, that are derived from plants are used? What if the seed was from corn that has been genetically altered? Would these examples of corn still be considered organic produce? These issues have been debated for some time, and despite many public hearings on the issue, the USDA has not been able to satisfactorily define organically produced food.

In 1990, Congress passed the **Organic Food Production Act** to (1) establish national standards governing the marketing of products labeled as organically grown; (2) assure consumers that organically produced products meet a consistent standard; and (3) facilitate commerce in organically produced fresh and processed foods. The act is administered through the Agricultural Marketing Service (AMS) of the United States Department of Agriculture (USDA). Although the agency does not define what "organically produced" means, it does allow other entities to certify products as being produced organically.

Presently, dozens of organizations set rules and qualifications for labeling food and fiber as organically produced (Figure 21–2). For example, an organization called California

Figure 21–2 *Many organizations set the regulations for foods that are certified as organic.*

Certified Organic Farmers sets the regulations for organically grown products in California. Many other states have similar organizations. They determine whether or not the producers and products meet the qualifications set forth by their organization. For example, the organization may specify what, if any, type of insect control is used, what types of fertilizers are used, or how the produce is processed. If the qualifications are met, the product may be labeled as "certified organically grown."

Initially, the organic industry began very slowly; however, since about 1970, the industry has achieved steady growth. Today it is a thriving and growing industry with almost 1 1/2 million acres of organically produced crops in the United States, and the acreage increases each year.

The United States Department of Agriculture (USDA) lists six forces driving the growth in the organic agriculture industry.

1. **Environmental awareness—**As the population of the country grows, people are more aware of the impact industry has on the environment. About the time the movement began, evidence that residual pesticides such as DDT were harmful to the environment was coming to the population's attention. The reasoning was (and still is) that by lessening the use of

pesticides on crops, the impact chemical residues have on the environment would decrease. Also, the concept that pesticides destroyed beneficial insects such as bees, lady beetles, and praying mantises came to the public attention. These factors have perpetuated the idea that buying organically grown products helps in maintaining a healthy environment.

2. **The relationship between diet and health**—Billions of dollars in research have gone into research on how diet affects the well being of the human body. Conclusive evidence shows that people who eat a well-balanced diet and exercise routinely live longer, healthier, more productive lives (Figure 21–3). Many of the diseases and health problems that have always plagued people can be prevented or controlled with the proper diet.

   Many people make the assumption that organically grown foods are more wholesome and nutritious than foods grown with conventional methods using inorganic fertilizers and pesticides. As the cost of health care increases and the median age of the citizenry increases, more attention will be placed on preventive measures, and organically grown food will likely continue to play a role in this movement.

3. **The declining costs of organic food production**—As people become more aware of the availability of organically grown products, the more they buy them. As the demand increases, the production also increases. This is the economic principle known as **supply and demand.** When demand increases, larger quantities are usually produced, and competition among producers expands. Generally more competition leads to lower price as larger, more established food companies enter the organic food business. Not only are more products available, but the accessibility of these items also helps to increase consumer awareness of the products (Figure 21–4).

   Another factor in the declining cost of organically grown products is the withdrawal of many conventional pesticides from the market. Producers who once used older, more residual pesticides now have to rely more on other methods of pest protection, and this drives up the cost of production. In many cases pest-protection plans may involve the techniques

Figure 21–3 *People recognize that there is a direct relationship between proper nutrition and good health.* Courtesy of USDA-ARS/Keith Weller.

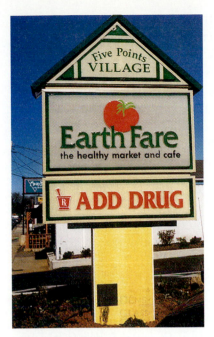

Figure 21–4 *There is an increasing number of establishments that specialize in organically produced food.*

used in organic production so producers may make the decision to grow products that can be labeled as organic in order to get better prices for their produce.

4. **Mainstreaming of organic consumers, products, and retailers—**When the movement toward organically grown products first began, it was considered by many to be the ideas of a small, almost radical group of people. Today, the concepts are embraced by a larger group of people and are accepted by a larger percent of the population. Remember that the current population is much more health conscious than generations in the past. Better marketing techniques, better packaging, and more publicity have all contributed to the popularity of organically grown products. Large grocery chains are now advertising the availability of organically grown foods in their stores. This all adds up to better awareness by the public about these foods and also leads to a wider variety of organically grown foods.

These foods and products have also improved in quality (Figure 21–5). At one time organically grown foods appeared to be of poor quality and did not have the visual appeal of the conventionally grown foods.

For example, organically grown fruit often had insect damage or were not as large, colorful, or bright as the more ordinary foods available at the supermarket. Organically grown fruits today look

Figure 21–5 *Organically produced foods and products have improved in quality. Courtesy of USDA-ARS/Scott Bauer*

much better because of the more advanced techniques used in pest control. Also much of the organically grown food is processed. Because defects can be removed before processing, the consumer never sees the blemishes in the product. Managers of large grocery stores demand that the products they sell be of high quality in order to maintain their reputation for selling only the very best products. Better quality has led to more consumer acceptance.

5. **Worldwide harmonization of organic standards—** The United States is not the only country to move toward the production and consumption of organically grown products. In fact many of the more advanced countries are ahead of the United States. According to the International Trading Center, retail sales of organic products are more than $10 billion each year. For several years, many of the European countries have been reluctant or have refused to buy products such as beef from the United States. Their concern has been over the use of growth hormones in cattle fed for market (Figure 21–6).

Figure 21–6 *Europeans are concerned over the growth stimulants used to produce U.S. beef. Courtesy of PhotoDisc, Adam Crowley.*

This concern has hurt the conventional beef industry. As a result, some beef, poultry, and pork producers are growing animals that qualify as "organically grown."

Several international organizations have established standards that define and regulate products that can be sold using an organic label. As communications and interchange improve among nations, the organic movement will likely increase.

6. **Capital investments from the financial community—** Any time a marketable product increases in popularity, people who have money to invest are attracted. The organic products industry is now large enough to cause investors to place assets into production, marketing, and sales. This is a tremendous boost to the movement because financing means that the organic industry can have the needed capital to increase all aspects of the industry.

THE PRODUCTION PROCESS

Methods used by organic producers are quite a bit different from those used by conventional producers. The vast majority of all the food and fiber crops produced in the United States are grown using **commercial fertilizers** and pest controls. Organic producers use different production schemes that make use of natural methods of fertilizing and pest control rather than the conventional pesticides and fertilizers.

Fertilizers

Producers of organically grown foods and fiber use a variety of methods to provide plant nutrients. The idea is to use only materials derived from nature and to not use fertilizers made purely from chemicals. One of the basic methods is to use **cover crop** plants, such as annual winter grasses or legumes, that are planted each year when crops are harvested.

The cover crops grow during the winter and early spring up until the time when new crops are to be planted and are then turned under before the next crop is planted to provide organic material for the new plants. The cover crop decays, and through the natural process of decomposition releases nutrients into the soil. In addition, such cover plants as vetch and clover, are legumes that have nodules on the roots that host nitrogen-fixing bacteria (Figure 21–7). These bacteria take nitrogen from the air and convert it to a form of nitrogen that can be used by plants. The process of nitrogen fixation improves the producing ability of the soil.

Figure 21–7 *Organic producers make use of cover crops such as clover to provide soil nutrients.* Courtesy of PhotoDisc, C. Borland/Photo Link.

Another method used by organic producers is that of **composting.** Materials such as leaves, grass clippings, and various other plant matter are put in a bin designed to hasten the decomposition process. As plant material decays, it is converted into a substance called **humus** that not only contains nitrogen, but also aids in the water- and nutrient-holding capacity of the soil.

Other materials, such as peat moss or decayed sawdust (called **soil amendments**), improve the soil by adding organic material (Figure 21–8).

Nonorganic amendments such as lime or gypsum may be used to adjust the Ph level of the soil to suit the particular crop being grown.

Producers of organically grown products can obtain materials for use as fertilizers from several sources. One is the fish-

Figure 21–8 *Mulch is used to enrich the soil. This machine is used to turn organic material so that it will decompose more quickly. Courtesy of USDA-ARS.*

ing industry. When fish are harvested and dressed, a large portion of the fish carcasses remain when the edible portions are removed. This combination of heads, skin, entrails, fins, and bones is ground and blended into a slurry that is used as a nutrient-rich fertilizer. **Fish slurry** is high in nitrogen content and contains phosphorus and potassium—all of which are major nutrients required by plants.

Perhaps the most important source of organic fertilizer is animal manure (Figure 21–9). All phases of the animal industry produce millions of tons of fecal material each year, which has to be disposed of properly. Because manure is very high in nutrient content and is valuable as fertilizer, the dual purpose of getting rid of the waste and fertilizing crops is achieved.

Unless the proper steps are taken, major problems may exist with using manure. One problem is that manure causes quite an odor problem when it is spread onto a field near where people live. Producers who grow crops near a residential area and fertilize with raw manure can come under heavy criticism from people in the area.

A second problem with using manure is that the fecal matter contains high levels of *E coli* and other forms of bacteria that may be harmful to humans. These bacteria can and do cause severe health problems and even death among humans.

A third problem with using manure is that most raw manure contains a considerable amount of weed seed. When

Figure 21–9 Animal manure is a major source of fertilizers for organic producers. *Courtesy of PhotoDisc.*

grain is grown, almost inevitably some amount of weeds will grow up and mature in the field alongside the crop. When the crop is harvested, the seeds from the weed are also harvested. Many weed seeds are too small to separate out, and go through the processing with the grain. Because most of the weed seeds are not harmful to livestock, they are left in the grain and are fed as part of the processed livestock feed. The weed seed then is ingested by the animal, goes through the digestive system of the animal, and is passed out with the feces. When the feces is used as manure fertilizer, the weed seeds are inadvertently planted and may germinate in the soil along with the crop.

All three of these problems can be eliminated by using the proper steps in the process of composting. When the manure is composted, it is piled together and allowed to age for at least two months prior to use. If properly composted, the internal temperature of the manure should reach temperatures of 210°F to 140°F. This degree of heat should be enough to kill most harmful bacteria and weed seed. In addition, the composting process helps alleviate problems with odor.

Insect Control

Perhaps the largest problem facing organic producers is that of controlling insects. When plants are growing and weather conditions permit, populations of insects can build up rapidly and destroy a crop very quickly. Conventional producers use an arsenal of chemical weapons to control insects. However, organic producers are against using insecticides to kill pests, and resort to more natural methods such as **crop rotation** to help keep insects in control.

The principle behind crop rotation as an insect control method is that many insect pests attack only particular plants. By planting fields with different crops each year, populations of insects have less chance of building because the plants they feed on are not grown two years in a row. For example, horn worms may attack tomatoes but may not bother squash

Figure 21–10 Horn worms may attack tomatoes but not squash. Crop rotation can help disrupt the life cycle of this pest. *Courtesy of PhotoDisc, Paul Beard.*

(Figure 21–10). By planting squash in a field that had tomatoes the previous year, the horn worms are deprived of food. This helps break up the insects' life cycle.

Another method is planting **border plants** around the outside of the field where the crop is grown. The borders serve to provide an alternative crop for the insects, and the idea is that the insects attack the plants around the border and not the crop. Border plants are usually perennials that don't have to be planted each year and must be plants that insects prefer to the crop plants.

Perhaps the best weapon organic producers use in controlling insects is the use of **predator insects.** In nature many insects are nourished by eating other insects. A high population of predator insects such as praying mantises, lacewings,

BIO BRIEF

Managing Manure Nitrogen to Curb Odors

People often think of livestock manure as waste, but scientists believe such animal byproducts should be considered production resources. They are trying to find new ways to use these resources from cattle feedlots and hog farms in farming operations to address environmental concerns and residential complaints about odors, says Vincent H. Varel. He is an Agricultural Research Service microbiologist at the Roman L. Hruska U.S. Meat Animal Research Center in Clay Center, Nebraska.

Manure from beef cattle feedlots could be valuable for use as nitrogen fertilizer, says Varel. But unfortunately, half to three-fourths of that nitrogen never reaches the field. Most of the loss occurs through a process called hydrolysis. Microbes in animal manure and soil produce the enzyme urease that converts the urea in urine into ammonia, which escapes into the air.

This same process happens in urea-based commercial fertilizers. To prevent ammonia loss, fertilizer manufacturers routinely add urease inhibitors—actually, chemical relatives of urea—to block hydrolysis; thus preserving nitrogen until it's taken up by plants.

In laboratory experiments, Varel also blocked this nitrogen loss from manure by adding urease-inhibiting compounds. He mixed either feedlot cattle manure or swine manure with cattle urine into a slurry. Like modern waste-handling systems, the slurry contained naturally occurring urease-producing microbes. Varel added as little as 10 milligrams (35 millionths of an ounce) of urease-inhibiting cyclohexylphosphoric triamide (CHPT) per liter of manure slurry mixture. In untreated samples, the urease hydrolyzed nearly all the urea within one day. But in treated samples, hydrolysis was completely prevented for at least four days.

Another urease inhibitor, phenyl phosphorodiamidate (PPDA), produced similar results. Varel found that adding more urease inhibitor each week provided longer term control. For example, 100 milligrams of PPDA added weekly to cattle waste preserved 70 percent of the urea for 28 days. Adding just 10 milligrams weekly preserved 38 percent.

lady beetles, and parasitic wasps are used to help control harmful insects (Figure 21–11).

Many of the predator insects are grown commercially for the purpose of selling to organic producers. Modern garden-supply firms offer predatory insects for sale through their retail catalogs. As soon as insect pests appear, the organic producer releases large numbers of the predatory insects to eat

The researchers are trying to determine why the larger amount of inhibitor didn't preserve even more urea. An understanding may lead to scaled-up, farm-size applications. "Our laboratory studies with CHPT and PPDA just started us on the learning curve," Varel says.

Blocking ammonia production in feedlots not only preserves nitrogen, but also reduces odors. Courtesy of USDA-ARS.

He tested a third urease inhibitor, n-(n-butyl) thiophosphoric triamide (NBPT). This compound is currently being used as a nitrogen preservative in no-till cropping systems. Preliminary experiments showed NPBT works even better in the feedlot than in the laboratory. Under feedlot conditions, Varel explains, NBPT is exposed to air that converts the compound into a more effective urease inhibitor. The researchers spread NBPT over a feedlot surface once each week. As cattle urinate, the chemical binds to urea and blocks ammonia production, thus preserving nitrogen.

Varel says urease inhibitors will reduce ammonia emissions, which contribute to odors. However, other odor-reducing compounds will be needed to more fully control a variety of unpleasant-smelling volatile compounds from manure. Encapsulating these mixtures in starch or other protective materials could ensure the slow release of the active compounds and require fewer applications to cattle feedlots, manure slurry tanks, and covered lagoons used on livestock farms.

Source: Agricultural Research Magazine.

the harmful insects. Specific predators are used for specific pests. For example, lacewings only feed on thrips and aphids, so if these insects are a problem, the producer introduces large numbers of lacewings. The larvae of parasite wasps eat a variety of insects and are used as a method to kill more than one type of insect pest. The wasp lays eggs on or in the harmful insect, and upon hatching, the larvae feeds on the insect.

Figure 21–11 *Predator insects such as the praying mantis help control insects for organic producers. Courtesy of PhotoDisc, Alan Dappe.*

ORGANIC ANIMAL AGRICULTURE

Although most of the organic production is centered around plants, animals are also grown and sold as organically raised. Most often livestock products labeled as "organically produced" applies to animals that have been raised without pesticides to control internal and external parasites. The animals must not have been subjected to the use of growth stimulants or hormones. Also, animal products labeled as organically produced are generally from animals raised in open-air, non-crowded conditions that include humane treatment of the animals (Figure 21–12).

Sometimes the certification may require that the animal be fed organically produced feed and raised by other stringent nutritional requirements.

For example, the California Certified Organic Farmers prohibit the following:

1. Nonorganic feed

2. Feed containing plastic pellets, urea, synthetic preservatives, or other synthetic materials.

3. Feed containing antibiotics, hormones, drugs, or synthetic growth promoters.

4. Intentional refeeding of manure.

Figure 21–12 *Organically grown animal products are usually from animals that were not raised in confinement.*

5. Maintaining animals on poorly balanced diets, including raising anemic animals.

6. Feed containing animal by-products for ruminants.

7. Feed that has been solvent extracted.

Only products from animals that meet all the production requirements may be labeled as organically produced. This includes products such as wool and hair, meat, milk and milk products, eggs, and any other animal-derived product. The labeling also extends to the use of irradiation for preserving meats. Although this method of preservation is allowed by the USDA, meats treated with radiation may not be labeled as organically produced.

CRITICISMS OF ORGANIC PRODUCTION

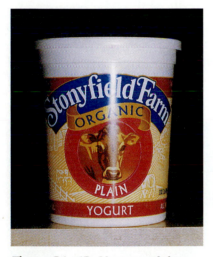

Figure 21–13 No research has shown conclusive evidence that organic foods are safer or more nutritious.

Since the concept of organic production began, controversy has surrounded the issue. Critics maintain that the public is misinformed about the safety and nutritional value of organically grown products. They point out that these products are much more expensive than conventionally produced food and fiber and that they are generally of poorer quality. Critics point out that a lot of research has been done comparing the safety and nutritional value of organic foods to conventionally grown foods. To date, no conclusive evidence has suggested that organically produced foods are safer or more nutritious (Figure 21–13).

In fact, organic production has several disadvantages. One of the primary reasons for organically grown food is to avoid the toxins from the trace amounts of insecticides that could possibly be on conventionally produced fruits and vegetables. Plants have self-defense mechanisms that produce **natural toxins** to help ward off insects. If attacked by pests, the plants release these substances as a result of the stress caused by the insects or disease. These natural toxins may be many times the amount contained in plants that were protected by pesticides.

It is difficult to produce fruits and vegetables that have the quality we are accustomed to without the use of chemical pesticides. Natural methods of controlling pests can be effective but when used along cannot offer the protection that will ensure that the food does not contain insect damage. This damage on fruits and vegetables provides a natural harbor for bacteria and other organisms that cause the food to spoil. Most scientists agree that the greatest danger we have from our food supply is that of food spoilage and the ingestion of bacteria, not the presence of pesticides.

Another problem with organically produced food is that often manure is used as fertilizer. Unless manure has been treated, it contains bacteria that can be harmful to humans. Critics point out that few regulations govern the use of manure as fertilizer, and contact with *E coli* bacteria negates any advantage offered by using manure. Also, they point out that nitrogen, phosphorus, and potassium, which are the major nutrients needed by plants, are the same whether they come from organic or inorganic sources.

Critics contend that organic production requires more land and more labor to produce the same amount of products grown using conventional means. In fact, some scholars argue that intensive farming methods are much more "environmentally friendly" than organic production. Projections are that around or prior to the middle of the 21st century, the world's population will double (Figure 21–14). This means that the world's production of food and fiber will have to double through either one of two methods: 1) to clear new land and increase the number of acres under cultivation; or 2)

Figure 21–14 *Projections are that within the next 50 years, the world's population will double. All these people will have to be fed.*
Courtesy of PhotoDisc, Doug Menuez.

grow more food and fiber on the land currently under production. The concern raised over organic production is that the use of these practices will necessitate the clearing of more land. As more land is cleared, the loss of wildlife habitat will increase. Most of the new land being cleared is through the "slash and burn" methods used in the tropical areas (Figure 21–15). A large portion of the land cleared is poor quality and highly susceptible to erosion. However, through the use of more intensive farming, less new land will have to be put into production.

The fact that organic production is based on the use of organic materials is also a weakness of the system if it is used as the main method of producing our food and fiber. If all the animal manure and all the available beneficial organic materials were used, there would not begin to be enough to fertilize all of the needed crop land. Obviously, fertilizer inputs will be needed, and the argument is made that organic sources alone cannot supply anything more than a small fraction of what is needed.

Figure 21–15 *Most of the new land that is cleared for agricultural purposes is in the rainforests.*
Courtesy of PhotoDisc, C. Sherburne/Photo Link.

SUMMARY

Despite the controversy over the issue, the production and consumption of organically grown products continues to grow. This relatively new branch of the agricultural industry is here to stay and will receive increasing attention in the years to come. As with all other aspects of the market, it is driven by consumer demand, and as that demand increases, production will also increase. The challenge will be to keep consumers well informed and up to date on research findings related to organically raised products.

CHAPTER 21 REVIEW

1. Search the Internet and find materials on organic agriculture. Prepare and present to the class your position on the merits or problems with organic agriculture. Debate others in your class who take the opposite view.

1. Go to the grocery store and make a list of the products that are labeled as organically grown. Also note the number of similar products that are not labeled as organic. What percent of the products are labeled as organically grown?

Define the Following Terms

1. organic
2. organic fertilizers
3. Organic Food Production Act
4. supply and demand
5. commercial fertilizers
6. cover crop
7. composting

8. humus
9. soil amendments
10. fish slurry
11. crop rotation
12. border plants
13. predator insects
14. natural toxins

True/False

1. Organic farming started in the 1950s as a way to increase food production.

2. The definition of organic farming is a major holdback for organically produced products.

3. The Organic Food Production Act addresses three areas and is administered through the Agricultural Marketing Service.

4. Only the USDA can decide what is "organically produced."

5. The organic movement began as a backlash to the perception that foods were contaminated by pesticides.

6. The United States Department of Agriculture (USDA) contends that the growth in organic agriculture industry will slow in the years to come.

7. Organic producers use cover crops to provide organic material for the new plants.

8. Organic producers use composting as a way of obtaining fertilizer.

9. The largest problem facing organic producers is that of controlling insects.

10. A way of controlling insects in organic production is with the use of insecticides.

11. Organically produced animals may require the animal to be fed organically produced feed.

12. Organically produced products are less expensive than conventionally produced products.

13. It is difficult to produce organic products with the quality to which we are accustomed.

14. The United States is the only country to produce organically grown products.

15. Organic producers use only inputs that are nonmanufactured or natural in origin.

Fill in the Blank

1. Today there are almost _____ million acres of organically produced crops in the United States.

2. The pesticide _____ is known to have harmful effects on the environment and has helped to pave the way for organic farming.

3. It is assumed that foods that are organically grown are more _____ and _____ than foods grown in conventional methods.

4. The economic principle known as _____ is evident in costs of organic food production.

5. _____, _____, and _____ have all contributed to the popularity of organically grown products.

6. Hosts that take nitrogen from the air and convert it to a form that can be used by plants are known as _____ _____ _____.

7. Materials known as _____ are added to the soil to improve the soil organically.

8. The most important source of organic fertilizer is _____ _____.

9. _____ _____ are perennials planted around the outside of a field where a crop is grown.

10. _____ is a process of preserving meats that disqualifies it from being labeled organically grown.

11. Critics argue that organic production requires more _____ than conventional farming.

Discussion Questions

1. Discuss the problems with using animal manure as a fertilizer.
2. Discuss the principle behind crop rotation as an insect control method.
3. List the three criteria for an animal to be considered organically grown.
4. Discuss the definitions of organic farming and the holdbacks because of them.
5. Discuss the Organic Food Production Act of 1990.

CHAPTER 22

NEW DIRECTIONS IN AGRICULTURE

KEY TERMS

crude oil
ethanol
gasohol
canola
kenaf
bast fibers
biomass
gasification
kelp
alginic acid
Global Positioning System
(GPS)

STUDENT OBJECTIVES

After studying this chapter, you should be able to:

✦ Analyze the trend toward fewer workers in production agriculture.

✦ Discuss the changes predicted for agriculture.

✦ Discuss how genetic engineering will affect agriculture.

✦ Explain the concept of renewable resources.

✦ Summarize the ways in which plants can be converted into energy.

✦ Discuss how new crops and animals can have an impact on agriculture.

✦ Explain how new uses can be found for currently used crops.

✦ Summarize the potential of the ocean as a site for production agriculture.

✦ Explain how developments in engineering and electronics will affect agriculture.

✦ Explain how computers will impact decision making.

✦ Analyze the concept of sustainable agriculture.

If you examine the statistics on the number of people involved in production agriculture for the past 50 years, you might arrive at the conclusion that agriculture is dying out. Fifty years ago in many states, more than 70 percent of

Figure 22–1 *Today, less than 2 percent of the population make their living producing food and fiber.* Courtesy of James Strawser, The University of Georgia.

Figure 22–2 *A large and growing number of people are needed to transport, process, and preserve food and fiber.* Courtesy of James Strawser, The University of Georgia.

the population worked on farms. Today less than 2 percent make their living producing food and fiber (Figure 22–1). Will this trend continue until we no longer have anyone producing crops and animals? Of course, this is impossible because humans will always have to have food, shelter, and clothing, and agriculture provides these necessities. The reason for the trend is that agriculture has become more efficient. It takes fewer and fewer people to produce the agricultural commodities that the nation's population requires. At the same time, the support necessary for this industry continues to grow. Someone must produce the fertilizers, seeds, medications, and supplies necessary for production. A common mistake is to consider agriculture and farming to be synonymous. Farming is a part of agriculture, but farmers represent only a small portion of the people involved in agriculture. A large number of people are needed to transport, process, and preserve food and fiber, and the list will continue to grow (Figure 22–2). This large group of support people is one reason that more food and fiber can be produced by a smaller number of farmers.

In the future there will be dramatic change in agriculture just as there will be in all aspects of our lives. Even though our agricultural system is marvelously efficient, it will have to become even more efficient to meet the demands of a growing population. Even though the population growth rate in the United States is expected to slow down, other countries in the world continue to have dramatic increases in population. Developed countries like the United States will have the challenge of producing enough food to feed these people. Not only will the larger number of people be a challenge to feed and clothe, but sprawling metropolitan areas will take an increasing amount of fertile farmland. To further compound the problem, agriculture will be faced with protecting a fragile and shrinking environment. Most of the population has little understanding about how agriculture operates. Agriculturists will have to address such groups as animal rights activists who will continue to target intensive animal production and environmentalists who object to disturbing the natural setting of the earth.

The Council for Agricultural Science and Technology (CAST) has identified several changes that may occur in early in the 21st century.

- Demographic changes will increase the demand for geriatric foods, smaller packages, snack food, ethnic food, and take-out and home-delivered food. The demand for poultry, cheese, soups, bakery products, and fruits will increase while the demand for fresh milk, beef, eggs, and processed vegetables will decrease.

- Changing attitudes will probably further reduce the role of the traditional family meal and the foods usually purchased for this purpose There will be increased purchases of minimally packaged, minimally processed foods with fewer nonnutritive additives.

- The growth in the demand for food in the United States will slow to about 1 percent per year by the year 2000. The reasons are the slowing of population growth, more modest increases in real income, and slower declines in the relative price of food.

In the future, agriculture will be challenged to overcome all of these problems, but these problems can be solved. Agricultural research has made agriculture what it is today and will create the agriculture we will need in the future. Although it is difficult to predict what agriculture will be like in the future, we can speculate on promising lines of research and thought.

Plants and animals that today's agriculturists produce are the result of years of development of plants and animals that

Figure 22–3 *Only about 3 percent of the known plants have been explored and evaluated for their potential use.* Courtesy of Progressive Farmer.

were once found in the wild. Two great potentials of agriculture are the abilities to develop organisms that now live in the wild and to develop new breeds and varieties through crossbreeding. For example, of all the known seed-bearing plants that grow in the world, less than 1 percent have been grown and used. Only about 3 percent have ever been explored and properly evaluated as to their potential use (Figure 22–3). There is an old saying that, "Necessity is the mother of invention," and this is true in agriculture. As we develop new needs, more plants will be evaluated and developed for use.

GENETIC ENGINEERING

Chapter 5 centered around the exciting possibilities of new discoveries and developments in the area of genetic engineering. This new technology will greatly add to the potential of new crops and animals to produce. It will also solve many of the problems with disease and pests that now face agricultural producers. Just think of all of the exciting possibilities of creating designer animals and plants. For example, scientists are now on the verge of being able to produce cotton that will come in a wide variety of colors. This should not only create new demand for cotton but also bypass the expensive steps necessary for dyeing the fabrics.

Perhaps animals could be produced that would have the capability to convert wood fiber (cellulose) to carbohydrates (Figure 22–4). The cost of producing meat animals would greatly lowered if animals could digest and utilize wood fibers.

In the future, these types of organisms will be commonplace to agricultural producers. Remember that just a few

Figure 22–4 *Perhaps animals could be produced that could convert wood fibers to cellulose.* Courtesy of James Strawser, The University of Georgia.

years ago common practices such as embryo transfer were considered to be science fiction. As the practice of genetic engineering becomes widespread, there will be a growing need for educating the public about the new methods. There will likely be public concern over this technology, and this concern will have to be addressed.

RENEWABLE RESOURCES

During the 1970s, the Oil Producing and Exporting Countries (OPEC) placed an embargo on **crude oil** sold to the United States. This brought to our attention just how dependent we are on petroleum products. For the first time in decades, there was a real shortage of gasoline, heating oil, and other petroleum products. Long lines appeared at gas stations as people waited to buy scarce gasoline. People came to the realization that crude oil is a natural resource that is finite and that sooner or later, we will run out of it (Figure 22–5). Vast reserves of

Figure 22–5 *One day our limited supply of crude oil will run out.*
Courtesy of James Strawser, The University of Georgia.

oil still are underground, but the energy needs of the world will one day exhaust the world's supply. Before that happens we must find substitutes for the petroleum that produces such a wide array of products, from gasoline and diesel fuel to plastics and cosmetics. Agriculture is an obvious answer. If agriculture can produce materials that will substitute for crude oil, we can have an inexhaustible supply of raw materials.

Alcohol made from corn and other grains can be made to run internal combustion engines as efficiently as gasoline with less pollution. The technology for this process is hundreds of

Figure 22–6 *Ethanol already replaces over 40 million barrels of crude oil each year.* Courtesy of University of Illinois.

Figure 22–7 *Canola has the potential of producing superior oil for automotive use. The oil comes from the seed.* Courtesy of James Strawser, The University of Georgia.

years old and is really a simple procedure. Many gallons of moonshine whiskey have been made from corn and other grains using crude homemade distilleries. This is the same type of alcohol, known as **ethanol**, that is currently available in many states as a gasoline additive. This mixture, called **gasohol**, now replaces more than 40 million barrels of crude oil each year (Figure 22–6).

A very promising new oil seed crop is **canola**, sometimes known as rape (from the Greek word "rhapys," which means turnip). This plant, a member of the mustard family, promises to have a wide variety of uses (Figure 22–7). Because of its rich content of erucic acid, oil from the seed has excellent potential as a lubricant. Canola is now being used on a limited basis as a lubricant and has been used to make plastic. This oil also has the potential to be used as a fuel that takes the place of diesel fuel. The emissions from an engine using canola oil have less particulate, sulfur, and carbon dioxide content than the emissions from diesel fuel.

One real advantage in using canola oil as an engine lubricant is that it is biodegradable. A major problem with petroleum-based motor oils is disposing of the used oil. Petroleum-based oil can cause severe environmental problems not only from the toxic materials in the oil, but also because petroleum remains in the soil for a long time. However, like any vegetable oil, canola oil breaks down soon after it is exposed to the elements and would pose little threat to the environment.

The high erucic acid content once made the oil unfit for human consumption. However, scientists developed varieties low in erucic acid and high in oleic acid. These new varieties with the altered fatty acid content produce seeds containing oil that can be used in human diets. As a food oil, canola has the lowest saturated fat content (6 percent) and the highest monosaturated fat content (58 percent) of any currently used vegetable oil.

An added benefit of canola is that the cake that remains after the oil has been pressed out makes an excellent feed for livestock. Scientists have also discovered that when the whole plant is plowed under, it may act as a natural pesticide for the control of nematodes and other soil pests.

What is so exciting about the future of plants like canola is that research with them is relatively new. Most of the crops we now grow have been researched and developed over a period of many years. Just think of the advances made in crops like corn since the time it was first grown. New species such as canola are particularly exciting because they lend themselves to biotechnology and gene transfer.

Another relatively new crop is one called **kenaf** (*Hibiscus cannabinus*). Kenaf is a tall-growing plant native to India and Indonesia (Figure 22–8). It grows particularly well in the lower parts of the South where there the climate is hot and humid. Kenaf is an annual plant that has two types of usable fibers: spongy core fibers and outer **bast fibers**. Bast fibers are a part of the plant's phloem (tubes that carry nutrients through the plant). The fibers in this plant have a wide variety of uses. The bast fibers are used to make newsprint and other types of paper. The advantage of using the bast fibers of kenaf over the conventional wood for making paper is that it grows so fast. In one year's time, the plant can reach a height of 15 feet; whereas pines used for pulpwood take 15 to 20 years before harvest. Also, the fibers from kenaf make a strong high-quality paper that requires less ink and does not yellow as much as conventional paper.

The bast fibers are also used to make mats that are used to seed lawns on steep hillsides. Erosion is always a problem on slopes, and grass seeds wash off before they can germinate. If the seeds are embedded in a fiber mat, and the mat is staked down, the seeds remain in place until they germinate. The long fibers of the kenaf plant are idea for making these mats.

The core fibers have even a wider use and are more valuable. They are used in applications ranging from packing material and animal bedding to components of potting soil and oil-absorption material.

Figure 22–8 *Kenaf is a plant that produces fiber with many uses.*
Courtesy of John Woodruff, Crop Science Department, The University of Georgia.

BIO BRIEF

Biodegradable Plant-Based Hydraulic Fluid

A commercial-grade, biodegradable hydraulic fluid to power heavy equipment is just around the corner, thanks to a new process that creates a key component from vegetable oil. Agricultural Research Service scientists at the National Center for Agricultural Utilization Research in Peoria, Illinois, have made hydraulic fluid that contains estolides from oil seeds, such as meadowfoam and high-oleic soybean oil.

A class of long-chain esters, estolides are the basic ingredient in many hydraulic fluids. These fluids, under pressure, transmit power to moving parts of many machines, including cars, bulldozers, tractors, and most heavy equipment used to build roads and structures.

The scientists began by making a plant-based estolide from an oilseed crop called meadowfoam. Researchers at the Agricultural Research Service have developed new uses for the crop. Grown primarily in the Northwest, meadowfoam is an ingredient in cosmetics and other facial-care products.

"We found that it also showed promise as a base stock in hydraulic fluid," says Terry A. Isbell, an ARS chemist who helped develop meadowfoam. But, poor low-temperature properties and cost were prohibitive. However, "we used the technology developed for meadowfoam estolides to make estolides from other vegetable oils," Isbell says. "We found that oils that are particularly high in oleic acid, such as sunflower, safflower, and some soybean, would serve as a good source of starting material for the formation of estolides."

Petroleum-based hydraulic fluids and lubricant base stocks do not degrade well, Recently, construction equipment manufacturers began seeking a biodegradable alternative in response to tighter environmental regulations. In tests, about 30 percent of a petroleum-based hydraulic fluid degraded in 28 days, compared to 80 percent for vegetable based estolides.

The scientists' challenge: making estolides in large enough quantities to be economically feasible for commercial manufacturers like Caterpillar, a heavy equipment manufacturer with headquarters in Peoria. Caterpillar is testing the new biodegradable hydraulic fluid in cooperation with ARS and Lambent Technologies of Chicago, Illinois.

"The initial yield of estolides in our tests with vegetable oil was very small because our process wasn't very efficient," says Isbell. "Estolides have been made for a long time but never in large enough quantities to be practical on a commercial scale."

Serendipity helped overcome this problem. "One day, Beth Stiner, a lab technician formerly with ARS, conducted an experiment mixing vegetable oils with sulfuric acid. The result was a very high yield of estolides. We've adapted that reaction for our work," says Isbell.

Researchers made estolides by breaking down vegetable oils into their two main components: fatty acids and glycerin. In doing this, they discovered that sulfuric acid acted as a catalyst to form the estolides.

Estolides form when two fatty acids—the building blocks of vegetable oils—link together. ARS researchers used a blend of fatty acids that could be obtained from high-oleic oils. Oleic acid is commonly used in formulating food products that seem to show a potential for lowering blood cholesterol in humans. It also displays chemical properties scientists want in formulating biodegradable hydraulic fluids.

Source: Agricultural Research Magazine.

Plant-based hydraulic fluid could replace conventional fluid in a wide range of industrial and farm equipment. Here, a specialized machine built for soil sampling uses hydraulics to raise or lower its height, adjusts its width, and drive the wheels.
Courtesy of USDA-ARS.

BIOMASS

Another new area for plant agriculture is that of generating **biomass**. Biomass is defined as any organic matter that is a renewable resource and can be converted to energy. This area holds tremendous potential because of the large amounts of plant material that can be produced from previously unwanted plants. For example, fast-growing trees, such as locusts and sweet gums that have limited commercial use, could be grown for use as fuel if ways could be found to make more efficient use of the biomass. One way to do this would be to revive an old energy source—steam power. Using all of the available modern technology, highly efficient steam power units can be constructed that make use of chipped up biomass for fuel. Steam generators can be designed to fit almost any electrical need, and biomass could provide an inexpensive way of powering them.

Another way biomass can be used for fuel is through **gasification**. In this process, the biomass is heated in a reactor, and the organic material is converted to a gas that can be used in much the same way as natural gas. The gas burns cleaner and is more easily transported.

NEW USES FOR OLD CROPS

Some new crops that have been around for many years could receive new emphasis through different uses. One such plant is tobacco. Since the 1960s, when research showed a link between smoking and cancer, the growing of tobacco has been controversial. Most government buildings are now smoke-free, and the trend is to ban smoking in all public places. This will have a profound impact on the people who make their living producing tobacco.

An alternative use for the controversial plant may be just around the corner. Researchers have discovered that certain varieties of wild tobacco contain sugar-based fatty acid compounds that act as a natural insecticide (Figure 22–9). Nicotine from the plant was once used extensively as an insecticide, but the newly discovered compound does not contain nicotine. The pesticide appears to be environmentally safe and effective in controlling white flies and aphids. If this compound proves

Figure 22–9 *Tobacco has the potential of producing a natural insecticide. Courtesy of James Strawser, The University of Georgia.*

to be effective and safe, a whole new use for tobacco could be developed that could help producers survive diminished demand for tobacco for human consumption.

Another plant that has been grown as a food for centuries is soybeans. In the 1980s, the American Newspaper Publishers Association searched for a substitute for the oil-based ink traditionally used. Publishers use about 400 million pounds of ink a day to print newspapers, and of that amount, about 300 million pounds is oil. The steeply rising price of petroleum was running up the cost of printing. A good replacement was found in ink that used soy oil instead of petroleum oil (Figure 22–10). In addition to being a renewable resource, soy oil has several advantages. It is more environmentally friendly than petroleum-based ink because it is biodegradable and safer to use. Soy oil-based ink does not have the volatile hydrocarbons emitted by conventional inks, and this reduces health concerns associated with printing with petroleum-based inks. Soy ink makes printed colors more vivid because the soy oil itself is colorless.

Figure 22–10 *Soybeans are being used as a source of ink. Courtesy of Lynn Betts, USDA Natural Resources Conservation Service, Iowa.*

NEW AGRICULTURAL ANIMALS

Just as new plants are being introduced into agriculture, so are new animals. Like the plants, they are not really new but are simply used in a new way. A good example is the ostrich. As a nation, we consume a lot of poultry, and most of this is from broilers that weigh 3 to 4 pounds. But think of a bird raised for meat that weighs in excess of 300 pounds! The ostrich has been raised commercially in South Africa and is now beginning to be produced in this country. The meat from the ostrich is high quality with relatively little fat content (Figure 22–11). The eggs from the birds could be a tremendous food source if researchers could develop birds that lay eggs daily.

Figure 22–11 *The mean from the ostrich is high in protein and low in fat.* Courtesy of James Strawser, The University of Georgia.

Agriculturists will continue to search for animals that will fill a need. Emus, buffaloes, alligators, rabbits, and bullfrogs are all animals that have potential as food sources. One problem may be the reluctance of the public to eat meat from animals other than the traditional meat animals. However, consider that once people considered tomatoes to be poisonous, and no one would eat them. Consumer education will be the key.

FARMING THE OCEAN

Many scientists think we have barely scratched the surface in realizing the potential for producing aquatic animals (Figure 22–12). The ocean covers a majority of the Earth's surface, and the potential for producing food is tremendous. Although we have harvested wild fish and other aquatic

Figure 22–12 *Many scientists think we have barely scratched the surface in realizing the potential for producing aquatic animals. Courtesy of* Progressive Farmer.

animals for centuries, efforts to actually farm the ocean have been small. With all of the vast amounts of water, it is just a matter of time before research gives us the means to use this resource to grow food in a controlled manner. The effects could be just as great as when people began to develop the wild plants from which they gathered food. Just think of how much more corn is grown in a modern field compared to the amount gathered in wild fields. The same could be true for the animals gathered from the ocean.

In many places in the world the ocean is filled with plant life. Off the coast of many countries grow giant forests of algae called **kelp**. These plants begin as tiny plantlets, and some species grow to a height of almost 200 feet. During the growth process these plants may put on more than 2 feet of new growth per day.

Several uses have been found for this plant. One abstract, **alginic acid**, is used in making tires, as an ingredient in ice cream, and in the manufacturing of paints. Also, kelp is high in vitamins and minerals and is used for food in Asia. Considering that this plant grows well in an area that has received little attention from agriculture and the many uses that can be made of it, the potential is high as an agricultural product.

HIGH-TECH ENGINEERING AND ELECTRONICS

The biological aspects of agricultural mechanization will be of increased importance in years to come. Machines were the key that began the agricultural revolution at the turn of the 20th century. Machines were designed to deal with crops that had been planted, cultivated, and harvested by hand for hundreds of years. The designs centered around the characteristics of the plants for which the machines were to be used. Today, with our ability to manipulate the genetic makeup of plants, plants can be designed that lend themselves better to mechanization. This ability, coupled with new computer-generated engineering and the science fiction-like world of electronics, will enable us to do things we once only dreamed of. For example, agricultural engineers have worked for years on a machine that can pick strawberries (Figure 22–13). The berries grow low on the ground, cling closely to the plant, and are very fragile. If strawberry plants can be developed that grow taller, have berries that grow on longer stems and that are not so fragile, the engineer's job becomes simpler. Modern robotics will be used in designing picking mechanisms that closely resemble the human hand in function. It will be so close that even fragile produce like berries and tomatoes can be mechanically harvested. Problems such as these will bring the fields of biology, electronics, and mechanical engineering closer together.

Fields may some day be completely planted, cultivated, and harvested by remote control (Figure 22–14). Equipment used in several different fields and environments may be controlled all at one time in a central location by one person. This will make production less labor-intensive and operations more scheduled and uniform.

Electronics are already having a tremendous effect on agriculture, and this impact will grow larger in the future.

Figure 22–13 *A challenge for agricultural engineers is to build a machine to pick strawberries.*
Courtesy of University of Illinois.

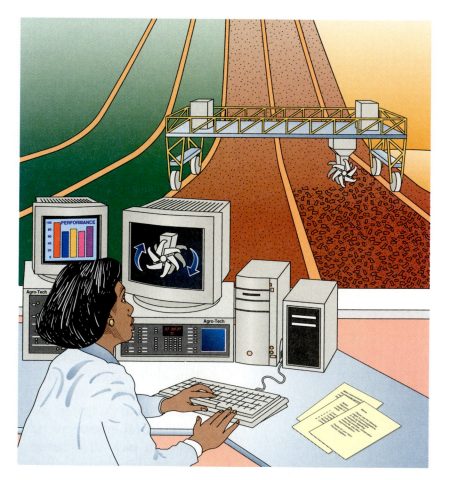

Figure 22–14 *Fields may one day be completely remote controlled.*

Soon young calves will be implanted with an electronic identification (Figure 22–15). This tiny apparatus will be injected under the skin at birth and remain with the animal until after it is slaughtered and the meat is marketed. A scanner will pick up information from the device at any time an operator wishes. The data will include an identification number, the name and location of the producer, pedigree data on the animal, and a production history, including any drug treatment or pest control measures. This will provide a wealth of information to research scientists as they collect data from each animal that goes through a slaughter plant. This will also allow closer monitoring of drug and pesticide residues from the animal. Producers will use the device to keep production records of feed consumed, medications given, health problems, and a wealth of other data.

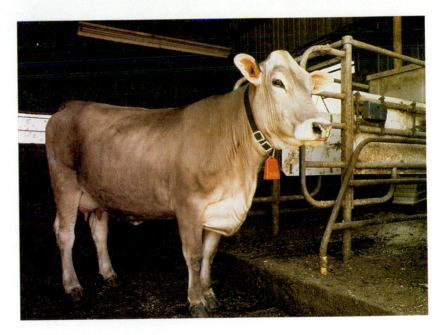

Figure 22–15 *Tags are now being used to electronically identify livestock. Someday they will be implanted with electronic identification chips that will stay with the animal from birth to after slaughter.* Courtesy of University of Illinois.

Computers have already revolutionized the way decisions are made on the farm. Animal sires and dams are selected based on computer databases. Marketing decisions are made based on computer predictions of commodity futures, and the commodities are actually bought and sold by computers. Computers tell producers when to irrigate, when to apply pesticides, and when to harvest.

As computers become more and more sophisticated and less expensive, a larger part of the decision-making process will be computer assisted (Figure 22–16). In addition, computers

Figure 22–16 *Computer units like this one that controls a modern feed mill will be more sophisticated and less expensive.*

will be a ready source of information. Through networks, producers can have access to an almost unbelievable amount of information on just about any topic.

Perhaps nowhere will computers have more impact on agriculture than through research. The development of simulation models will greatly shorten the time necessary for conducting tests on new methods. Also, data can be analyzed in greater quantity, more thoroughly, and in a shorter time. Also electronic banks of data will provide access to variables of almost any type. As new discoveries are made, connections with other discoveries can be found and new applications suggested through the use of computers.

PRECISION AGRICULTURE

One of the most exciting waves of the future is that of precision farming. A problem that has always existed occurs in large fields. The soil may be different in one part of the field from another part. Differing fertility levels, pH levels, soil type, drainage, and other factors may lead to inefficiency. Some parts of the field may get too much irrigation water, pesticides, or fertilizers, while other parts may not get enough.

In 1995, the department of defense allowed the full use of its **Global Positioning System (GPS)** to anyone (Figure 22–17). The system consists of 24 satellites that can use a small electronic device, called a GPS locator, to pinpoint any location on earth. There are a wide array of possibilities with the system. For example, a producer call install a GPS locator and a yield monitor on a combine, and the system can tell him/her the crop yield at any particular point in the field. This will help a producer find spots in the field where the yields are lower. Prescriptive measures then can be taken to bring the low yielding places up to the level of the higher yielding areas.

Figure 22–17 *The Global Positioning System (GPS) is used to pinpoint locations in fields that do not produce as well as the rest of the field.* Courtesy of John Deere Company.

Another use of the technology is that the producer can get a detailed map of the boundaries of his/her properties. Any irregularities, wet spots, trees, buildings, etc., can be mapped using GPS.

Satellites can also be used to photograph land and crops. Satellites do not take conventional photographs, but they measure wavelengths of light including those the human eye cannot detect. The different wave lengths can be converted to images that can be analyzed to determine the amount of moisture on a certain crop, the beginnings of an insect of disease infestation, or problems with fertilizer applications.

This technology is exciting because of the potential. In fact, the use of precision agriculture is beginning to receive widespread use. In the future, it will be used more as other methods are discovered for its use.

SUSTAINABLE AGRICULTURE

Agriculture has a bright future, and many things are left for speculating, predicting, and dreaming. But one thing is sure; agriculture will have to center all of its efforts around the concept of sustainability (Figure 22–18). The definition of sustainable agriculture is open to many interpretations. To the

Figure 22–18 In the future, agriculture production will have to be geared to the concept of sustainable agriculture. Courtesy of Lynne Betts, USDA Natural Resources Conservation Service, Iowa.

producer, it may mean making enough profit to sustain the family year after year. To the environmentalist, it may mean that production methods will have to be geared toward caring for the ecosystems of earth. To the economist, it may mean coordination among all of the different segments and phases of agriculture to ensure that the system continues to work. To world leaders, the concept may mean producing enough food to sustain the world's population and to alleviate hunger. Who is right? They all are. For agriculture to be sustained in an increasingly complex world with a rapidly growing population, all of these definitions of sustainable agriculture will have to be addressed.

The concept of sustainable agriculture centers around finding ways to produce food and fiber that allow the continuation of production year after year. When our country had vast expanses of land that could be cleared or simply moved onto and farmed, few people worried about using up the land because there was always more. Now it has become clear that the land and all of our natural resources must be cared for.

Many concepts such as low tillage cultivation, integrated pest management, water conservation, and pollution control are all a part of sustainable agriculture. The challenge for the future will be to coordinate these methods and concepts into production systems that will produce enough food and fiber to meet the needs of the world, and at the same time sustain the land and the environment for the production in the future. This is a tall order! The science of agriculture will be challenged as it never has been before.

SUMMARY

Agriculture is vital to the lives of all people. Without it, we would have few ways of clothing, feeding, and sheltering people. This industry will continue to be important as the population of the world increases. The challenge will be to find newer and more efficient ways to produce food and fiber.

CHAPTER 22 REVIEW

Student Learning Activities

1. Write an essay on what you consider to be the biggest problem facing agriculture and your vision of how the problem will be solved in the future. Share your ideas with the class.

2. Interview three producers and get their opinions of how things will change in the future in their areas of production. Share your findings with the class.

3. Locate new products from agriculture that were not in existence five years ago. Share your findings with the class.

Define the Following Terms

1. crude oil
2. ethanol
3. gasohol
4. canola
5. kenaf
6. bast fibers
7. biomass
8. gasification
9. kelp
10. alginic acid
11. Global Positioning System (GPS)

True/False

1. Canola oil is biodegradable.
2. Petroleum-based oil is not harmful to the environment.
3. Canola oil has the lowest saturated fat content of vegetable oils.
4. Large amounts of plant materials can be produced from previously unwanted plants.
5. The ocean does not have a great potential for producing food.
6. Kelp is high in vitamins and minerals.
7. Fields may some day be completely planted, cultivated, and harvested by remote control.
8. Presently, the erucic acid content in canola is too high for human consumption.
9. Nicotine was once used as an insecticide.
10. New animals are being introduced into agriculture.

Fill in the Blank

1. Kenaf is an annual plant that has two types of usable fibers: _____ and _____.
2. Fast-growing trees such as _____ and _____ could be grown as fuel if ways could be found to use the biomass more efficiently.
3. One way to make efficient use of the biomass is through _____.
4. Soy oil-based ink does not have the volatile _____ emitted by conventional inks.
5. _____ are having a tremendous effect on agriculture by storing and analyzing data.
6. Fifty years ago in many states, more than _____ percent of the population worked on farms.
7. The OPEC embargo reminded us how dependent we are on _____.
8. Because of its rich content of erucic acid, _____ oil is considered to be an excellent potential lubricant.
9. Researchers have discovered that certain varieties of wild _____ contain sugar-based fatty acid compounds that act as insecticides.
10. _____ animals could be a potential food source.

Multiple Choice

1. A promising new seed crop oil is
   A. petroleum.
   B. canola.
   C. soy.

2. The emissions from an engine using canola oil would have less particulate, sulfur, and carbon dioxide than emissions from
   A. diesel fuel.
   B. gasohol.
   C. ethanol.

3. Highly efficient steam power units can be constructed to make use of chipped up biomass for
   A. food.
   B. fertilizer.
   C. fuel.

4. An example of a crop that has been around for many years but could receive new emphasis through a different use is
   A. tobacco.
   B. corn.
   C. wheat.

5. Nicotine was once used extensively as
   A. an insecticide.
   B. a fertilizer.
   C. a medically prescribed drug.

6. The meat from ostriches is high quality with relatively little
   A. protein.
   B. fat.
   C. sodium.

7. A food that was once considered poisonous is
   A. wheat.
   B. corn.
   C. tomatoes.

8. Agricultural engineers have worked for years on a machine that can pick
   A. cotton.
   B. strawberries.
   C. neither A nor B.

9. Picking mechanisms that closely resemble the human hand in functions might be developed using
   A. remote controls.
   B. computers.
   C. robotics.

10. Perhaps nowhere will computers have more impact on agriculture than through
    A. research.
    B. robots.
    C. data screens.

Discussion

1. What are the advantages of canola oil?
2. What are the advantages of soy oil?
3. List some potential food animals.
4. What is electronic identification?
5. What information does electronic identification have on it?
6. What are the advantages of simulation models?
7. What is meant by sustainable agriculture?
8. What are some of the changes predicted by the CAST that may occur in agriculture after the year 2000?
9. How is genetic engineering beneficial to plants and animals?
10. What are some of the uses of bast fibers?

CHAPTER 23

CAREERS IN AGRICULTURAL SCIENCE

STUDENT OBJECTIVES

After studying this chapter, you should be able to:

✦ Analyze the career opportunities in agriculture.

✦ List the broad categories of jobs in agriculture.

✦ Discuss the education needed for different jobs in agriculture.

✦ Explain some of the characteristics of various jobs in agriculture.

✦ Explain the different levels of education in agriculture.

✦ Discuss agriculturally related student activities associated with college programs.

✦ List some important factors to consider when choosing a career.

Sooner or later you will have to decide what you want to do with your life. Everyone faces the choice of what career he or she wants to enter. There are many considerations in this decision. What do you enjoy doing? How much time and effort are you willing to spend in training and education to get the job you want? Are you willing to leave your home area for employment? How important is making a large salary?

Notice that the first question was "What do you enjoy doing?" This is perhaps the most important aspect of career choice because all of the other questions will have a direct bearing on how happy you are at what you do. A major portion of your life will be spent on the job, and the enjoyment of that job will greatly enhance your quality of life.

If you really enjoy science, and applied science in particular, then a career in agriculture might be right for you (Figure 23–1). At one time a career in agriculture meant farming for a living, but that was long before you were born and is no

Figure 23–1 *If you really enjoy science, a career in agriculture might be for you. Courtesy of USDA.*

longer true. Today's agriculture is a vast field with a lot of opportunity. If you have studied all the chapters in this book you must realize the diversity of the opportunities in agricultural science. In this chapter you will learn about some of the jobs that are involved with the science of agriculture and the education you will need. The financial rewards of any job are usually directly tied to the amount of preparation it takes to secure and succeed on that job. This may mean community college, a four-year college, graduate school, or on-the-job training. The choice is up to you, and you must decide among a complex assortment of job opportunities and qualifications.

OPPORTUNITIES IN PLANT SCIENCE

If you enjoy working with plants, hundreds of different types of job opportunities exist in this area. This category can be broken into the areas of horticulture, agronomy, landscaping, and turf management. With such a wide assortment of jobs, there is probably one suited to you.

Plant science deals with growing fruits and vegetables or ornamental plants. It may also deal with row crops such as corn, soybeans, or cotton. In this industry you might work in a greenhouse, outside, or in a laboratory (Figure 23–2). You may be involved with planting, potting, or propagating new plants from tissue cultures (Figure 23–3). Or, you could be engaged in protecting plants from pests such as weeds and insects. Plants that grow in a greenhouse have to be carefully monitored, and environmental controls have to be maintained. Retailers have to keep plants alive and flowers fresh until they are sold. **Elevator operators** have to analyze grain dry and store grain as it is brought in for sale.

If you wish to get a job working with plants as soon as you graduate from high school, you will need to start preparing early. Courses in agricultural education that emphasize plant science and greenhouse management will be of great benefit to you. Here, you will study the fundamentals of plant propagation and growth as well as how to care for plants in the greenhouse or in open fields. You might even get the opportunity to start your own business as a result of your Supervised Agricultural Experience Program (**SAEP**). You will also have the opportunity to test your skills in programs of the FFA. Some of the jobs you might obtain with a high school diploma include:

Figure 23–2 *A career in plant science might involve working in a greenhouse. Courtesy of James Strawser, The University of Georgia.*

Figure 23–3 *In a career in plant science you might propagate plants. Courtesy of James Strawser, The University of Georgia.*

Greenhouse caretaker

Plant propagator

Orchard caretaker

Pesticide applicator

Groundskeeper

Tractor operator

Combine harvester

Many community colleges have programs in horticulture. At these schools you will study biology and botany along with other related courses (Figure 23–4). In addition you will study advanced methods of crop production and greenhouse management. The program of study will also include courses in the use of computers. Plant science jobs that require a two-year degree include:

Greenhouse manager

Golf course manager

Pesticide dealer

Crop supply salesperson

Greenhouse supply salesperson

Figure 23–4 *For a career in plant science you will study several laboratory sciences. Courtesy of USDA-ARS, K-5040-01.*

For those who attend a four-year college or university, more technical careers are available. At the university, some of the areas you will study are chemistry, biology, botany, entomology, **agricultural economics**, soil science, agronomy, horticulture, and turf management. Most colleges have student organizations that sponsor activities centered around your area of study. These may include an agronomy club, a horticulture club, Collegiate FFA, or Collegiate 4-H. Possible job opportunities include:

Crop marketing specialist

Crop production specialist

Fertilizer technologist

Erosion control scientist

Soil conservationist

Seed technologist (Figure 23–5)

Turf grass specialist

If you have the ability and would like to continue your study of plants, you might consider a graduate degree in the plant sciences. With the Doctorate of Philosophy degree (PhD) you will be able to work with the world's most knowledgeable people. To obtain this degree you will take courses in the most advanced areas dealing with genetics, biochemistry, entomology, microbiology, pathology, and other subjects. In this program you will be able to narrow your focus on an area that appeals to you and concentrate on developing a deep understanding of that area. Careers with a graduate degree include:

Plant geneticist

Plant breeder

Biotechnologist

Genetic engineer

Plant pathologist

Plant physiologist

Statistician

Weed scientist

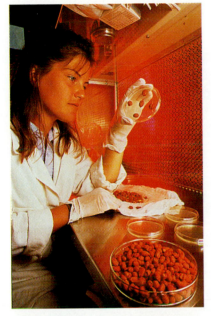

Figure 23–5 *As a plant scientist, you might study the structure and function of seed.* Courtesy of USDA-ARS.

CAREERS IN THE ANIMAL SCIENCES

If you enjoy studying about and working with animals, you might think about a career in animal science. Working with animals can be a most enjoyable job. There is something about caring for and working with animals that is richly rewarding. Jobs in this area may range from actually owning or raising animals to selling supplies for animal production. It may also include producing pharmaceuticals or equipment that aid in the management of animals (Figure 23–6).

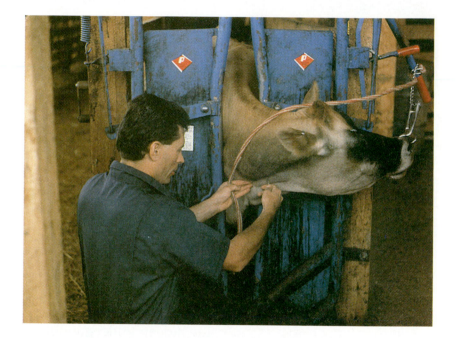

Figure 23–6 In the area of animal science, you might work with equipment and instruments that help in the management of animals. *Courtesy of Calvin Alford, The University of Georgia.*

If you choose to go to work immediately after graduating from high school, several jobs are available to you in animal science. You should take all of the courses in agriculture, agriscience, or agribusiness that your high school offers. In these courses you will learn the basics of how plants and animals live, grow, and reproduce. In addition you will learn responsibility in caring for animals and the essentials of properly caring for them. There will be opportunities for you to learn to work with animals in actual job situations through a Supervised Agricultural Experience Program (SAEP). You will also obtain leadership and personal development skills through the FFA Organization. You can participate in such activities as livestock judging, career development events, dairy products judging, poultry judging, livestock shows, and Proficiency Awards. Most of these activities are available to you whether you live on a farm or in the center of a city.

Some of the jobs that can be secured with a high school diploma include:

Herdsman

Feed mill worker

Milking machine operator

Sheep shearer

Pet groomer

Chick grader

Egg candler

Small animal producer

Slaughterhouse worker

Milk hauler

Poultry processing plant worker

Small animal breeder

Pet store worker

An **associate's degree** is a two-year degree from a community college or two-year institution. Many programs teach animal science in community colleges all across the country. If you enroll in one of these programs you will study practical courses in the fundamentals of producing and caring for animals. You will also study the sciences of chemistry, biology, and zoology as well as math and English. Many of the courses taken at a community college may be transferred to a university if you decide to continue your education.

If you enjoyed livestock judging in high school, you may continue your interest with competitive livestock evaluation at the community college level (Figure 23–7). You may also participate in student organizations, such as the Postsecondary Agriculture Students Organization (PAS), that participate in many activities involved with animal science.

The following are jobs available with an associate's degree.

Veterinarian assistant

Computer operator

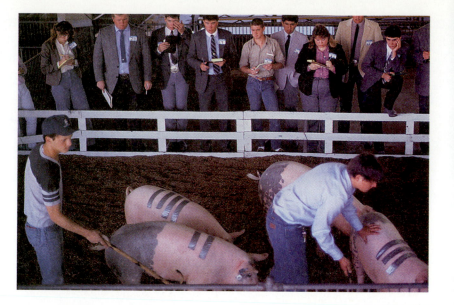

Figure 23–7 *As you study animal science you can judge livestock competitively on the high school, community college, and university levels.*

Farrier

Poultry vaccinator

Producer

Artificial insemination technician

Meat cutter (Figure 23–8)

Embryo implant technician

Wool grader

Animal buyer

Figure 23–8 *An associate's degree might lead to a career as a meats technician.* Courtesy of James Strawser, The University of Georgia.

A **bachelor's degree** requires four years of education at a college or university. In animal science, there is a broad array of choices in majors. These include animal science, dairy science, poultry science, and agricultural education.

You will study courses in science such as chemistry, biology, and zoology. At a large number of universities, the departments that teach these courses are housed in the College of Agriculture. In agricultural science you will study animal anatomy, nutrition, animal growth and development, and other subjects dealing with how animals live, grow, and reproduce (Figure 23–9).

You can also participate in such competitive events as livestock evaluation and meat evaluation. Student organizations include Block and Bridle, Collegiate FFA, and Collegiate 4-H.

Careers with a bachelor's degree include:

Farm or ranch manager

Producer

High school agriculture teacher

Extension agent

Agricultural journalist

Meat grader

Company representative

Field service technician

Hatchery manager

Dairy inspector

After you complete your bachelor's degree you may want to continue your education with a **master's degree** or **PhD** (doctor of philosophy). With a graduate degree such as the PhD, you will be able to conduct scientific research or continue with a degree in veterinary medicine (Figure 23–10). You will choose a specific area of animal science in which to concentrate your studies, and most of your course work will be in that area. For example, you might want to study in the area of nutrition or animal reproduction. Your course work will include study in statistics and research methodology so you will be able to understand, design, and conduct scientific research. A graduate degree requires good grades in college and a determination to study hard to reach your career objective.

Figure 23–9 *In college you will study the theoretical as well as the practical side of animal science.*
Courtesy of James Strawser, The University of Georgia.

BIO BRIEF

What Does a Research Scientist Do?

Just how does a research scientist conduct research? The process involves several steps and takes a lot of time. Although many types of research investigate a wide variety of topics, most research projects follow a similar scheme.

The first step is that the scientist defines a definite problem to investigate. The problem has to be defined so that the scientist knows exactly what is to be accomplished. When this is established, the scientist begins a literature search. This involves reading all the research study summaries, books, and journal articles that have been written on the topic. This helps to provide a basis for designing the project.

After the literature search has been completed, the scientist carefully plans how the research project is to be conducted. Remember from Chapter 1 that many factors have to be considered to make the results of the research usable. Questions such as "Where will the research be conducted?"; "What steps will I take?"; "What materials will be needed?"; and "What will be the timeline?" all have to be answered.

Next, the scientist has to secure funding for the project. Quality research is very expensive, and plans have to be made to provide money. If the scientist works for a corporation, the money will come from the research and development funds of the corporation. If the scientist works for a university, funding will be applied for on a competitive basis. Only so much money is available, and researchers have to compete for it. The scientist may write a proposal for a grant. Grants come from foundations that sponsor research or from corporations that have an interest in a particular area. The grant proposal contains the rationale for doing the research, the benefits of the research, how the research will be conducted, how the results will be used, a timeline, and a budget. The proposals with the most potential are the ones that get funded.

When funds have been secured, the project begins. All the procedures outlined in the plant are followed. The research study is completed, and the data are gathered. The data are analyzed using appropriate formulas and

Examples of jobs requiring a PhD or doctorate of veterinary medicine are

Veterinarian

Meat inspector

Animal geneticist

Animal nutritionist

Reproductive physiologist

Microbiologist

tests. Conclusions are drawn based on the analysis of the data.

When the conclusions have been made, the research is still not complete. The scientist must decide what to do with the findings of the research. Recommendations have to be made as to what impact the research will have and how it should be used. The final step is to publish the results. The research is described in a written report that details how the project was done and what the results were. The report then can be written in the form of a journal article for a research publication or an oral presentation at a research meeting. By publicizing the results, other scientists can have access to the research and use the findings to conduct other research. Of course, the practical applications can be put into use.

A research scientist needs skills in communication almost as much as expertise in research methodology and technical know-how. Research is of not use if the results cannot be communicated to others.

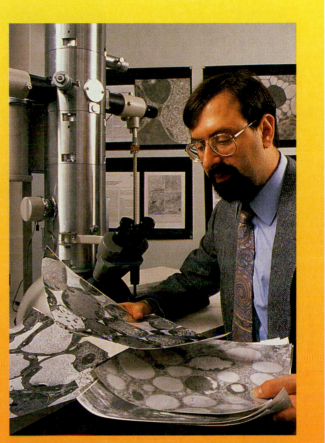

Collecting data is only part of a research scientist's duties. Courtesy of USDA-ARS.

Research scientist

College professor

CAREERS IN NATURAL RESOURCES

If you enjoy the beauty of nature and like to be outdoors, you might consider a career in natural resources. This area is expected to become increasingly more important. As the population

Figure 23–10 *With a degree in veterinary medicine, you will help prevent disease and treat animals in need of your care.* Courtesy of James Strawser, The University of Georgia.

gets larger, we will have to do a better job of protecting natural resources and the environment. This will require many specially trained and educated people. Our country has vast national forests, wilderness areas, and recreational areas that need care and protection (Figure 23–11). Every state in the nation has programs to support fish and wildlife. All aspects of agriculture must be monitored to ensure that the environment is properly protected.

Most of the careers available in natural resources and the environment will require a four-year degree. You will need a

Figure 23–11 *Our country has vast national forests, wilderness areas, and recreational areas that need professionals to manage them.* Courtesy of Jim Peterson, Evergreen Magazine.

sound background in botany, zoology, entomology, chemistry, pathology, and ecology. You will study courses that deal with the environment, the threats to it, and the solutions. In addition you will study governmental regulations and laws that deal with the environment (Figure 23–12). Examples of careers are

Chemist

Ecologist

Game warden

Range specialist

Soil conservationist

Forestry specialist

Laboratory technician

A graduate degree will give you the knowledge and skills to enter the following occupations:

Wildlife research scientist

Wildlife biologist

Forest entomologist

Wildlife parasitologist

Wildlife pathologist

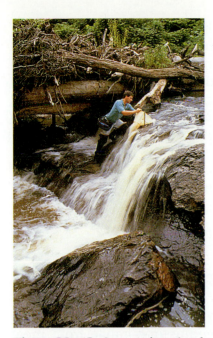

Figure 23–12 *A career in natural resources might include helping to safeguard the environment. Courtesy of USDA-ARS, K 5213-33.*

CAREERS IN FOOD SCIENCE

One of the largest industries in the nation is the processing and distribution of food. This industry includes all of the jobs involved from the time the food crop is harvested until it reaches the consumer. This includes preparing, processing, inspecting, packaging, labeling, maintaining equipment, advertising, marketing, storage, wholesaling, and retailing. Millions of people are employed in this industry. A young person who chooses a career in this field will have many opportunities.

As a high school graduate, you can look forward to a variety of jobs in the food industry. Someone must operate the machinery and equipment involved with the processing of produce, the slaughtering of animals, processing of meats, storing food, and providing help in the retail grocery stores. Courses in high school Agricultural Education will give you skills in record keeping, time management, and the people skills needed to succeed on the job. You will also gain an understanding of all the segments of the food industry and how these segments interconnect. Examples of jobs in the food industry that require a high school education include:

Production line worker

Equipment maintenance person

Grocery store clerk

Produce manager

Meat cutter

Warehouse foreman

Several two-year colleges have programs in food technology (Figure 23–13). These programs may range from meat cutting to equipment maintenance and repair. These areas are specialized, and the course work can be tailored to prepare you to enter a particular career. A degree at a two-year institution might give you the background to be employed in the following jobs:

Figure 23–13 *A two-year degree might lead to a career as a food technologist. A college degree in food science will involve taking courses in biochemistry.* Courtesy of James Strawser, The University of Georgia.

Food technician

Meat technologist

Refrigeration specialist

Processing supervisor

Restaurant cook

Restaurant manager

Colleges and universities educate students in a broad array of programs, ranging from food preservation to restaurant management. A four-year degree will give you opportunities in several areas:

Food chemist

Food inspector

Plant manager

Wholesale or retail manager

Food engineer

A graduate degree can prepare you for a career as a **food scientist**. Food scientists improve or develop new food products. Or, you might work toward discovering new and improved ways of preserving foods. Your coursework will be heavy in many branches of chemistry—analytical, biochemistry, and organic chemistry. You will also take in-depth courses in microbiology and theories of food preservation. A graduate degree will allow you to enter the following jobs:

Microbiologist

Food research scientist

Bacteriologist

CAREERS IN SOCIAL SCIENCE

A field that is growing in importance is **social science** in agriculture. Social science is the study of human society and how people interact with each other. This area may include economics, education, and politics. Most of these jobs will require at least a four-year degree. Coursework will be concentrated in economics, communications, and sociology. A major component will be acquiring computer skills to allow you to analyze data and communicate with people from all across the world. Several government agencies deal with agriculture, and many advisors are needed to help devise policy for the agricultural sector of the economy.

The field of agricultural economics needs many people to deal with the financial and human resource aspects of agriculture. You will help develop management schemes, or you might work in the futures market where you will be involved in the trading of agricultural commodities.

Rural sociology deals with the interactions of people in rural settings (Figure 23–14). In this field, you will study how people deal with change, how they communicate, or how their lifestyle is affected by different phenomena. If you like working with people, this might be an area for you.

Figure 23–14 *Rural sociology deals with the structure of groups and the interactions of people in a rural setting.*

The number of people who grow up on farms has diminished for many years, and this phenomenon will continue. This means that a large and growing segment of our society knows little about the agricultural industry. The field of agricultural education is an important part of that industry. The mission of agricultural education is to teach people about agriculture. People who communicate about agriculture through the media, those who operate the Cooperative Extension Service, university professors, and middle and high school agriculture teachers are all agricultural educators.

The shortage of middle and high school agriculture teachers is rapidly becoming a serious problem. Some states have difficulty each year in finding qualified people to fill these positions. If you enjoy working in all the areas of agriculture and you enjoy working with people, you may be suited to becoming a middle and high school Agriculture Education teacher. Your agriculture teacher is a good source of information on how to become a teacher in your state.

Other jobs in social science in agriculture are

County extension agent

Home economist

Agricultural economist

Agricultural journalist

Rural sociologist

Bank loan officer

Congressional aide

Agricultural aide for government officials

The following are several places where you may obtain information concerning careers in agriculture.

The Institute of Food Technologists
221 North LaSalle Street
Chicago, IL 60601

American Meat Science Association
444 North Michigan Avenue
Chicago, IL 60611

American Association of Cereal Chemists
3340 Pilot Knob Road
St. Paul, MN 55121

American Agricultural Economics Association
80 Heady Hall
Iowa State University
Ames, IA 50011-1070

Association of Official Seed Analysts
c/o Nebraska Crop Improvement Association
268 Plant Science IANR-UNL
University of Nebraska-Lincoln
Lincoln, NE 68583-0911

American Society of Agronomy
Crop Science Society of America
Soil Science Society of America
677 South Segoe Road
Madison, WI 53711

National Grain and Feed Association
1201 New York Avenue N.W., Ste. 830
Washington, DC 20005

The Poultry Science Association, Inc.
309 West Clark Street
Champaign, IL 61820

Animal Industry Foundation
1501 Wilson Boulevard, # 1100
Arlington, VA 22209

National Vocational Agriculture Teachers Association
P. O. Box 15440
Alexandria, VA 22309

National Aquaculture Association
Drawer 1569
Shepardstown, WV 25443

SUMMARY

As you decide what you want to do for the rest of your life, do not overlook the opportunities in agriculture. If you enjoy working with the industry that feeds, clothes, and shelters people, there will be a career waiting for you. The choice is yours.

CHAPTER 22 REVIEW

Student Learning Activities

1. Visit your guidance counselor and discuss your abilities and aptitudes. List all the things you enjoy doing and try to envision a career that will allow you to use your interests, abilities, and aptitudes. Decide on several possible careers and obtain more information using the sources listed in the text.

2. Spend the day with a person who works in a career you might like. Make a list of all the things you liked and disliked about the job. Share your experiences with the class.

Define the Following Terms

1. plant science
2. elevator operator
3. SAEP
4. agricultural economics
5. associate's degree
6. bachelor's degree
7. master's degree
8. PhD
9. food scientist
10. social science
11. rural sociology

True/False

1. If you really enjoy applied science, a career in agriculture might be right for you.
2. A career in agriculture means farming for a living.
3. The financial rewards of any job are usually directly tied to the amount of preparation it takes to secure and succeed on that job.
4. As a worker in the plant science industry, you might work in a greenhouse, outside, or in a laboratory.
5. Plants that grow in a greenhouse do not have to be carefully monitored.
6. There are more technical careers available for those who attend a four-year college or university.
7. Natural resources careers involve working outdoors.
8. Our country does not have a lot of acreage in national forests, wilderness, and recreational areas.
9. Most of the careers available in natural resources and the environment will require a four-year degree.
10. One of the largest industries in the nation is involved with the processing and distribution of plants.

Fill in the Blanks

1. _____, _____, _____, and _____ are all involved in working with plants.
2. If you enjoy learning about and working with animals you might think about a career in _____.
3. At a large number of universities the departments that teach science courses are housed in the College of _____.

4. Every state in the nation has programs aimed at supporting _____ and _____.

5. All aspects of agriculture must be monitored to ensure that the _____ is properly protected.

6. One of the largest industries in the nation is involved in the processing and distribution of _____.

7. Courses in agricultural _____ will give you skills in record keeping, time management, and the people skills you need to succeed on the job.

8. Several two-year colleges have programs in food _____.

9. To become a food scientist, you will probably have to earn a _____ degree.

10. An area of agriculture that deals with social interaction is called _____.

Discussion

1. What is involved in studying plant science?

2. List some of the jobs you might obtain with a high school diploma.

3. List some of the jobs you might obtain with a four-year degree.

4. What is involved in a career in animal science?

5. What is the benefit of the SAEP?

6. What are the benefits of the FFA?

7. What jobs are available with an associate's degree?

8. What are some choices of areas of study at a four-year institution?

9. What is studied in agricultural science?

10. What jobs are available in social science in agriculture?

GLOSSARY

A

Abomasum   The fourth, or true, stomach division of a ruminant animal.

Acid   Term applied to any substance with a pH less than 7.0.

Active immunity   The type of immunity in an animal that is permanent.

Adipose   Fat tissue.

Adrenal glands   A pair of ductless glands, each located on or adjacent to a kidney, which secrete adrenaline and cortin, whence cortisone.

Adrenaline   One of the hormones produced by the medulla of the adrenal glands.

Aerobic   Pertaining to organisms that grow only in the presence of oxygen.

Afferent neurons   A single nerve cell that conveys impulses toward a nerve center.

Aflatoxin   The highly toxic substance produced by some strains of the fungus *Aspergillus flavus*. It is found in feed grains.

Agricultural animal   An animal that is raised for the purpose of making a profit.

Agricultural products   Goods that are produced as a result of plants or animals that are raised by humans.

Agriculture   The broad industry engaged in the production of plants and animals for food and fiber, the provision of agricultural supplies and services, and the processing, marketing, and distribution of agricultural products.

Agriculturist   Any person involved with the production, processing, or distribution of agricultural products.

Algae   Comparatively simple plants containing photosynthetic pigments.

Alkaline   A chemical term referring to basic reaction where the pH reading is above 7, as distinguished from acidic reaction where the pH reading is below 7.

Allele   A matched pair of genes that control a characteristic.

Alluvial soil   Soil developed from transported and relatively recent water-deposited material (alluvium), characterized by little or no modification of the original material by soil-forming processes.

Alveoli   Small grapelike structures in the mammary system that take raw materials from the bloodstream and synthesize them into milk.

Amino acid   The basic building block of protein.

Amniotic fluid   The fluid that surrounds a fetus before birth.

Anabolism   The growth process by which tissues are built up.

Anaerobic   Pertaining to organisms that grow without the presence of oxygen.

Animal and Plant Health Inspection Service (APHIS)   A federal agency with the responsibility of inspecting plant and animal products that cross state lines or enter the country.

Animal welfare   A line of thinking that animals should be treated well and that their comfort and well being should be considered in their production.

Annuals   Plants that germinate, grow, reproduce, and die in one year.

Ante mortem   After death.

Anthropologist   One who studies humans in relation to distribution, origin, classification, and relationships of race, physical character, environmental and social relations, and culture.

Antibiotics   A group of drugs that are used to fight bacterial infections.

Antibodies   Substance produced by an animal's body that fights disease or foreign materials in the bloodstream or other places in an animal's body.

Antigen   Any substance that stimulates the production of antibodies in an animal's body.

Applied research   Research that has a practical application for the results.

Aquaculture   The production of animals that live predominantly in the water.

Aquatic animals   Animals that spend all or most of their lives in the water.

Aquifers   A geologic formation or structure that transmits water in sufficient quantity to supply the needs for water development, such as a well.

Arteries   Blood vessels that carry blood from the heart.

Artificial hormone   A manufactured substance that is used in the place of a naturally produced hormone.

Artificial insemination   The placing of sperm in the reproductive tract of the female by means other than that of the natural breeding process.

Artificial vagina   A tubelike device used to collect semen from a male animal.

Artificially acquired immunity   Immunity that comes about as a result of a vaccination.

Asexual reproduction   The production of young by only one parent.

Atriums   The upper chambers of the heart.

Auxin   A substance that promotes the growth of plants.

B

Basic research   Research with no immediate application, but yet it adds to the general understanding of a topic.

Bast fibers   Strong woody fibers obtained mostly from the phloem of plants.

Beef   The meat from cattle more than a year old.

Biennials   Plants that live for two years. They germinate one year and reproduce and die the next year.

Binomial classification   A system of scientific classification of living organisms that uses two names, the genus and the species. The names are usually derived from Latin.

Biological control   The use of natural means rather than chemical means to control pests.

Biomass   The amount of matter of biological origin in a given area.

Blastula   A mass of cells with a cavity that occurs from the dividing of a fertilized egg. The cells begin to differentiate at this stage.

Border plants   Plants established along the borders of a field to attract insects away from the crop.

Botanist   A person who studies plants for a career.

Bovine somatropin (BST)   A naturally occurring hormone that aids in stimulating the production of milk in cows.

Broiler   A chicken approximately eight weeks old weighing 2.5 pounds or more.

Bronchi   The tubes that connect the lungs to the windpipe.

Buffered   Soil-containing substances such as organic matter, clay, carbonates, or phosphates, which resist changes of soil pH.

Bull   A male bovine that has not been castrated.

By-product   A product that is created as the result of producing another product.

C

Cage operation   An operation in which hens are kept in cages all their lives as they produce eggs.

Callus cells   Undifferentiated or organized tissue that grows from a plant cell or piece of leaf when it is placed on media containing certain hormones.

Calyx   The outer, usually green, leaflike parts of a flower.

Cambium   The actively growing cells between the bark and the wood of a tree or shrub.

Canning   The process of preserving food by applying heat to a sealed container holding the food.

Canopy   The branches of a tree that have grown together to form a structure that shades the ground.

Capillaries   Extremely narrow, microscopic blood vessels.

Carbon cycle   The sequence of transformation undergone by the carbon utilized by organisms.

Carcass   That part of a meat animal that is left after the hide, hair, feet, head, and entrails have been removed.

Carcinogen   A substance that causes cancer.

Cardiac   Pertaining to the heart.

Carding   One of the first steps in the processing of wool. The fibers are separated from other fibers in the locks or bunches of wool.

Carnivorous   An animal whose diet consists mainly of other animals.

Carpel   One of the units composing a pistil or ovary.

Cartilage   Firm but pliant tissue in an animal's body that may turn to bone as the animal ages.

Castration   The removal of an animal's testicles.

Catabolism   The process of breaking down tissues from the complex to the simple as in the digestive process.

Catheter   A tube that is inserted into an animal's body to inject or withdraw fluid.

Cation exchange   The interchange among cations in soil solution and cations on the surface of clay, humus, or plant roots.

Cecum   The enlargement on the digestive tract of animals, such as the horse, that allows them to digest large amounts of roughages.

Cell   The basic building block of living tissue. It generally consists of a membrane wall, a nucleus, and cytoplasm.

Cell membrane   A cell wall.

Cell walls   The membranous coverings of cells secreted by the cytoplasm in growing plants.

Cellulose   An inert complex carbohydrate that makes up the bulk of the cell walls of plants.

Centromere   A small structure located on a chromosome that appears to form an attachment to the spindle fibers during cell division.

Cereal   Grain or a product made from grain.

Cervix   The organ that serves as an opening to the uterus.

Chalazae   A ropelike structure inside of an egg that holds the yolk in the center of the egg.

Cheese   A food product made from the solids in milk.

Chicle   A substance made from the sap of a tree that is used in the manufacture of chewing gum.

Chlorophyll   The green substance in plants that transforms light energy to chemical energy.

Chloroplasts   Minute objects within plant cells, which contain the green pigment chlorophyll.

Cholesterol   A fat-soluble substance found in the fat, liver, nervous system, and other areas of an animal's body. It plays an important role in the synthesis of bile, sex hormones, and vitamin D.

Chromatin   The complex of DNA and proteins that make up the chromosomes of eukaryotic cells.

Chromoplasts   A colored plastid usually containing red or yellow pigment.

Chromosomes   A linear arrangement of genes that determines the characteristics of an organism.

Chromotids   One strand of a doubled chromosome.

Chronological age   The actual age of an animal in days, weeks, months, or years.

Cleavage   The splitting of one cell into two parts.

Clone   An organism, produced by asexual means, with the same genetic makeup as another.

Cocoon   A covering that surrounds an insect pupa.

Codominant genes   Genes that are neither dominant nor recessive.

Cold-water fish   A fish that will not thrive in water temperatures greater than 70°F.

Colon   The large intestine.

Colostrum   The first milk that a mammal gives to the young following birth. It is rich in nutrients and imparts immunity from the mother to the offspring.

Combine   A machine that combines the jobs of cutting and threshing grains.

Commercial fertilizer   Plant nutrients derived from processing, involving chemical mixtures, which are sold to producers.

Commodity   A transportable resource product with commercial value.

Companion animals   Animals that are raised for use as pets or to assist humans.

Composting   The process of piling organic matter together so that it will undergo chemical changes that make the matter usable for fertilizer.

Compound leaf   A leaf composed, usually, of two or more leaflets.

Concentrate   A feed that is high in carbohydrates and low in fiber.

Conception   The uniting of the egg and sperm.

Confinement operation   A system of raising animals in a relatively small space.

Conifer   A tree or shrub that bears cones, as pines and firs.

Consumer   A person who buys or uses food, manufactured goods, or other products.

Contact herbicides   Chemicals that kill plants after they are sprayed on the plants.

Contagious disease   A disease that may be passed from one organism to another.

Control group   In a scientific experiment, a group of animals, plants, etc., that do not receive the treatment under study.

Copulation   The act of sexual union between two mating animals.

Corms   Enlarged fleshy base of a stem, bulblike but solid, in which food accumulates.

Corolla   The petals of a flower making up the inner floral envelope surrounding the sporophylls.

Corpus luteum   A swelling of tissue that develops on the ovary at the site where an ovum has been shed.

Cortex   The outer layer or region of any organ.

Cotyledons   The first leaf to be developed by the embryo in seed plants.

Council for Agricultural Science and Technology (CAST)   An organization of agricultural scientists that represent all the branches of agriculture. One of its main goals is to inform the public about agricultural issues.

Cover crop   Crops established for the purpose of adding organic matter to the soil. They are usually planted in the winter and plowed under in the spring.

Cow   A female bovine that has had a calf.

Cow-calf operation   A system of raising cattle; the main purpose is the production of calves that are sold at weaning.

Cowper's gland   A gland in the male reproductive tract that produces a fluid that is added to the ejaculate.

Crimp   The amount of waves in wool fiber.

Crop rotation   Planting a different crop than was planted the last year.

Cross-pollination   The uniting of plant gametes from different plants.

Crossbred   An animal that is the result of the mating of parents of different breeds.

Crowns   The upper part of a tree that bears branches and leaves.

Crustaceans   Aquatic animals with a rigid outer covering, jointed appendages, and gills.

Cud   A small wad of regurgitated feed in the mouth of a ruminant that is rechewed and swallowed.

Cultivars   Plants that have been developed by humans for their use.

Cultivation   The process of growing plants.

Curd   The coagulated part of milk, which results when the milk is clotted by adding rennet, by natural souring, or by adding a starter.

Cured   Meat that has been treated to retard spoilage.

Cuticle   A thin layer of cutin that covers the epidermis of plants above the ground, except where cork has replaced the epidermis.

Cuttings   Portions of plants that are removed for the purpose of growing new plants.

Cytokinesis   The dividing of the cytoplasm and organelles into two daughter cells during cell division.

Cytoplasm   The living material within a plant or animal cell excluding the nucleus.

D

Dam   The mother of an animal.

Daughter cells   Newly formed cells resulting from the division of another cell.

Deciduous   Plants that lose their leaves every year and grow new ones.

Dehydration   The removal of 95 percent or more of the water from any substance by exposure to high temperature.

Deltas   Areas of land that have been built up as the result of soil that has been moved by water.

Dental pad   A hard pad in the upper mouth of cattle and other animals that serves in place of upper teeth.

Deoxyribonucleic acid (DNA)   A genetic proteinlike nucleic acid on plant and animal genes and chromosomes that controls inheritance.

Diaphragm   The muscle between the thoracic and abdominal cavities of animals.

Dicots   Plants whose seed have more than one cotyledon or seed leaves, such as beans.

Differentiation   The development of different tissues from the division of cells.

Diffusion   The total movement of particles from a region of high concentration to one of low concentration.

Digestion   The changes that food undergoes within the digestive tract to prepare it for absorption and use in the body.

Diploid   Having one genome comprising two sets of chromosomes.

Disease   Any deviation in the normal health of plants or animls.

Dissolved oxygen   Oxygen in water that is available for the use of animals with gills (such as fish).

DNA   *See* deoxyribonucleic acid.

Docile   Having a quiet, gentle nature.

Docking   The removal of an animal's tail.

Domesticated   Animals that are raised under the care of humans.

Dominant gene   A gene that expresses its characteristics over the characteristics of the gene with which it is paired.

Donor cow   A cow of superior genetics from which an embryo is taken to implant in a cow of inferior genetics.

Dormant season   The part of the year when plants are not growing. They may also lose their leaves during this time.

Duodenum   The first portion of the small intestine.

E

***E. coli* bacteria**   Bacteria that normally inhabit the human colon.

Ecology   The totality or pattern of the interrelationship of organisms and their environment and the science that is concerned with that interrelationship.

Ecosystem   The entire system of life and its environmental and geographical factors that influence all life, including the plants, animals, and environmental factors.

Ectoderm   The outer layer of the three basic layers of the embryo, which gives rise to the skin, hair, and nervous system.

Efferent neurons   A single nerve cell that conveys impulses away from a nerve center.

Egg   The female sex cell; the female gamete.

Ejaculation   The expulsion of semen from the male reproductive system.

Electron microscope   A microscope in which a beam of electrons is focused using an electron lens to produce an enlarged image of a minute object on a fluorescent screen or photographic plate.

Embryo   An organism in the earliest stage of development.

Embryo transplant   The removal of an embryo from a female of superior genetics and placement of the embryo in the reproductive tract of a female of inferior genetics.

Endocrine system   The system of glands in an animal's body that secretes a substance that controls certain bodily processes.

Endoderm   The innermost layers of cells of an embryo that develops into internal organs.

Endoplasmatic reticulum   The structure of membranous channels within eukaryotic cells, which is the site of most protein and lipid biosynthesis.

Endosperm   The nutritive portion in some seeds that originates in the embryo sac but which is outside the embryo.

Energy   The capacity to do work.

Entomologists   Persons who study insects for a career.

Entomology   The study of insects.

Environment   The sum total of all the external conditions that may act upon an organism or community to influence its development or existence.

Environmental Protection Agency   The governmental agency that is charged with protecting the environment.

Enzyme   A protein that is produced by an animal's body that stimulates or speeds up various chemical reactions.

Eolian soils   Loosely designating soils that are derived from geologic deposits, which are windborne in origin.

Epicotyl   The part of the axis of an embryo above the region of attachment to the cotyledons.

Epididymis   A small tube leading from the testicle where sperm mature and are stored.

Epiglottis   A flap made of cartilage that covers the opening of the larynx during swallowing.

Erosion   The wearing away of the soil through the action of wind or water.

Esophagus   The tube leading from the mouth to the stomach.

Essential amino acid   Any of the amino acids that cannot be synthesized by an animal's body and must be supplied from the animal's diet.

Essential nutrients   Nutrients that are essential to the life of a plant or animal.

Estimated breeding value   In beef cattle, an estimate of the value of an animal as a parent.

Estrogen   A hormone that stimulates the female sex drive and controls the development of female characteristics.

Estrus   The period of sexual excitement (heat) when the female will accept the male.

Estrus cycle   The reproductive cycle of female animals measured from the beginning of one heat period until the beginning of the next.

Estrus synchronization   Using synthetic hormones to make a group of females come into heat (estrus) at the same time.

Ethology   The science of animal behavior.

Eukaryotic cells   Cells that contain a membrane-bound nucleus and other membrane-bound organelles.

Evergreen   A plant that stays green all year long.

Ewe   A female sheep.

Exoskeleton   A skeleton that is on the outside of an animal. Insects have exoskeletons.

Expected progeny difference   An estimate of the expected performance of an animal's offspring.

Export   A product that is shipped to another country.

Extender   A substance added to semen to increase the volume.

External parasite   A parasite that lives in the hair or on the skin of an animal.

F

F1 generation   The first generation of offspring from purebred parents of different breeds or varieties.

Fallopian tubes   The tubes leading from the ovaries to the uterus.

Farrow   To give birth to a litter of pigs.

Farrowing crate   A crate or cage in which a sow is placed at the time of farrowing to protect the newborn pigs.

Fatty acids   An organic molecule comprised of a long chain of carbon atoms with a carboxylic acid.

FDA   *See* Food and Drug Administration.

Feedlot   A pen in which cattle are placed for fattening prior to slaughter.

Felting   The property of wool fibers to interlock when rubbed together under conditions of heat, moisture, and pressure.

Fermentation   The processing of food by the use of yeasts, molds, or bacteria.

Fertile   Capable of producing viable offspring.

Fertilization   The union of the sperm and egg.

Fertilization membrane   A membrane surrounding an egg that is formed after the egg is fertilized to prevent another sperm from entering.

Fertilized egg   An egg that has united with a sperm.

Fertilizers   Any organic or inorganic material added to soil or water to provide plant nutrients and to increase the growth, yield, quantity, or nutritive value of the plants grown therein.

Fetus   The growing animal within the uterus.

Fingerling   A small fish that is of sufficient size to use for stocking.

Finish   The degree of fat on an animal that is ready for slaughter.

Fish slurry   A material made from the entrails, heads, bones, etc., of fish. It is a rich source of plant nutrients.

Flagella   Whiplike appendages of certain single-celled aquatic animals and plants, including some bacteria, the rapid movement of which produces motion.

Flax   A plant that is grown for fibers that are woven into a cloth called linen.

Floodplains   The area along a stream that is covered by periodic floods.

Follicle   A small blisterlike structure that develops on the ovary that contains the developing ovum.

Follicle stimulating hormone   The naturally occurring hormone that stimulates the development of the follicle on the ovary.

Food and Drug Administration (FDA)   The federal agency that is charged with ensuring that marketed foods and drugs are safe for human consumption.

Food chain   The linkage of predator and prey. A rabbit eats grass, and an eagle eats the rabbit.

Forage   Livestock feed that consists mainly of the leaves and stalks of plants.

Forestry   The process of cultivating and managing trees for commercial use.

Freeze drying   A method of preserving food that dries food through the use of freezing under a tight vacuum.

Fructose   The sugar found in fruit.

Fry   Small, newly hatched fish.

Fumigate   To kill pathogens and insects by the use of certain poisonous liquids or solids that form a vapor.

Fungicides   Substances used to kill fungi.

G

Galactose   The sugar in milk.

Gamete   The sex cell, either an egg or sperm.

Gastrointestinal tract   The digestive system, made up of the stomach and intestines.

Gelding   A male horse that has been castrated.

Genes   A unit of inheritance, which is composed of DNA.

Genetic base   The breeding animals available for a producer to use.

Genetic code   The sequencing of genes that determine the makeup of an organism.

Genetic defect   An impairment of an animal that was passed by the parents to the offspring.

Genetic engineering   The alternation of the genetic components of organisms by human intervention.

Geneticists   People who study inheritance for a career.

Genetics   The science that deals with the processes of inheritance in plants and animals.

Genotype   The genetic makeup of an organism.

Geotropic   A term referring to the way shoots and roots of plants respond to the stimulus of gravity.

Geriatric   Pertaining to elderly humans.

Germination   The process of a plant emerging from a seed.

Gestation   The length of time from conception to birth.

Gills   The organs in an aquatic animal that allow them to take oxygen from the water.

Gilt   A female pig that has not given birth.

Glacier   A slow-moving mass of ice.

Global Positioning System   A system using a series of satellites and a locator to pinpoint any location on earth.

Glucose   A common sugar that serves as the building blocks for many complex carbohydrates.

Golgi apparatus   The site of processing and separation of membrane components and secretory materials of the cell.

Gossypol   A substance in cottonseed that is toxic to certain animals.

Grafting   The process of joining parts of different plants together to form a new plant.

Grease wool   Wool that has been shorn from a sheep and has not been cleaned.

Greenhouse   A structure covered in glass, plastic, or fiberglass that is heated or cooled to provide the proper environment for growing plants.

Growing season   The period from the last killing frost in the spring to the first killing frost in the fall.

Growth   An increase in the cell size or cell numbers of an animal.

Grub   The larva stage of some insects, particularly beetles.

Guard cells   One of two epidermal cells in a plant leaf or needle that enclose a stoma.

H

Ham   A cut of pork that consists of the hindquarter from the hock to the hip; especially one that has been cured and smoked.

Haploid   An organism or cell with one set of chromosomes.

Hardwood   A tree that is deciduous.

Heifer   A female bovine that has not produced a calf.

Helix   Strands consisting of molecules of DNA that are shaped like a corkscrew.

Hemoglogin   The red pigment in the red blood cells of people and animals that carries oxygen from the lungs to other parts of the body.

Herbaceous   Not woody, dying back to the ground each year, such as rhubarb and asparagus.

Herbicide   A substance used to kill plants.

Herbivorous   Describing an animal that eats plants as the main part of its diet.

Heritability   The portion of the differences in animals that is transmitted from parent to offspring.

Heterosis   The amount of superiority in a crossbred animal compared with the average of their purebred parents. Also called hybrid vigor.

Heterozygous   An animal who carries genes for two different characters.

Homeostatis   Maintenance of a constant internal environment by a combination of body mechanisms.

Homogenized milk   Milk that has been blended to dissolve the fat molecules so that the fat (cream) will not become separated from the rest of the milk.

Homozygous   Possessing identical genes with respect to any given pair or series of alleles.

Hormone   A chemical substance secreted by various glands in an animal's body that produces a certain effect.

Host   An animal on which another organism depends for its existence.

Humus   Organic material in the soil.

Hybrid   An animal produced from the mating of parents of different breeds.

Hybrid vigor   *See* Heterosis.

Hydrologic cycle   The complete cycle through which water passes, commencing as atmospheric water vapor, passing into liquid and solid forms as precipitation, into the ground surface, and finally again returning in the form of atmospheric water vapor by means of evaporation and transpiration.

Hydrolysis   Chemical reaction in which a compound reacts with water to produce a weak acid, weak base, or both.

Hypocotyl   The short stem of an embryo seed plant, the portion of the axis of the embryo seedling between the attachment of the cotyledons and the radicle.

Hypothalamus   Portion of the forebrain of vertebrates that controls hormonal activities and behaviors such as feeding, drinking, aggression, and fear responses.

Hypothesis   A theory by a scientist as to the cause or effect of a phenomenon. This is rested by experimentation or other types of research.

I

Immunity   Resistance to catching a disease.

Import   A good that is brought into this country from another country.

Inbreeding   The mating of animals that are closely related.

Incubation   The process of the development of a fertilized poultry egg into a newly hatched bird. The eggs must have the proper heat, humidity, and length of time.

Infectious disease   A disease that is contagious.

Infundibulum   The enlarged funnel-shaped structure on the end of the fallopian tube that functions in collecting the ova during ovulation.

Ingest   To take in substances through eating, drinking, breathing, or absorption through the skin.

Inorganic   Depicting substances that do not contain carbon. Usually derived from nonliving sources.

Insect   An animal of the class Insecta. They have three body parts and six legs.

Insulin   The hormone from a part of the pancreas that promotes the utilization of sugar in the organism and prevents its accumulation in the blood.

Integrated pest management   A system of controlling pests that includes a variety of methods.

Interferon   A protein released by cells when they are attacked by viruses that increase the resistance of uninfected cells.

Intermediate host   An animal other than the primary host that a parasite uses to support part of its life cycle.

Internal parasite   A parasite that lives inside the body of the host animal.

Internode   The portion of a stem or other structure between two nodes.

Irradiation   The process of treating a food or feed with ultraviolet light to increase the vitamin D content.

Irrigation   The artificial watering of crops.

K

Keratin   A complex protein distinguished by high insolubility, which is contained in substances such as hair, horns, claws, and feathers.

Kiln   A large oven used to dry lumber.

Knock-down herbicides   Chemicals that will kill any plant they contact.

L

Laboratory animal   An animal that is raised for the purpose of being used for laboratory experimentation.

Lactation   The process of an animal giving milk.

Lactose   A sugar obtained from milk.

Lacustrine soil   Sands deposited on the bottom of a lake, or lacustrine terrace formed along the margin of a lake.

Lagoon   A body of water used for the decomposition of animal wastes.

Larva   The immature stage of an insect from hatching to the pupal stage.

Law of dominance   The law pertaining to the way in which a dominant gene in a pair hides the effect of the other gene.

Law of independent assortment   The law pertaining to the way the distribution of alleles for one trait into the gametes does not affect the distribution of alleles for other traits.

Law of segregation   The law pertaining to the way that each gamete receives only one of an organism's pair of genes.

Layer   A chicken raised primarily for egg production.

Layering   The method of propagating woody plants by covering portions of their stems or branches with moist soil or sphagnum moss so that they take root while still attached to the parent plant.

Leaching   The removal of nutrients from the soil by water actions.

Lean-to-fat ratio   The amount of lean meat in a carcass compared to the amount of fat.

Leather   Material made from the hides of animals.

Legumes   Plants that produce nitrogen through a symbiotic relationship with rhizobia bacteria.

Leucoplasts   A colorless plastid in the cytoplasm of interior plant tissues that is potentially capable of developing into a chromoplast.

Libido   The sexual drive of an animal.

Life cycle   The changes in the form of life an organism goes through in its life.

Ligaments   The tough, dense fibrous bands of tissue that connect bones or support viscera.

Lignin   A complex, highly indigestible material associated with cellulose material and the other fibrous parts of a plant.

Linen   Cloth made from the fibers of the flax plant.

Lipid   A fat or fatty tissue.

Live weight   The weight of an animal before slaughter.

Lutinizing hormone   The hormone that stimulates ovulation.

Lymphocytes   A kind of white blood cell produced by the lymph glands and certain other tissues. It is associated with the production of antibodies.

Lysomes   A membrane-bound organelle in which hydrolytic enzymes separate.

M

Macronutrients   Nutrients that are needed by a plant in relatively large quantities.

Manure   The excrement from animals.

Marbling   The desired distribution of fat in the muscular tissue of meat that gives it a spotted appearance. Marbling is used in the quality grading of a carcass.

Mare   A female horse that has produced a foal.

Marrow   The substance in the center of bones that produces blood cells.

Mastication   The act of chewing food.

Mastitis   A disease involving the inflammation of the udder of milk-producing females.

Maturity   The point in an animal's life when it is old enough to reproduce. Also refers to the age of an animal or carcass.

Meat   The edible flesh of an animal.

Meat animal   An animal that is raised primarily for the meat in its carcass.

Mechanical cultivation   The preparing and tillage of soil using mechanical means.

Meiosis   Cell division that results in the production of eggs and sperm.

Membrane   The thin protoplasmic tissue connecting, covering, or lining a structure, such as a cell of a plant or animal.

Meristem   Plant tissue capable of cell division and, therefore, responsible for growth.

Mesoderm   The central layer of cells in a developing embryo that gives rise to the circulatory system and certain other organs.

Mesophyll   The parenchyma tissue between the upper and lower epidermis of a leaf.

Metabolism   The chemical changes in cells, organs, and the entire body that provide energy for the animal.

Metamorphosis   The process by which organisms change in form and structure in their lives, such as insects do.

Microbe   Minute plant or animal life. Some cause disease; others are beneficial.

Microfilaments   Strands consisting primarily of the protein actink, sometimes in conjunction with a second protein, myosin.

Micromanipulator   A very small instrument used to dissect cells and embryos in the cloning process.

Micronutrients   Nutrients that are required in relatively small amounts.

Microtubules   A hollow, cylindrical strand found in eukaryotic cells, made up of the protein tubulin.

Minerals   A chemical compound or element of inorganic origin.

Mitochondria   A double-membraned organelle that is the site of reactions of aerobic metabolism.

Mitosis   Cell division involving the formation of chromosomes.

Molds   Fungi distinguished by the formation of a network of filaments or threads, or by spore masses; usually saprophytes.

Molt   The process of poultry casting off old feathers before a new growth occurs.

Monoscots   Plants having a single cotyledon or seed leaf, such as corn.

Monogastric   Refers to an animal that has only one stomach compartment such as swine.

Moraine   The geological deposition formed on the margins of glaciers or beneath moving ice sheets.

Morula   A spherical mass of cells that develops into an embryo.

Mother breeds   Those breeds of animals that make the best mothers, such as the Yorkshire and Landrace breeds of swine.

Motility   Active movement of the sperm in the male's semen.

Mucous membrane   A form of tissue in the body openings and digestive tract that secretes a viscous watery substance called mucus.

Mule   A cross between a horse and a donkey. The mother is a mare, and the father is a jack.

Muscling   The degree and thickness of muscle on an animal's body.

Mutation   An accident of heredity in which an offspring has different characteristics than the genetic code intended.

Mutton   The flesh of a sheep older than one year of age.

Mutualism   Animals that live together and benefit from each other.

Mycotoxins   Chemical substances produced by fungi that may result in illness and death of animals and humans when food or feed containing them is eaten.

Myofibrils   A cylindrical part of the muscle cells, composed of a series of sarcomeres.

N

Naiad   The aquatic young of a mayfly, dragonfly, damselfly, or stone fly.

Natural fibers   Fibrous materials produced from natural sources, such as cotton and flax.

Natural resource   The elements of supply inherent to an area that can be used to satisfy needs of people, including air, soil, water, native vegetation, minerals, wildlife, etc.

Naturally acquired immunity   Immunity to a disease that is acquired by the animal's having had a disease.

Natural toxins   Poisons released by plants as a natural defense against insects.

Nectar   The sweet substances in flowers. Bees make honey from nectar.

Nematodes   Microscopic, wormlike, transparent organisms that can attack plant roots or stems to cause stunted or unhealthy growth.

Neural tube   A structure derived from the ectoderm during early embryonic development, which becomes the brain and spinal chord.

Node   The place upon a stem that normally bears a leaf or whorl of leaves.

Nodules   A root tubercle or lump formation on certain leguminous plants produced by the invation of symbiotic, nitrogen-fixing bacteria.

Nonpoint pollution   Pollution that results from many sources instead of a single source.

Noninfectious disease   A disease that cannot be transmitted from one animal to another.

Noxious weeds   Weeds that have been declared as especially troublesome.

Nucleus   The central portion of the cell that contains the genetic material.

Nutrient   A substance that aids in the support of life.

Nutritional disease   A disease that is caused by not enough or too much of a certain nutrient in an animal's diet.

Nymph   A stage in the development of some insects that immediately precedes the adult stage.

O

Offspring   The young produced by animals.

Omasum   The third compartment of the ruminant stomach. A lot of the grinding of the feed occurs here.

Omnivorous   Describing an animal that eats both plants and other animals.

Oogenesis   The process of egg production in the female.

Organelles   Membrane-bound structures found in the cytoplasm of a cell that perform a specific function.

Organic   Containing carbon or being of living origin.

Organically grown foods   Food that is grown without the use of chemical fertilizers or pesticides.

Organic fertilizer   Plant nutrients derived from organic sources such as compost, manure, etc.

Organic Food Production Act   An act passed by Congress in 1990 aimed at helping to regulate organically produced fresh and processed foods.

Organism   Any living being, plant or animal.

Ornamental plants   Plants produced for their beauty.

Osmosis   The flow of a fluid through a semipermeable membrane separating two solutions, which permits the passage of the solvent but not the dissolved substance.

Ossification   The process of forming bone.

Ovary   The female organ that produces the egg and certain hormones.

Ovulation   The process of releasing eggs from the ovarian follicles.

Ovum   An egg.

Oxidation   Any chemical change that involves the addition of oxygen.

Oxytocin   The hormone released from the posterior pituitary of the female, which causes contractions of the uterus at the time of breeding and parturition.

P

Palatability   The degree to which a feed or food is liked or accepted by an animal or human.

Palatable   Describing a feed or food that is preferred over another.

Palisade mesophyll   A leaf tissue composed of slightly elongated cells containing chloroplasts; located just beneath the upper leaf epidermis and above the spongy mesophyll in broad-leafed plants and in some conifers.

Palmately veined   Containing three or more veins, nerves, lobes, or leaflets radiating fanwise from a common basal point of attachment.

Pancreas   A gland below and behind the stomach that secretes pancreatic juice.

Papillae   Any small nipple-like projections.

Parasitism   The condition when one organism lives on or in another organism at that organism's expense.

Parent cell   The original cell that divides to produce daughter cells.

Parent material   The horizon of weathered rock or partly weathered soil material from which the soil is formed. Horizon C of a soil profile.

Passive immunity   Immunity that is temporary.

Pasteurization   The process of heat treating milk to kill microbes.

Pathogens   A living, microscopic, disease-producing agent, such as a bacterium or a virus.

Pathologist   One who studies the science that deals with diseases and the effects that diseases have on the structure and function of tissues.

Pathology   The science that deals with diseases and the effects that diseases have on the structure and function of tissues.

Peds   Units of soil structure such as an aggregate, crumb, prism, block, or granule formed by natural processes (in contrast with a clod, which is formed artificially by compression of a wet clay soil).

Penis   The male organ of copulation.

Per capita consumption   The average quantity of something, such as food or energy, consumed per person within a time period, usually a year; calculated by dividing the total amount available for consumption by the population.

Periosteum   The outer membrane or covering of bone.

Pesticides   Substances that are used to kill pests.

Petiole   The stem of any leaf.

Petrie dish   A small shallow dish of thin glass with a loose cover used for cultures in bacteriology.

ph   A measure of the acidity or alkalinity of a substance.

Phagocyte   An animal cell capable of ingesting microorganisms or other foreign bodies.

Pharmaceuticals   Substances that are used to enhance the health of humans or animals.

Pharynx   The cavity that connects the mouth and nasal cavity to the throat; a passage common to the digestive and respiratory tracts.

Phenotype   The observed characteristic of an animal without regard to its genetic makeup.

Pheromones   Substances secreted to the outside of the body by an individual organism, which causes a specific reaction by another organism of the same species.

Phloem   Inner bark; the principle tissue concerned with the translocation of elaborated food produced in the leaves, or other areas, downward in the branches, stems, and roots.

Photoperiodism   The reaction of plants to periods of daily exposure to light, which is generally expressed in formation of blossoms, tubers, fleshy roots, runners, etc.

Photosynthesis   The process by which green plants, using chlorophyll and the energy of sunlight, produce carbohydrates from water and carbon dioxide.

Physiological age   The age of an animal that is determined by an examination of the carcass.

Pickling   A process of food preservation that uses a solution, such as vinegar, that is too acidic for microbes to grow.

Pigment   The naturally occurring color in the hair, skin, etc., of a plant or animal.

Pinnate Constructed somewhat like a feather, with the parts (e.g. veins, lobes, branches) arranged along both sides of an axis, as in pinnate venation.

Pistil The female element of a flower; composed of a stigma, style, and ovary.

Pituitary gland A small gland at the base of the brain that secretes hormones that stimulate growth and other functions.

Placenta The membranous tissue that envelopes a fetus in the uterus.

Plastids A body in a plant cell that contains photosynthetic pigments.

Platelets Parts of cells formed from megakaryocytes in bone marrow.

Plumule In a germinating seed plant, the primary bud that develops into the primary stem.

Plywood A wood product that is made by gluing thin strips of wood together under pressure.

Polar nuclei In the ovule, the two haploid nuclei contained in the central cell.

Polled An animal that is naturally hornless.

Pollen The substance of a flower that carries the sperm.

Pollution The presence of substances in a body of water, soil, or air to impair the usefulness or render it offensive to the senses of sight, taste, or smell.

Polyploid An organism with more than two sets of the basic or haploid number of chromosomes; e.g. triploid, tetraploid, pentaploid, and so on.

Pony Horses that weigh 500 to 900 pounds at maturity.

Pork The edible flesh of a pig.

Post mortem After death.

Postnatal After the birth of an animal.

Poultry Any or all of the domesticated fowls that are raised for their meat, eggs, or feathers.

Predator An animal that kills and eats other animals.

Predator insects Insects that eat other insects.

Prenatal Before birth.

Preservatives Chemical additives placed in food to help preserve the food.

Primary nutrients Nitrogen, phosphorus, and potassium.

Produce Usually refers to fruits and vegetables grown for the market.

Progeny The offspring of animals.

Progesterone A hormone produced by the ovary that functions in preparing the uterus for pregnancy and maintaining it if pregnancy occurs.

Prokaryotic cell Cells of the kingdom Monera, which do not have a membrane-bound nucleus and lack other membrane-bound organelles.

Prolactin A hormone that stimulates the production of milk.

Propagate To create new plants from old plants.

Protoplasm   The material of plant and animal tissues in which all life activities occur.

Protozoa   One-celled animals.

Pruning   Selectively cutting limbs or other parts from a plant to shape it.

Pubescence   Having hair-like projections on the surface.

Pulmonary   Relating to or associated with the lungs.

Pulp   A substance made from wood fiber that is used in the manufacture of paper.

Pupa   The stage in an insect's life between the larva stage and the adult stage.

Purebred   An animal that belongs to one of the recognized breeds and has only that breed in its ancestry.

Q

Quality grade   The grade given to a beef carcass that indicates the eating quality of the meat.

R

Ram   A male sheep that has not been castrated.

Rancid   The putrified state of foods.

Ration   The feed allowed for an animal in a 24-hour period.

Recessive gene   A gene that is masked by another gene that is dominant.

Recipient cow   A genetically inferior cow in which an embryo from a genetically superior cow is placed.

Recycle   To save used products or the remains of used products for use in the manufacture of new products.

Refrigerated truck   A truck that contains its own refrigeration unit used for transporting meat or other perishable products.

Renewable resources   Resources such as trees that can be replaced.

Rennin   An enzyme extracted from the stomach of cattle used in the cheesemaking process.

Replication   The systematic laying out of several test rows or plats to test soil for fertility and other environmental factors, and also, the process of genetic material duplicating itself.

Reproduction   The process of creating new life from organisms.

Respiration   The act of breathing; the drawing of air into the lungs and its exhalation.

Reticulum   The second compartment of a ruminant's stomach.

Retort   A device used to place cans or jars under pressure to kill pathogens.

Rhizobia   Bacteria that live in the nodules on the roots of certain plants that fix nitrogen.

Rhizomes   Horizontal underground stems or branches of a plant that send off shoots above and roots below and are often tuber-shaped.

Rhizosphere   The region of the soil where roots grow.

Roe   A mass of fish eggs.

Roughage   A feed low in carbohydrates and high in fiber content.

Rumen   The largest compartment of the stomach system of a ruminant. This is where a large amount of bacterial fermentation of feed occurs.

Ruminant   Any of a class of animals having multicompartmented stomachs that are capable of digesting large amounts of roughages.

Runoff   Water that flows across the ground after a rain.

S

Safflower   Family Compositae, and herb, resembling a thistle, cultivated for its orange-colored flower heads, which yield a drug and red dye, and for its seed, a source of a rapid-drying oil used in food and medicine and by the paint industry.

Salmonella   A large group of bacteria, some of which cause food poisoning.

Salting   A method of preserving food that uses salt to prevent the growth of microbes.

Sap wood   That portion of the wood of a tree more recently formed, which contains living cells intermingled with nonliving, woody tissue.

Saturated fats   A fat whose carbon atoms are associated with the maximum number of hydrogen atoms; no double bonds exist.

Scarification   The scratching or modification of the surface of a seed coat to increase water absorption.

Scientific method   The process used in research that uses observation, hypothesis, experimentation, and conclusion.

Scion   An unrooted portion of a plant having one or more buds, used for grafting or budding onto rootstock.

Scouring   The cleaning of wool.

Scrotum   The pouch that contains the testicles.

Secondary nutrients   Calcium, magnesium, and sulfur.

Seedlings   The early growth stage of a plant grown from seed as it emerges above the ground surface.

Seeds   The embryo of a plant; also kernels of corn, wheat, and others, which botanically are seedlike fruits and include the ovary wall.

Selective breeding   Choosing the best animals and using those animals for breeding purposes.

Semen   A fluid substance produced by the male reproductive system containing the sperm and secretions of the accessory glands.

Semipermeable membrane   A membrane that permits the diffusion of some components and not others. Usually water is allowed to pass, but solids are not.

Sepals   One of the separate units of a calyx, usually green and foliaceous.

Septum   Any dividing membrane or other layer in plants and animals.

Sericulture   The growing of silkworms.

Serum   The clear portion of any animal fluid.

Sessile leaves   Leaves, flowers, fruits, and so on attached directly by the base without a stem or stalk, such as leaves of grasses, sedges, and certain other plants.

Sex character   The physical characteristics that distinguish males from females.

Sexual reproduction   Reproduction that requires the uniting of an egg and a sperm.

Shrubs   A plant that has persistent, woody stems and a relatively low growth habit and that generally produces several basal shoots instead of a single bole.

Silk   The long, silky styles with stigmas of the corn plant. Also the soft, fine, and shiny fiber that is produced by the silkworm.

Silt   Small, mineral, soil particles, ranging in diameter from 0.05 to 0.002 mm. Also a textural class of soils that contains 80 percent or more of silt and less than 12 percent clay.

Simple leaf   Leaf blades consisting of one unit.

Sire   The father of an animal.

Sire breeds   Those breeds of agricultural animals that are used as sires in a cross-breeding program.

Softwood   Wood from a conifer.

Soil   The mineral and organic surface of the earth capable of supporting upland plants.

Soil amendment   Any material added to the soil that improves it.

Soil horizon   A natural layer of a vertical section of the soil profile.

Solum   The upper part of the soil profile above the parent material, in which the processes of soil formation have taken place or are taking place.

Sow   A female pig that has had a litter of pigs.

Spawning   The depositing of fish eggs.

Sperm   The male reproductive cell that unites with an egg.

Spermatogenesis   The development of the sperm cell.

Sphincter muscle   A ring shaped muscle that closes an orifice.

Spindle   The fine threads of achromatic protoplasm arranged in a fusiform mass within the cell during mitosis.

Spongy mesophyll   Irregularly shaped cells below the Palisade parenchyma, often with large spaces between them.

Spores   The one- to many-celled units of a fern, fungus, bacterium, or protozoan that has entered the resting state and is capable of growth and reproduction when conditions become favorable.

Stamen   The organ of a flower that bears the pollen, consisting of the stalk and the anther.

Staphylococci   A bacteria that causes swollen joints, diarrhea, and depression in poultry.

Steer   A male bovine that has been castrated before sexual maturity.

Sterile   Being unable to produce an offspring; containing no life.

Sternum   The breastbone of an animal.

Stimuli   Any agent that causes a response.

Stolons   An aboveground horizontal stem that propagates vegetatively by forming new roots and shoots at the nodes.

Stoma   An opening in the epidermal layer of plant tissues, which leads to intercellular spaces.

Streptococci   Bacteria responsible for various diseases in humans and animals.

Style   In the pistil of a flower, the part between the ovary and the stigma.

Subsoil   The soil beneath the topsoil. It is generally very low in organic matter.

Succession   The replacement of one plant community by another.

Supply and demand   An economic principle that says when demand of a product increases, the supply decreases. This leads to more production of the product, which leads to a higher supply. Price is affected by the shift.

Sustainable agriculture   Agricultural practices aimed at maintaining yields of plants and animals over a period of time.

Synapsis   The association of alike chromosomes that is characteristic of the first meiotic prophase and is thought to be the mechanism for genetic crossing-over.

T

Taxonomy   The science of classification of organisms and other objects and their arrangement into systematic groups such as species, genus, family, and order.

Tendon   The strong tissue connecting a muscle to a bone.

Testa   The outer coat of a seed.

Testicles   The male organs that produce sperm and certain hormones.

Testosterone   The male hormone responsible for the male sex drive and the development of the male sex characteristics.

Thermoclines   The layers in a body of water in which the drop in temperature equals or exceeds 1°C for each meter of water depth.

Thyroxin   An iodine-containing amino acid that is used to treat thyroid disorders.

Tissue culture   The process or technique of making plant or animal tissue grow in a culture medium outside the organism.

Topsoil   Surface soils and subsurface soils that presumably are fertile soils, rich in organic matter or humus debris.

Toxin   A poison.

Trachea   The tube by which air passes to and from the lungs.

Transgenetic animals   Animals that are the results of crossing genes from different types of animals such as a sheep and a goat.

Treatment group   In a scientific experiment, the group of animals or plants that receives the treatment that is being researched.

Trunicate   Appearing as if cut off nearly or quite straight across at the end.

Tubers   Thickened or swollen underground branches or stolons with numerous buds.

Tumor   A swelling; a new growth of cells or tissues governed by factors independent of the laws of growth of the host.

Turgid   Swollen, or slightly drawn, membrane or covering expanded by pressure from within.

U

U. S. Forestry Service   A part of the USDA charged with the overseeing and care of national forests.

Umbilical cord   The part of the fetal membranes that connects to the navel of the fetus; it carries the blood vessels that transport fetal blood to and from the placenta.

United States Department of Agriculture (USDA)   The federal bureaucracy that conducts research, inspections, and regulations pertaining to U. S. agriculture.

Unsaturated fats   Fats of vegetable origin, such as olive oil or cotton seed oil, that have more than one double bond in their carbon chain.

Urethra   The tube that carries urine from the bladder and serves as a duct for the passage of the male's semen.

USDA   See United States Department of Agriculture.

Uterus   The female reproductive organ in which the fetus develops before birth.

V

Vaccination   The process of injecting an animal with certain microorganisms in an effort to make the animal immune to specific diseases.

Vaccine   A substance that contains live, modified, or dead organisms that is injected into an animal to make it immune to a specific disease.

Vagina   The canal in the female reaching from the uterus to the vulva.

Vas deferens   The tubes connecting the epididymis of the testicles to the urethra.

Vascular bundles   In plant anatomy, a unit containing both the phloem and xylem.

Veal   The meat from calves slaughtered before they are three months of age.

Vegetative propagation   Increasing the number of plants by methods such as cuttings, grafting, or layering.

Veins   One of the systems of branching tubes that carry blood back to the heart.

Veneers   Layers of wood of superior value or excellent grain to be glued to an inferior wood.

Ventricles   The pumping chambers of the heart.

Vertebrate   An animal having a backbone.

Viable   Capable of living.

Villi   Microscopic, hairlike projections of the lining of the digestive tact.

Virgin forest   A mature or overmature forest, which has grown entirely uninfluenced by human activity.

Virus   A self-reproducing agent that is considerably smaller than a bacterium and can multiply only within the living cells of a suitable host.

Vulva   The external reproductive organ of the female.

W

Warm-blooded   An animal whose body temperature is warmer than its surroundings.

Warm-water fish   A fish that does not thrive in water colder than 70°F.

Wean   To make a young animal cease to depend on its mother's milk.

Weed   Any plant that is growing where it is not wanted.

Wetland   A parcel of land that stays wet for most of the year.

Withdrawal period   The length of time that must transpire between the time an animal was given a certain drug and the time the animal's milk can be used or the animal is slaughtered.

X

Xylem   The woody part of the tree that conducts water and nutrients up the stems from the roots.

Y

Yeasts   A yellowish substance composed of microscopic, unicellular fungi of family *Saccharomycetaceae,* which induces fermentation in juices, worts, doughs, etc.

Yield grade   A grade in meat animals that refers to the ratio of lean meat produced in a carcass.

Yogurt   A semi-solid, fermented milk product.

Yolk   The yellowish part of a fowl's egg that contains the germinal disk. Also the substances such as wool grease in a fleece.

Z

Zygote   A fertilized egg.

INDEX